Smart Sensors, Measurement and Instrumentation

Volume 25

Series editor

Subhas Chandra Mukhopadhyay
Department of Engineering, Faculty of Science and Engineering
Macquarie University
Sydney, NSW
Australia
e-mail: subhas.mukhopadhyay@mq.edu.au

More information about this series at http://www.springer.com/series/10617

Boby George · Joyanta Kumar Roy
V. Jagadeesh Kumar
Subhas Chandra Mukhopadhyay
Editors

Advanced Interfacing Techniques for Sensors

Measurement Circuits and Systems for Intelligent Sensors

Springer

Editors
Boby George
Department of Electrical Engineering
IIT Madras
Chennai, Tamil Nadu
India

V. Jagadeesh Kumar
Department of Electrical Engineering
IIT Madras
Chennai, Tamil Nadu
India

Joyanta Kumar Roy
ECE Department
MCKV Institute of Engineering
Liluah, West Bengal
India

Subhas Chandra Mukhopadhyay
Department of Engineering
Macquarie University
Sydney, NSW
Australia

ISSN 2194-8402 ISSN 2194-8410 (electronic)
Smart Sensors, Measurement and Instrumentation
ISBN 978-3-319-55368-9 ISBN 978-3-319-55369-6 (eBook)
DOI 10.1007/978-3-319-55369-6

Library of Congress Control Number: 2017934448

Printed on acid-free paper

This Springer imprint is published by Springer Nature
The registered company is Springer International Publishing AG
The registered company address is: Gewerbestrasse 11, 6330 Cham, Switzerland

Preface

Sensors play a key role in almost all the scientific advancements in the world. Successful operation of the engineering systems, from automotive engines to International space stations, is almost impossible to imagine without efficient and reliable sensors. Sensors are inevitable, for the operation and maintenance of various industries including automation, automotive, shipping, transportation, power, manufacturing, wood, and paper. With the advanced sensing techniques and processing/manufacturing technologies available today, the variety of sensors that are/can be realized and manufactured to sense vital parameters of any system which are of tremendous interest are plenty compared to the situation prevailed a decade back.

Output from any sensor or sensing element is converted into a usable and standardized form using a unit, typically, called as signal conditioning unit. In some cases, these are referred as interfacing or read-out circuits. This unit is present immediately after the sensing element, in a measurement system. In most of the cases, the raw signal from the sensing element can be as bad as noise signal. Unless processed correctly by a suitable interfacing unit, the sensor system will fail to work or perform unsatisfactorily. In addition, the interfacing unit plays an important role in exploiting the full capabilities of the sensing element. Thus, in general, the interfacing instrumentation is an integral part of any sensor system and a deep understanding of those schemes is a requirement for a scientist or instrumentation engineer who selects the appropriate interfacing circuit for a sensor that is of interest. This understanding will also help the reader to develop his/her own interfacing schemes with more useful features for an existing sensor or a new interfacing scheme that may be required for a new or an existing sensor.

The interfacing scheme can be an analogue electronics solution, a digital-based solution, or a mixed signal solution. Certain operations can be achieved effectively using analogue systems while a mixed signal approach is preferred in some other applications. There are interfacing schemes available, purely, using digital schemes/systems. Each one has its own advantages and disadvantages. This book will help the reader to acquire basics as well as advanced knowledge about the interfacing schemes for various sensors that are widely used. The book will also discuss the important signal processing techniques, certain sensing applications, future trends that are worth to understand and follow for a sensing and instrumentation scientist. The book contains invited chapters from experts working in the field of sensors and instrumentation domain, presenting various aspects of sensors, advanced interfacing techniques for sensors and interesting applications. The book is organized as discussed below.

The book begins with a chapter that deals with the fundamentals of the measuring/sensing systems. This includes types of sensor systems, important performance characteristics, propagation of errors, statistical analysis of random error and transient and steady-state response of sensors. The above-mentioned topics presented in Chapter "Sensors and Their Characteristics" provide the necessary background information for those who are relatively new to the field. It helps others in clearing most of the possible ambiguity or doubts that may come out, in relation to some of the fundamentals of the measurement, while reading other chapters of the book.

Chapters "Advanced Interfacing Techniques for the Capacitive Sensors" and "A Simple Embedded Sensor: Excitation and Interfacing" deal with impedance sensors and interfacing circuits. Chapter "Advanced Interfacing Techniques for the Capacitive Sensors" introduces a few important sensors. Later, the advantages of the impedance sensors such as capacitive or inductive sensors over the resistive sensors are presented. The chapter also presents the design and development of a low-cost portable sensing system, which measures impedance of a capacitive sensor. In addition, some of the possible applications of the capacitive sensors are presented in Chapter "Advanced Interfacing Techniques for the Capacitive Sensors". Chapter "A Simple Embedded Sensor: Excitation and Interfacing" concentrates more on different types of capacitive sensors and the interface electronics for instrumentation applications. Some of the advanced electronic circuits that combines the well-established bridge methods of measurement together with certain oscillator configuration for accurate interfacing of the perfect and the lossy capacitive sensors are discussed in Chapter "A Simple Embedded Sensor: Excitation and Interfacing".

An interesting way to design low-cost but efficient digital measurement systems, for interfacing resistive sensors is presented in Chapter "Advanced Techniques for Directly Interfacing Resistive Sensors to Digital Systems". The technique involves the use of relatively advanced techniques for direct connection of resistive sensors to the digital systems. The digital system measures the resistance with the help of an embedded digital timer that digitizes the charging/discharging time of an RC circuit formed by the resistive sensor and a known capacitor. The uncertainty sources involved in measurement and the performance, in some applications are discussed. Next, the chapter deals with the direct connection of resistive sensor arrays to field-programmable gate arrays, where different resistors of the array are measured in parallel through a set of timers that are running simultaneously.

Combination of sensors and IoT can provide very efficient sensing solutions that we, otherwise, would not have imagined. While the techniques for wireless internet connectivity for the sensor unit is relatively a solved problem, powering the node and keeping the power requirement to ultra-low values remain challenging in many applications. Chapter "Interfaces for Autarkic Wireless Sensors and Actuators in the Internet of Things" presents various aspects of those challenges and possible solutions through ISO/IEC/IEEE 21450-2010 smart transducer standard.

Output from some of the sensors are very low and many times special techniques are required to improve the SNR at the final output. Chapter "Lock-In Amplifier Architectures for Sub-ppm Resolution Measurements" presents some of the effective schemes to recover sensor signals from noise. The fundamental limitations of the widely used lock-in amplifier technique are discussed in this chapter, followed by possible ways to enhance the performance to achieve high resolution measurements.

From here onwards, the reader will find the application-oriented chapters of the book. Chapters "Biomedical Sensors and Their Interfacing" and "Interfacing and Pre-processing Techniques with Olfactory and Taste Sensors" discuss some of the requirements in the healthcare domain and possible approaches to support or help the caregiver, to realize un-obstructive monitoring, etc. Chapter "Biomedical Sensors and Their Interfacing" presents schemes for continuous monitoring of patients in their normal activity. This includes real-time recognition of abnormal 'events' such as fall and unconsciousness. This is achieved with the help of autonomous sensor systems with necessary processors having the required computational capability. Chapter "Interfacing and Pre-processing Techniques with Olfactory and Taste Sensors", in the similar lines, presents the challenges in un-obstructive monitoring of important parameters and possible solution through smart sensing systems. The sensor system discussed is compatible with IEEE1451 standard.

In Chapters "Harnessing Vision and Touch for Compliant Robotic Interaction with Soft or Rigid Objects" and "IEEE1451 Smart Sensors Architectures for Vital Signs and Motor Activity Monitoring", interfacing challenges and solutions of couple of industrial applications are presented. Chapter "Harnessing Vision and Touch for Compliant Robotic Interaction with Soft or Rigid Objects" concentrates on the details of electronic nose and tongue. Basic principle, sensing mechanism, interfacing and pre-processing techniques, etc. employed for typical electronic nose and tongue are presented. Number of groups across the world are working in this domain. Chapter "IEEE1451 Smart Sensors Architectures for Vital Signs and Motor Activity Monitoring" presents the sensing issues and interfacing techniques to enable safe interaction of commercial-grade robot manipulators with objects exhibiting rigid or soft surfaces. The solution involves combination of vision and touch sensing techniques. Details of the proposed solutions and results of the experimental studies validating the usefulness are presented in the chapter.

We would like to express our deep appreciation to the distinguished authors of the chapters. The book has taken an interesting shape with valuable content, due to the expertise, professionalism and kindness of the contributed authors of this book. We are expecting that the readers will find the book very interesting, informative and enriching.

Chennai, India	Boby George
Liluah, India	Joyanta Kumar Roy
Chennai, India	V. Jagadeesh Kumar
Sydney, Australia	Subhas Chandra Mukhopadhyay

Contents

About the Editors

Dr. Boby George received the M. Tech. and Ph.D. degrees in Electrical Engineering from the Indian Institute of Technology (IIT) Madras, Chennai, India, in 2003 and 2007, respectively. He was a Postdoctoral Fellow with the Institute of Electrical Measurement and Measurement Signal Processing, Technical University of Graz, Graz, Austria, from 2007 to 2010. He joined the faculty of the Department of Electrical Engineering, IIT Madras in 2010. Currently, he is working as an Associate Professor there. He serves as one of the Associate Editors of the IEEE Sensors journal. His areas of interests include measurements, sensors and instrumentation.

Prof. Joyanta Kumar Roy has been working in electronics and automation engineering since 1984 as Company Director, Consulting Engineering, Developer, Researcher and Educationist. He graduated from the Department of Physics from University of Calcutta, India, and received Master of Science degree in Physics in 1977. He started his carrier as entrepreneur in the year 1984 founded a small manufacturing enterprise named System Advance Technologies Pvt. Ltd., dealing turn-key execution of SCADA, Automation and industrial instrumentation system. In 2004, he obtained Ph.D. (Technology) degree in Applied Physics from University of Calcutta, India, and executed number of projects, related to control, automation and instrumentation in several engineering sectors. After long association with industry, he started his academic carrier from 2005.

He worked with many educational institutes as principal, dean and professor. He is also serving technical consultancy to leading consulting company like GKW Consult GmbH and many more. He significantly contributed to industrial automation in water sector, laboratory instrumentation in development of low-cost sensors and sensing system for process control industry. His industry group developed low-cost indigenous substitute of high-cost sensors and sensing systems for power industry, which are working satisfactorily till date. His industry group was introduced Distributed SCADA system for plant and water quality surveillance 1st time in Asia. His research group developed low-cost non-contact liquid-level transmitter, temperature transmitter, pressure transmitter, vortex flow transmitter, mass flow metre, etc. He has contributed more than 150 scientific and technical publications in the form of book, book chapters, journal papers, conference papers, manuals and engineering design of industrial project. He is a technical speaker and has given invited talks in number of International and national level conferences. He organized many technical events, national and international events and worked as TPC member. He is a senior member IEEE, Chairman and EC member IET (UK) Kolkata Network, Fellow of IWWWA and Fellow of IETE. Presently, he is working as Editor of S2IS and regular reviewer of research articles. His present research interest includes development of smart measurement and control system for water production and distribution, multifunction sensor, IoT based m-health, technology-assisted living, smart home and city.

Prof. V. Jagadeesh Kumar obtained his B.E. (ECE) (1978), from College of Engineering Guindy, M.Tech. (1980) and Ph.D. (1986) from IIT Madras. He is presently working as a Professor of Electrical Engineering at IIT Madras. He is heading the Central Electronics Center, IIT Madras and also is the Dean of Academic Courses at IIT Madras. He received the Young Scientist award from the Department of Science and Technology in 1988 and the DAAD fellowship award in 1997. He had guided 7 Ph.D. scholars and 11 M.S. (Research) scholars. Has published 50 Journal articles (mostly in IEEE Journals) and presented 90 papers in international conferences. He has obtained 6 patents. He had worked at King's College London (1988), Asian Institute of Technology, Bangkok (1996), University of Braunschweig (1998) and University of Aachen (1999, 2007, 2011, 2013). His areas of interests are: Measurements, Instrumentation, Biomedical Engineering and Signal processing.

Prof. Subhas Chandra Mukhopadhyay has been working as an engineer and scientist from 1989 in industry and academic institutions. He has significantly contributed in the area of smart sensors and instrumentation for monitoring health of inhabitant especially for the elderly people. So far, he has contributed in the form of scientific contribution over 400 scientific items (6 authored books, 118 authored scientific journal papers, 34 book chapters, 215 conference contribution, 26 edited books, 15 edited conference proceedings and 16 journal special issues). He contributes regularly as a highly acclaimed speaker in different international conferences in the form of keynote/invited/tutorial talk. His contribution as an author and an editor for international journals has been recognized in the form of numerous international awards.

His group has developed smart MEMS sensors for detection of endotoxin (Lipopolysaccharide, LPS) and detection of contaminated phthalates in water and juices towards safe food for healthy living. The group is also involved in the development of a novel, a non-invasive, real-time and label-free smart assay technique for the prognostic detection of bone loss by electrochemical impedance spectroscopy (EIS) and has done preliminary work using blood sample. Prof. Mukhopadhyay's research on the development of different sensors and instrumentation-based smart home will provide a safe, sound and secured living environment for any individual especially for the elderly people living alone at home. The developed medical instrumentation will measure the physiological parameters as well as smart sensors and instrumentation-based monitoring household appliances will be seamlessly upload measured data to a secured website through the Internet of Things (IoT). The data will determine the wellness of the individual in real time and also calculate the pattern of the trend of the change of lifestyle. The developed system can be extremely useful for generating warning message during any unforeseen accident when there is a dire need of attention. All the research and development activities help to cut down the rising health care costs by increasing the health monitoring and improving the efficiency of the monitoring system in developed countries.

Sensors and Their Characteristics

V. Jagadeesh Kumar

Abstract This chapter introduces sensor as a device (sort of black box approach) and describes various characteristics that define such a sensor and most importantly how to interpret its performance in actual use employing these characteristics. Depending on the actual sensor and its intended application, some of these characteristics become relevant and some are not. For example, the dynamic characteristics of a sensor (load cell) used in a stand alone weighing machine is not at all relevant and just the static characteristics will do. On the other hand if the same sensor is used to weigh objects on a conveyor belt, then the dynamic characteristics of the sensor along with the static characteristics have to be considered. Hence the reader must exercise caution and not blindly insist that a sensor must possess all the characteristics described herein.

1 Introduction to Sensors

What is a sensor? Dictionary defines (the meaning of) the word "*sensor*" as:

> **Sensor** A device that responds to a physical stimulus (as heat, light, sound, pressure, magnetism, displacement, velocity or acceleration) and provides an output for either measurement of that physical quantity or operating a control.

For example, the *U-tube manometer*, a differential pressure sensor shown in Fig. 1 translates a pressure difference $(P_1 - P_2)$, not an easy quantity to measure directly, into a height of a fluid which is easily measureable. In Fig. 1 the pressure P_1 is assumed to be higher than pressure P_2. Traditionally, the brachial arterial *blood pressure* is measured using such a device with mercury as the fluid. The blood pressure measuring device that includes the mercury manometer, shown in Fig. 2, is called the '*sphygmomanometer*'.

V. Jagadeesh Kumar (✉)
IIT Madras, Chennai, India
e-mail: vjk@iitm.ac.in

B. George et al. (eds.), *Advanced Interfacing Techniques for Sensors*,
Smart Sensors, Measurement and Instrumentation 25,
DOI 10.1007/978-3-319-55369-6_1

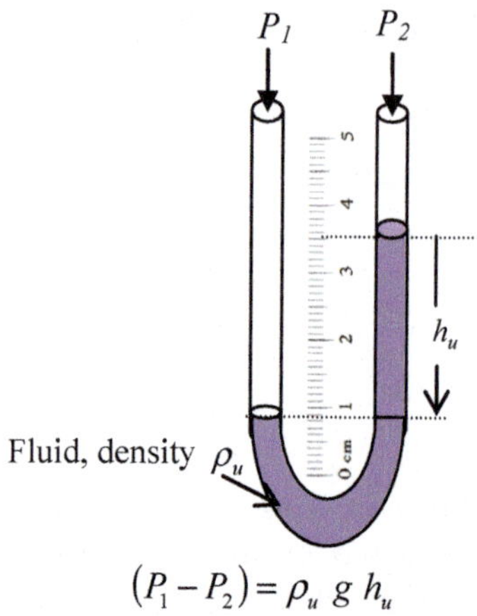

Fig. 1 The U tube manometer

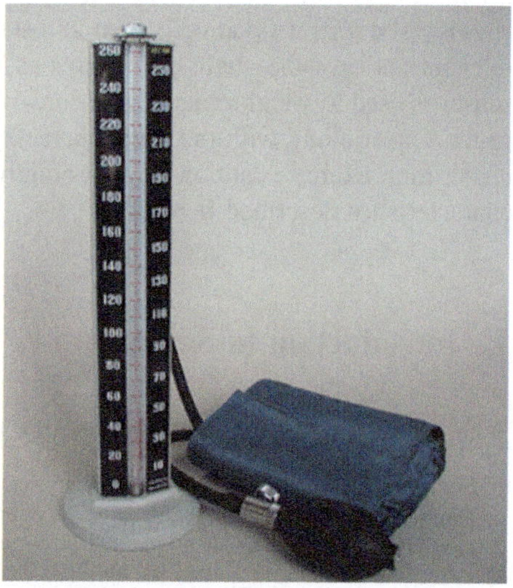

Fig. 2 A mercury sphygmomanometer

From the measurement of h_u, with the knowledge on the acceleration due to gravity g at the measurement site, and the density of the fluid ρ_u, we can compute the pressure difference $(P_1 - P_2)$ as:

$$(P_1 - P_2) = \rho_u g \, h_u \qquad (1)$$

If the density of the fluid ρ_u is expressed in kg/m^3 (kilogram per cubic meter), the acceleration due to gravity g in m/s^2 (meter per second per second) and h_u in meter then $(P_1 - P_2)$ in Eq. (1) will have the units of pascal (Pa). Exercise 1 illustrates this aspect.

Exercise 1 The liquid used in a U-tube manometer is mercury possessing a density of 13,534 kg/m^3. Determine the minimum height required of the straight section of the U-tube if the maximum differential pressure to be measured is 33.32 kPa. Assume the value of g to be 9.81 m/s^2.

Solution for Exercise 1:

$$(P_1 - P_2) = \rho_u g\, h_u \rightarrow 33.32 \times 10^3 = 13534 \times 9.81 \times h_U$$

$$h_u = \frac{33.32 \times 10^3}{13{,}534 \times 9.81} = 0.25005\ \text{m} \approx 250\ \text{mm}.$$

Figure 3 provides an *engineering model* of the differential pressure sensor of Fig. 1 that explains how different parameters, some internal to the sensor and some external, combine to provide the final output of that sensor. From the model it is abundantly clear that the output of a sensor is dependent on, other than the input pressure difference $(P_1 - P_2)$, intrinsic (density ρ_u of the fluid) and extrinsic (gravity g) parameters associated with that sensor.

2 Sensors and Transducers

Modern measurement, instrumentation and control systems invariably are implemented with electrical or electronic devices due to the fact that electrical or electronic instruments provide ease of operation and or advanced signal processing power. Thus today, a sensor is required to provide an electrical output that is not only easily readable, but can be processed and more importantly stored for future reference. Thus majority of sensors designed, developed and marketed invariably

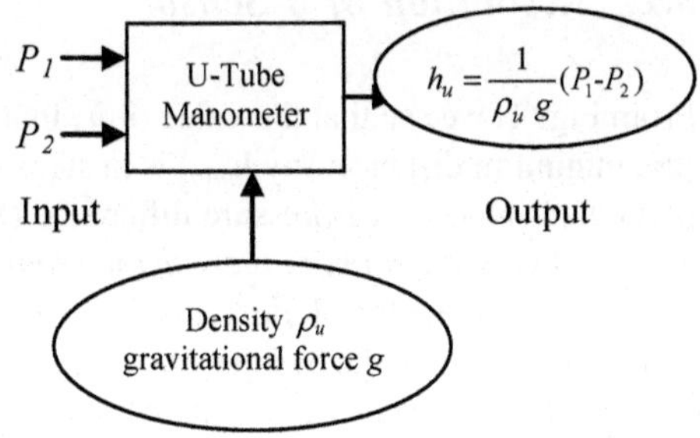

Fig. 3 Generalized model of the pressure sensor of Fig. 1

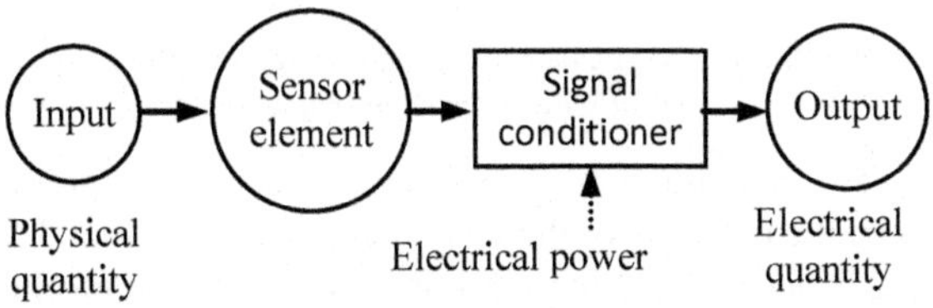

Fig. 4 Block schematic of a typical sensor/transducer

provide an electrical output. In general, to obtain an electrical output from a sensor, the sensing element of the sensor will have to be operated upon by an electrical or electronic *signal conditioning circuit* as shown in Fig. 4. When a sensor employs a signal conditioning circuitry then it is customary to call the sensor as a '*transducer*'. The dictionary defines the word transducer as:

Transducer A device that converts one type of energy or signal into another.

However, in usage today, a *transducer* is understood to be "*A device that converts variations in a physical quantity into an electrical signal*" and thus points to a sensor. On the other hand a device (such as motors, solenoids…) that converts an electrical stimulus (electrical energy/signal) into a physical quantity is called an *actuator*. Today the words '*sensor*' and '*transducer*' are used to represent similar or same devices and are taken as *synonyms*. Since most of the requirement today for sensing requires a sensor or transducer to provide an electrical output, in this book, we deal mainly with those types of sensors (transducers) that provide an electrical output.

3 Important Characteristics of Sensors

In this section we will learn about different parameters that characterize a sensor. Learning and understanding the characteristics of a sensor helps us

(i) Optimize the design of a sensor to provide required input output characteristics and
(ii) Properly and optimally choose a sensor for a particular application.

3.1 Resolution of a Sensor

From Fig. 1, we see that the value of h_u that we can measure without any ambiguity (the minimum distance, say h_{min}) is in steps of 1 mm, and that would translate into a particular value of the pressure difference ($P_1 - P_2$) depending on the values of ρ_u and g. Thus the *minimum incremental step value that can be read unambiguously* using the manometer of Fig. 1 is $(P_1 - P_2)_{\min} = \rho_u g\, h_{\min}$. This minimum value is called (termed as) the resolution of the sensor. Thus resolution is the step size between two adjacent values that can be clearly read using a sensor. We can define the '*Resolution*', an important characteristic of a sensor, as:

Resolution The smallest incremental input quantity a sensor can read (resolve or respond to and provide a readable output) without any ambiguity.

The resolution of a sensor can also be expressed as a simple fraction. For example the scale of the manometer shown in Fig. 1 has 50 divisions of 1 mm each and hence the resolution is 1 in 50 or 1/50. The resolution determines the number of *significant digits* that can be written in the result obtained from a reading of a sensor. For example, if the height of the mercury in the manometer of Exercise 1 is measured with a scale 250 mm long and graduated in millimeters, then its resolution is 1/250. If for a particular pressure difference being read, the difference in height between the left and right columns is 100 mm then the pressure difference can be calculated (at sea level) as:

$$(P_1 - P_2) = \rho_U g\, h_U = 13,534 \times 9.81 \times 100/1000 = 13276.854\ \text{Pa}$$

Though the computation as given above is "*mathematically*" correct but gives a wrong impression on the actual measurement that was performed. When we measured the difference in height as 100 mm, we are certain that the difference in height is 100 mm and not 101 mm nor 99 mm. This would result in a measurement (100 mm) having only three *significant digits*. Hence the result of the computation of the pressure difference should also be expressed to three significant digits by performing a *rounding off* as:

$$(P_1 - P_2) = \rho_U g\, h_U = 13,534 \times 9.81 \times 100/1000 = 13276.854 = 13.3\ \text{kPa}$$

If, instead of 100 mm, say we got the reading as 101 mm or 99 mm then the pressure difference computed would be:

$$(P_1 - P_2) = \rho_U g\, h_U = 13,534 \times 9.81 \times 101/1000 = 13409.62254\ \text{Pa} = 13.4\ \text{kPa or}$$
$$(P_1 - P_2) = \rho_U g\, h_U = 13,534 \times 9.81 \times 99/1000 = 13144.08546\ \text{Pa} = 13.1\ \text{kPa respectively.}$$

Note: When the height is 99 mm, the number of significant digits is two and hence the result displayed also should have only two significant digits. Here we see that a step change of one division (one millimeter) results in a change of 0.1 kPa (when the reading changes from 100 to 101 mm) to 0.3 kPa (reading changes to 99 mm from 100 mm) on the computed pressure. In other words the resolution of 1 mm on the manometer scale results in a resolution in the range of 0.1–0.3 kPa on the pressure measurement. Thus when we deal with computations with measured values obtained not only with sensors but with any other device or instrument, we must not get carried away with the large number of digits our calculator or computer can spew as a result of that calculation. Instead, we should ensure that correct information about the measurement is disseminated to others without any ambiguity. We have seen above, that to inform others of the number of significant digits in our measurement, we need to *round off*. The rules to be followed for rounding off are:

(i) If the digits to be dropped are $< 50,000\cdots$ then the digits are dropped and the rest of the digits retained as it is. (*vide* Exercise 2)

Exercise 2 Round of 12.34999999 to three significant digits. The result is 12.3.

(ii) If the digits to be dropped are $> 50,000\cdots$ then the digits are dropped and the least significant digit of the digits to be retained is incremented by 1 (*vide* Exercise 3)

Exercise 3 Round of 12.3500000001 to three significant digits. The result is 12.4.

(iii) If the digits to be dropped are exactly $50,000\cdots$ and

(a) if the least significant digit of the digits to be retained is even then the digits ($50,000\cdots$) are dropped and the set of significant digits retained as it is. (*vide* Exercise 4)

Exercise 4 Round of 12.4500000000 to three significant digits. The result is 12.4.

(b) if the least significant digit of the digits to be retained is odd then the least significant digit is incremented by 1. (*vide* Exercise 5)

Exercise 5 Round of 12.3500000000 to three significant digits. The result is 12.4.

Note: Often it would happen that a reading (in the present case, the difference in heights of the left and right columns of the manometer of Exercise 1) can be such that it is between two markings on the scale. While the right measurement practice is to take the nearest marking as the reading, we may quite often '*judge*' the reading to be a fraction. For example, we may note down the reading in Fig. 5 as 100.5 mm (some may even argue that the reading is 100.6 mm!) and we may be justified in doing so. But the fact that we have guessed the fraction and the actual reading has only three significant digits must be brought out when we present the reading as:

$$(P_1 - P_2) = \rho_u g\, h_u = 13,534 \times 9.81 \times 100.5/1000 = 13343.23827\ \text{Pa} = 13.\underline{3}4\,\text{kPa}$$

Here we place an underscore to inform the world that the result has only three significant digits and the additional digit has come about by judicious approximation. Beware! Rounding off must be carried out only after all computations were accomplished and should never be done at the initial and or intermediate steps!

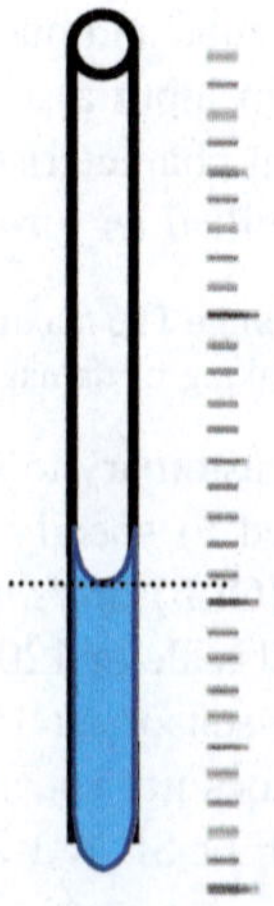

Fig. 5 Reading—fraction of the resolution

3.2 *Range (Full Scale) of a Sensor*

From Fig. 1, we can also deduce that the maximum h_w, *say* h_{max}, we can read is 50 mm (after which the fluid levels are out of the range of the scale). Hence using this particular manometer we can measure a maximum pressure difference of

$$(P_1 - P_2)_{\max} = \rho_u g\, h_{\max}. \tag{2}$$

The maximum value that can be read using a sensor, another important characteristic of a sensor, termed as the '*operating range*' or simply '*range*'; also called the '*Full Scale*' or '*Full scale value*' (FS or FSV) of the sensor is defined as.

Operating range (Full Scale) The maximum value that can be read using a sensor.

Note: Some times we may come across a sensor that can sense both positive and negative values. For example the manometer of Fig. 1 can sense both positive and negative pressure differences. We take one of the ports, say P_1, as the reference and then refer to the condition when P_2 is $>P_1$ as positive pressure and when P_2 is $<P_1$ then we term the pressure difference as negative. It is perfectly fine if we take P_2 as the reference and thus say the pressure difference is positive when P_1 is $>P_2$. In either case, the range is expressed as $\pm P_{max}$, where P_{max} is the maximum pressure that can be measured in either polarity.

3.3 *Safe Range (Overload Factor) of a Sensor*

If we keep increasing the pressure P_1 applied to the manometer of Fig. 1, then the pressure difference $(P_1 - P_2)$ will rise over and above the full scale and the fluid will rise (Since $P_1 > P_2$, the fluid will rise on the right hand side tube) and flow out of the U-tube and may make the manometer unusable for future measurements. The maximum input a sensor can accept and not get destroyed as a sensor, is another important characteristic of a sensor, termed as the '*safe range*'. The safe range is also specified as '*proof range*'. Thus safe range of a sensor is defined as:

Safe range The maximum value of the input that can be applied to a sensor without the fear of breaking or damaging it.

It is customary to specify the safe range as a factor of full scale. In that case, the term used to specify the safe range of operation of a sensor is either "*Overload factor*" (OLF) or "*safety factor*" (SF). For example if a particular pressure sensor has a full scale of 120 kPa and its safe range is 180 kPa, then the SF or OLF of that pressure sensor is: 180/120 = 1.5 (times the FSV). If the details on a particular sensor does not specifically mention the safe range, then it is universally assumed that OLF or SF = 1.2 (times the full scale).

3.4 Accuracy of a Sensor

The *accuracy* of a sensor relates to its ability to provide at its output, the correct or true value of the quantity being sensed. As we have already learnt, the output of a sensor may depend on many parameters, some intrinsic to the sensor and some on the surrounding environment. If either the input or output parameter changes then the output will change, even though the input itself has not changed. If the output is different from the true (expected) output, then we call the difference as an '*error*' in the output. Smaller the error, the sensor is deemed to be more *accurate*. If an error occurs due to the sensor, the we term it as "*instrument error*" An error can occur at the output due to the measuring system itself and such an error is termed as '*systematic error*'.

Systematic Error

The '*Instrument error*' (an indicator of a sensor's *accuracy*) that may be present in a sensor's output (particular reading), another important characteristic of a sensor, is defined as:

$$\textit{Error: Measured Value} - \textit{True Value}.$$

To compare different measurements and or different sensors or different measuring instruments, the error is always expressed as a percentage as given in Eq. (3).

$$\%\,\textbf{Error} = \frac{\textit{Measured Value} - \textit{True Value}}{\textit{True Value}}\,100\%. \tag{3}$$

Instrument error can occur at the output of a sensor due to one or more of the reasons listed below.

(a) When a sensor is manufactured, after fabrication, the sensor is *calibrated*. Calibration of a sensor is achieved by giving known inputs to the sensor, operate the sensor under standard testing conditions and then adjust (vary) the sensor's parameters (that affect the output) so that the sensor output reads correctly. During calibration, the external conditions are set as per standards. For example, when pressure sensors are calibrated, the operating temperature during calibration is set as per IEC60770 (IEC stands for International Electro Technical Commission, an independent global organization that prescribes materials, methods and procedures for calibration of measuring apparatus such as sensors). Quite often it may happen that even after adjusting, the output of the sensor may be a little less or more than the expected output. The difference between the actual output from the sensor and the expected output is then marked as '*error*' (more recently, use of the term *uncertainty* instead of error is advocated by organizations such as IEC, SI and NIST. However, the continued use of the terms "*error*" and "*accuracy*" is not forbidden, yet). Even if the error is zero when the sensor is being calibrated (not easily achievable, but we can assume such a possibility), the operating conditions may be different when the sensor is used for measurement or during the course of time the parameters of the sensor may vary due to ageing and or wear and tear, leading to an error in the sensor's output. Deviations in the parameters of the sensor that influence the output of that sensor can also introduce an error at the sensor's output. For example, in the U-tube manometer, the output is dependent on the density of the liquid, which is intrinsic to the sensor. The output is also dependent on the external factor, namely, *g*, the acceleration due to gravity at the measurement site. These factors can change from the expected or assumed values. In general, the output of a sensor may not always indicate the correct or true value of the parameter being sensed. In such a case, the sensor is said to read erroneously. Such an error is termed as '*instrument errors*'. Examples are:

(i) As per the standard, a pressure sensor is calibrated with the standard operating temperature (condition) of 25 °C. The sensor may be used where the room temperature is 35 °C, resulting in an error.

(ii) The U-tube manometer described above is normally calibrated at sea level, where the acceleration due to gravity is 9.81 m/s^2. If the manometer is used atop the mount Everest, then *g* at that measurement site would be different leading to an error. (*g* atop mount Everest is 0.972 times the *g* at sea level.)

(iii) The elasticity of the spring in a mass spring accelerometer can change due to ageing and or usage. Hence the characteristic of a mass-spring type accelerometer put in use over a long period of time will slowly change, leading to errors.

(b) **Error due to loading**: An error can also occur at the output of a sensor due to the very act of sensing or measuring. For example, when we measure the temperature using a thermometer, the specimen whose temperature we are trying to measure may be altered due to the insertion of the thermometer. This is due to the fact that

Fig. 6 Measuring the temperature of an ant!

sensing element of the thermometer has to attain the temperature of the test specimen and thus 'sense' the actual temperature. In doing so, the sensor extracts thermal energy from the specimen, whose temperature is being measured and thus may alter the temperature of the specimen itself. Figure 6 caricatures an extreme situation illustrating this point. It is easily seen from Fig. 6 that the thermal capacity of the ant is inadequate to heat the mercury blob inside the thermometer and hence may not even register a reading. Worst by placing the thermometer on the ant, we may even kill the ant!! The error created by such a situation is called '*the error due to loading*'. The scenario depicted in Fig. 6 is the 'loading' of the measurand (the input) by a sensor. "*Loading*" can also occur at the output terminals of a sensor, as illustrated next.

Figure 7 portrays a '*light intensity sensor*' that converts the intensity of the input light into an output voltage. Such a typical light sensor possesses appreciable series resistance R_{se}. If we connect a voltmeter possessing an internal resistance of R_{vm} as indicated in Fig. 7 to measure the voltage V_{li} generated by the input light intensity, we will only measure $V_{vm} = \frac{R_{vm}}{R_{vm} + R_{se}} V_{li}$ and thus the measured value will not indicate the actual intensity of light being

Fig. 7 Light intensity sensor

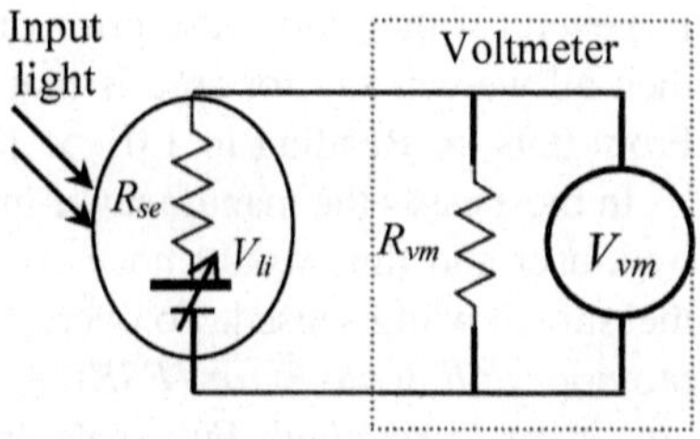

sensed by the sensor. The error introduced in this case is due to '*loading*' effect on the output of the sensor.

The error introduced in this measurement computed utilizing Eq. (3) is:

$$Error = \frac{\frac{R_{vm}}{R_{vm}+R_{se}}V_{li} - V_{li}}{V_{li}} 100\% = \frac{-R_{se}}{R_{vm}+R_{se}} 100\% \quad (4)$$

The error in a sensor's output either due to '*instrument error*' or due to '*loading*' (either at the input or at the output or both) is termed as '*Systematic Error*' (meaning: Error due to the measurement system). Once we know the parameters of the measurement system, we can apply corrections to the measured value and thus can '*compensate the systematic error*' in that measurement. For example, in the measurement of light intensity as given in Fig. 7, if we know the values of R_{se} and R_{vm} then we can multiply the measured value V_{vm} by $\frac{R_{vm}+R_{se}}{R_{vm}}$ and obtain:

$$V_{corrected} = V_{vm} * \frac{R_{vm}+R_{se}}{R_{vm}} = \frac{R_{vm}}{R_{vm}+R_{se}} \frac{R_{vm}+R_{se}}{R_{vm}} V_{li} = V_{li} \quad (5)$$

Thus by applying correction to the reading of the voltmeter, we can get the correct value of the intensity, namely, V_{li}. Thus in general '*systematic error (instrument error or loading error or both) is calculable and thus can be compensated*'.

Since manufacturers of sensors do not have a priori knowledge on the exact values of the intrinsic and extrinsic factors at the point where the sensor is to be operated, they assume extreme values of possible deviations from expected values that may occur in *normal usage* and provide the "*worst case error*" as a specification for a sensor.

For example, the density ρ of the liquid of a U-tube manometer is expected to vary between 990 and 1010 kg/m^3 in usage (may be due to evaporation, contamination etc.). The manufacturer will then calibrate (mark the readings on the scale) assuming the nominal density to be 1000 kg/m^3 and will express the worst possible error of the manometer to be between +1.0% (when ρ_U = 990 kg/m^3) and −1.0% (ρ = 1010 kg/m^3). It may so happen that when a particular reading is taken with this manometer, ρ_U = 1000 kg/m^3 and hence that particular reading has no error.

In general if the worst case error of a sensor is indicated as ±1.0% of its reading, then all we can say for sure is that the true value may lie anywhere in the range: From 0.99 × Reading to 1.01 × Reading.

In most cases the manufacturer may not also know the actual reading being taken by a user and thus would not have a priori knowledge on the "true value" of the measurand being sensed. To overcome this problem, most manufacturers of sensors provide '*Full Scale Error (FSE)*' as a typical specification of errors of a sensor and not '*error in reading*'. Full scale error is defined as:

$$\%\mathit{Full\ Scale\ Error} = \frac{(\text{Measured Value} - \text{True Value})_{\text{max}}}{\text{Full Scale Value}} \times 100\%. \qquad (6)$$

$(\text{Measured Value} - \text{True Value})_{\text{max}}$ in Eq. (6) is the maximum error that can occur at a particular reading and the reading can be anywhere within the range of the sensor. The worst case error in a particular reading from a sensor, whose full scale error is known, can be computed using the identity:

$$\%\,\mathit{Error(in\ reading)} = (\%\,\text{Full Scale Error}) \times \frac{\text{Full Scale Value}}{\text{Measured Value(reading)}}. \qquad (7)$$

It must be noted here that the worst case error of a reading when full scale error alone is specified will increase as the reading obtained moves down the scale towards the minimum value on that range, as illustrated by the Exercise 6 given next.

Exercise 6 A 100 °C full scale thermometer is specified as having a full scale error of ±1.0%. Then the worst case error of this thermometer is:

±1.0% = ±1 °C when the reading obtained is 100 °C,
At 50 °C the % error in the reading can be calculated using Eq. (5) as
$\pm 1\% \times \frac{100}{50} = \pm 2\%$. (*note*: the worst case error is ±2% of 50 °C = ±1 °C)
Similarly the worst case error in the reading will be
±5.0% when the reading obtained is 20 °C (±1 °C in 20 °C),
±10% when the reading obtained is 10 °C (±1 °C in 10 °C) and
±100% when the reading obtained is 1 °C! (±1 °C in 1 °C).

From the foregone calculations in Exercise 6, we learn that whenever we use a sensor whose full scale error is specified, it is prudent to obtain a reading close to the full scale value of that sensor in order to achieve minimum errors in the measurement. Alternatively, one should choose the range of a sensor for a particular measurement such that the reading obtained is as close to the full scale as possible.

Propagation of Systematic Error

Quite often we may use measured values (with systematic errors in them) from sensors and compute another quantity. The error in the computed quantity depends on the error in the measurement and the computational process. Let A and B are measured quantities having worst case errors of $a\%$ and $b\%$ respectively and C is computed as $C = A + B$, then

Exercise 7 The measured values from two different sensors are: $A = 100$ mV with a worst case systematic error of $\pm 1\%$ and $B = 95$ mV with a worst case error of $\pm 1.5\%$. determine the error $c\%$ in C if $C = A + B$.

Using Eq. (6) we can calculate c as:

$$c\% = \left(\frac{A}{C}a + \frac{B}{C}b\right)\% = \left(\frac{100}{195}1 + \frac{95}{195}1.5\right) = 1.24\%$$

we can compute the worst case error in C as given next. The value of C will lie in the range:

$$\left[\left(A - \frac{aA}{100}\right) + \left(B - \frac{bB}{100}\right)\right] : \left[\left(A + \frac{aA}{100}\right) + \left(B + \frac{bB}{100}\right)\right].$$

Thus the worst case error ΔC in C is: $\Delta C = \left(\frac{aA}{100}\right) + \left(\frac{bB}{100}\right)$

Hence the worst case error $c\%$ in C is:

$$c\,\% = \frac{\Delta C}{C}100 = \frac{100}{C}\left[\left(\frac{aA}{100}\right) + \left(\frac{bB}{100}\right)\right] = \left(\frac{A}{C}a + \frac{B}{C}b\right)\% \tag{8}$$

Equation (8) indicates that the error in A (or B) influences the error in C to the extant A (or B) influences C. Exercise 7 illustrates the propagation of error.

If $C = A - B$, then the error in C will be:

The extreme values of C will lie in the range:

$$\left[\left(A - \frac{aA}{100}\right) - \left(B + \frac{bB}{100}\right)\right] : \left[\left(A + \frac{aA}{100}\right) - \left(B - \frac{bB}{100}\right)\right].$$

Thus the worst case error ΔC in C is: $\Delta C = \left(\frac{aA}{100}\right) + \left(\frac{aA}{100}\right)$

Hence the worst case error $c\%$ in C is:

$$c\,\% = \frac{\Delta C}{C}100 = \frac{100}{C}\left[\left(\frac{aA}{100}\right) + \left(\frac{bB}{100}\right)\right] = \left(\frac{A}{C}a + \frac{B}{C}b\right)\% \tag{9}$$

Though Eqs. (8) and (9) look identical, the resultant error in the latter case, that is $C = (A - B)$, can be very much higher than a or b as illustrated by Exercise 8.

From Exercise 8, we see that if we compute the difference of two measured quantities that are close to each other, then the error in the result can become very high. Thus if a situation arises that we need to determine the difference between two physical quantities then we must first get the difference and measure that difference rather than compute the difference from the measured values!

It would be nice if we can derive a formula to ascertain the error in a computed result using the errors in measured values. To do that, let us take that we compute C as a function of measured quantities A, B, X, Y... as expressed in Eq. (10)

$$C = f(A, B, X, Y\ldots) \tag{10}$$

Then an incremental deviation (error) ΔC in C can be expressed as:

$$\Delta C = \left(\frac{\partial f}{\partial A}\Delta A\right) + \left(\frac{\partial f}{\partial B}\Delta B\right) + \left(\frac{\partial f}{\partial X}\Delta X\right) + \left(\frac{\partial f}{\partial Y}\Delta Y\right) + \cdots \tag{11}$$

where $\Delta A, \Delta B, \Delta X, \Delta Y\ldots$ are incremental values (errors) in A, B, X, Y... and $\frac{\partial f}{\partial A}$ is the partial derivative of the function f (that determines C) with respect to the measured variable A, $\frac{\partial f}{\partial B}$ is the partial derivative of the function f with respect to the measured variable B and so on. The error $c\%$ in C can be derived as:

$$c\% = \frac{\Delta C}{C}100 = \frac{A}{C}\left(\frac{\partial f}{\partial A}\frac{\Delta A}{A}100\right) + \frac{B}{C}\left(\frac{\partial f}{\partial B}\frac{\Delta B}{B}100\right) + \frac{X}{C}\left(\frac{\partial f}{\partial X}\frac{\Delta X}{X}100\right) + \frac{Y}{C}\left(\frac{\partial f}{\partial Y}\frac{\Delta Y}{Y}100\right)\cdots$$

Resulting in:

$$c\% = \frac{A}{C}\left(\frac{\partial f}{\partial A}\right)a + \frac{B}{C}\left(\frac{\partial f}{\partial B}\right)b + \frac{X}{C}\left(\frac{\partial f}{\partial X}\right)x + \frac{Y}{C}\left(\frac{\partial f}{\partial Y}\right)y + \cdots\% \tag{12}$$

Exercise 8 The measured values from two different sensors are: $A = 100$ mV with a worst case systematic error of $\pm 1\%$ and $B = 95$ mV with a worst case error of $\pm 1.5\%$. determine the error $c\%$ in C if $C = A - B$.

Using Eq. (9) we can calculate c as:

$$c\% = \left(\frac{A}{C}a + \frac{B}{C}b\right)\% = \left(\frac{100}{5}1 + \frac{95}{5}1.5\right) = 48.5\%!$$

Exercise 10 Derive the error in C in Exercise 9 without using the formula as given in Eq. (12).

$$C=\left(A\pm\frac{aA}{100}\right)\times\left(B\pm\frac{bB}{100}\right)=AB\pm A\frac{bB}{100}\pm B\frac{aA}{100}\pm\left(\frac{aA}{100}\frac{bB}{100}\right)$$

$$\therefore\quad \Delta C=\pm A\frac{bB}{100}\pm B\frac{aA}{100}\pm\left(\frac{aA}{100}\frac{bB}{100}\right) \tag{13}$$

$$c\%=\frac{\Delta C}{C}100=\left(\pm a\pm b\pm\left(\frac{ab}{100}\right)\right)\%$$

Substituting the values of a and b from Exercise 9 in Eq. (10) gives us:

$$c\%==\pm 1\pm 1.5\pm\left(\frac{1\times 1.5}{100}\right)=2.515\%$$

Let us look at an example given in Exercise 9.

Note: Eq. (12) is derived assuming that the errors are small, hence valid only if the errors are small. Let us look at the example 9 differently.

Let us redo the calculation of error in Exercise 9 utilizing the long-hand method as illustrated in Exercise 10. From Exercise 10, we see that the error calculated using the long hand method as given in Eq. (13) is slightly different from the one portrayed in the formula of Eq. (12). While the formula states that if two measured quantities A and B having errors $\pm a\%$ and $\pm b\%$ are multiplied to get C (=AB) then the error in C, namely c% is the addition of $a\%$ and $b\%$ (2.5%). However, the *exact* worst case error as given by Exercise 10 is [$\pm a \pm b \pm$ ab/100]% (2.515%). Since the errors are

Exercise 9 The measured values from two different sensors are: $A = 100$ with a worst case systematic error of $\pm 1\%$ and $B = 95$ with a worst case error of $\pm 1.5\%$. determine the error $c\%$ in C if $C = A\ B$.

Using Eq. (12) we can calculate c as:

$$c\%=\frac{A}{C}\left(\frac{\partial f}{\partial A}\right)a+\frac{B}{C}\left(\frac{\partial f}{\partial B}\right)b=\frac{AB}{C}a+\frac{BA}{C}b=(a+b)=\pm 1\pm 1.5=\pm 2.5\%$$

Thus if we multiply two measured quantities, the percentage errors in the measured values simply add up.

small the difference between the calculation of error using the formula of Eq. (12) and the exact value is quite small. Readers should also note that the 'error' used in these calculations are 'error in the reading'. Since most manufacturers of sensors provide full-scale error, the full scale error must be converted to 'error in reading' using Eq. (7) before attempting to use Eq. (12) or Eq. (13).

Random Error

An error in a sensor's output can also arise due to external/internal influence (especially by the environment in which the sensor is operating) on the input or output of a sensor. The external influence can also be an interference (electrostatic or electromagnetic) from another source. Many times we witness these interferences. For example when a cellular phone (mobile) is activated, it introduces unwanted sounds in a public address system near by. If you are listening to radio (especially AM radio) and some one switches on or off an electrical equipment near by, once again you hear the interference of the switching transients. Apart from the interference from other equipment, the output of a sensor may also be affected by 'noise'. Noise can be generated internally to the sensor or can come from the environment in which a sensor is operating. For example, if a sensor employs a resistor (either as the sensing element or as a part of the signal conditioning electronics) thermal noise will be introduced in the output of the sensor. Each resistor (whether the resistor is a sensor element or not) is a noise generator and the noise voltage depends on the value of the resistor, its operating temperature and the bandwidth of the sensor system. The thermal noise inherent to a resistor, in general, is random in nature as the noise is generated due to the random motion of free charge carriers (electrons if the material of the resistor is of metal and can be electrons are holes if the resistor is made of *n*-type or *p*-type semiconductor) inside the resistor. The rms value of the thermal noise voltage e_n generated in a resistor of value R ohms at a temperature of T kelvin is:

$$e_n = \sqrt{4kTR\Delta f}\,\mathrm{V} \tag{14}$$

The Δf in Eq. (10) is the bandwidth of interest. Error introduced by either interference or noise is termed as '*random error*'. When a random error is present in the output of a sensor, then even for the same input to the sensor we may obtain different readings at different times at its output. Let us look at an example.

A resistive sensor element whose value R_x changes with the measurand x as $R_x = R_0(1 \pm K_T x)$. To obtain a measurable output voltage, this sensor resistance is excited with a dc current of I_{dc} A. Then the output voltage V_o, we expect is $V_o = I_{dc}R_0(1 \pm K_T x)\mathrm{V}$. However, the practical output voltage will be $V_o = I_{dc}R_0(1 \pm K_T x) + i_r R_0(1 \pm K_T x) + \tilde{e}_{\mathrm{n}}\mathrm{V}$, where i_r is the "*ripple*" current (interference, ac in nature) in I_{dc} and $\tilde{e}_{\mathrm{n}}$ is the '*thermal noise*' voltage that is inherent in any resistor. Thus the terms $i_r R_0(1 \pm K_T x)$ and $\tilde{e}_{\mathrm{n}}$ in the output V_o will give rise to an error. It should be noted that this error is time dependent and is unpredictable. When random errors are present in the output of a sensor, the only option is to take multiple readings of the output within a measurement window and compute the average of the multiple readings acquired from the sensor.

Let us look at an example of a resistive sensor of 400 kΩ value having at its terminals a 100 μV translated from the parameter it is sensing, as indicated in Fig. 8. The terminal voltage V_T will be the addition of the voltage being sensed (100 μV) and the noise voltage generated by the 400 kΩ resistor as indicated in Fig. 9. It is evident from Fig. 9, that the value measured will depend on the

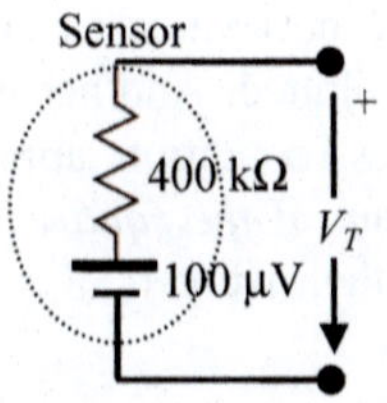

Fig. 8 Error due to noise

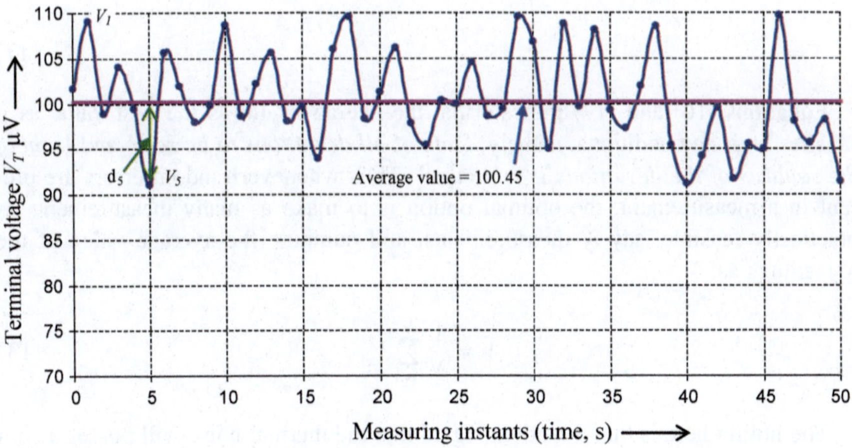

Fig. 9 Terminal voltage V_T as a function of time

particular time, the measurement is taken. Thus due to the thermal noise of the 400 kΩ resistor, the measured value will not reflect the true value, and will be different when measured at different points of time. The error arising out of such a phenomenon is called the "*Random error*". For example if we take the measurement at time = 5 s, then the reading obtained will be 91 μV resulting in a deviation of −9 μV from the expected value. The deviations at each measured value can be determined as:

$$d_j = V_j - V_E \tag{15}$$

From Fig. 9, we also see that the deviation can be either positive or negative. One way of determining the expected value from the measured values is to ensure that the sum of all deviations is zero.

$$\sum_{j=1}^{N} d_j = \sum_{j=1}^{N} Vj - V_E = 0 \quad \Rightarrow V_E = \frac{1}{N}\sum_{j=1}^{N} V_j = \bar{V} \tag{16}$$

Thus the average value is the best value as the sum of deviations is zero. In Fig. 9, the average value is also plotted. Another way of determining the expected value is through the "least squares" condition, applying which, we need to find the expected value such that the '*sum of the squares of the deviations is minimum*'. Now if we apply this condition then we get:

$$\frac{d}{dV_E}\sum_{j=1}^{N} d_j^2 = 0 \Rightarrow \frac{d}{dV_E}\sum_{j=1}^{N} (Vj - V_E)^2 = 0 \quad \Rightarrow \frac{d}{dV_E}\sum_{j=1}^{N}\left(V_j^2 + V_E^2 - 2V_jV_E\right) = 0$$

$$\sum_{j=1}^{N}\left(2V_j - 2V_E\right) = 0 \quad \Rightarrow \quad V_E = \frac{1}{N}\sum_{j=1}^{N} V_j = \bar{V} \tag{17}$$

Equations (16) and (17) indicate that the average value is the best value as it satisfies both he conditions, namely, '*sum of all deviations to be zero*' and '*sum of the squares of the deviations is minimum*'. Thus whenever random errors are present in a measurement, the optimal option is to make as many measurements as practically feasible, say N measurements, and compute the average value of the N readings as:

$$\bar{V} = \frac{1}{N}\sum_{j=1}^{N} V_j. \tag{18}$$

The notion here is that a random signal like the thermal noise will posses a zero average value and hence will be eliminated when we take the average of N readings. The average value of the 50 readings shown in Fig. 9 is 100.45 μV. Thus we see that the average value is not the true value (which is 100 μV) but close to the true value, hence we call the average value as the '*best value*'. We have seen that systematic errors determine how accurate is a measurement and the reading obtained is expressed as it is and the accuracy of that measurement is indicated by expressing a '*percentage error*'. When random errors are present in a measurement, multiple readings are taken, the average value is expressed as the best value and the differences between individual readings and the average value obtained and expressed as deviations. The difference between a measurement and the average value is called the deviation and not called as 'error' so as to differentiate between 'systematic' and 'random' errors. It would be unwieldy and impractical to retain all the measured values and or computed deviations. A concise way of indicating the deviations that are inherent in a measurement affected by random errors is to just indicate the average value and a representative value that provides information on the deviations.

$$\bar{V}\pm_{d_-}^{d_+}. \tag{19}$$

The representative values d_+ (positive deviation, when V_j is more than $\bar{V}$) and d_- (negative deviation, $V_j < \bar{V}$) can be chosen in three possible ways as given below.

(i) **The peak deviation**: Here we ascertain the maximum values of deviation in either polarity and indicate those maximum values. In all manufactured items, the dimensions are indicated in this way. For example if cylindrical shafts of diameter 10 mm are manufactured, the deviations in a set of 100 shafts (with the maximum diameter is 100.1 mm and the minimum diameter is 99.2 mm then the result may be expressed as:

$$100\,\text{mm}^{+100\,\mu\text{m}}_{-80\,\mu\text{m}} \tag{20}$$

The above can also be written taking the worst case scenario as:

$$100\,\text{mm} \pm 100\,\mu\text{m} \tag{21}$$

In some situations (where fitting operation with another fabricated part is involved), it may turn out that deviation in one direction can not be 'tolerated'. For example if a washer is to be fitted with a shaft, the diameter of the shaft can never be more than the inner diameter of the washer. In such a case, the shaft specification has to be:

$$100\,\text{mm}^{+0\,\mu\text{m}}_{-100\,\mu\text{m}} \tag{22}$$

and the washer (inner hole) diameter has to be specified as:

$$100\,\text{mm}^{+100\,\mu\text{m}}_{-0\,\mu\text{m}} \tag{23}$$

Since we can 'tolerate' these deviations in the shaft outer diameter and the washer's inner diameter, we also call them as "Tolerances". From the foregone discussions, it is evident that the tolerance (when expressed in terms of peak deviations) can be (i) Uni-polar tolerance and (ii) Bi-polar tolerance. In the uni-polar tolerance, the deviation can only be in one polarity (either positive or negative) and should be zero in the other polarity (negative or positive). On the other hand, when bi-polar tolerance is specified, the deviations can be either positive or negative.

(ii) **The average deviation**: When average deviation is specified then we indicate the measurement as:

$$\bar{V} \pm d_A, \tag{24}$$

where

$$d_A = \frac{1}{N} \sum_{j=1}^{N} |d_j|. \tag{25}$$

The average deviation is not very popular and is seldom used.

(iii) **Standard deviation**: If we take a look at Fig. 9, we see that the average value is like a dc voltage and the deviations are like an ac voltage riding on the dc (in fact it is so in the example that resulted in that figure, but in many cases it may not be so), then we can represent the deviation as a dc + ac. Since ac (voltages or currents) is always represented as a root mean square (rms) value, the result in a measurement can be written as the average value + rms value σ_{rms} of the deviation, where

$$\sigma_{rms} = \sqrt{\frac{1}{N} \sum_{j=1}^{N} d_j^2}. \tag{26}$$

The rms deviation though looks logical, when we look at the deviations as 'variable' in a statistical sense then we only have $(N - 1)$ independent deviations $(d_1, d_2, d_3, \ldots d_{N-1})$ with N readings as the Nth deviation d_N will depend on all the other $(N - 1)$ deviations. Then what is mathematically correct and called as unbiased or standard deviation is:

$$\sigma = \sqrt{\frac{1}{(N-1)} \sum_{j=1}^{N} d_j^2}. \tag{27}$$

When rms or standard deviation is specified the we indicate the measurement as:

$$\bar{V}^{\pm \sigma_{rms}} \text{ or } \bar{V}^{\pm \sigma}. \tag{28}$$

The standard (or rms deviation) is called the '*Precision index*' of the set of measurements. Though indicating the result in such a concise way is nice, but such an act throws away all the measurements (in many instances painstakingly collected). If we can get more insights from all those N measurements made, it not only justifies our labor but also provides additional information on the measurements made. Towards achieving this goal, we can now represent the measurements made in another form as given in Fig. 10. Here the number of readings within a range, say (100 ± 1.0) μV, is plotted. Since this picture presents the 'history' of all the measurements made, the graph of Fig. 10 is called the '*Histogram*'. Figure 10 gives actual number of readings obtained in N measurements and thus would depend on the total number of readings obtained. If we want to compare two different sets of

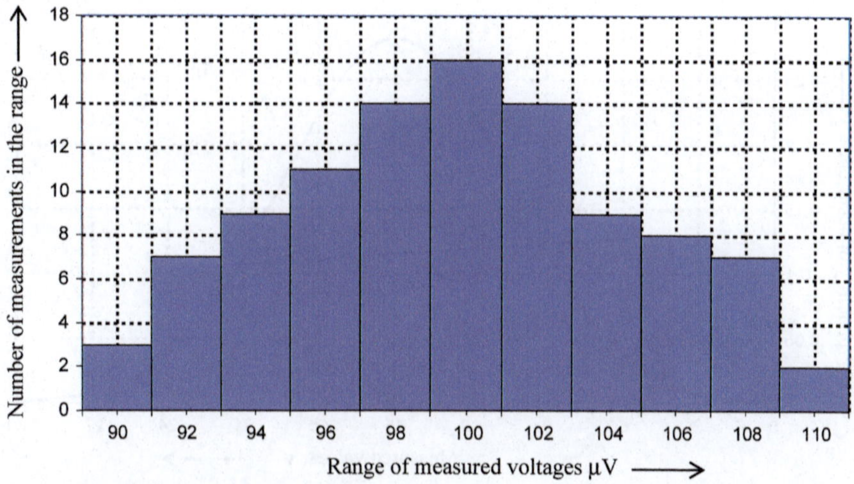

Fig. 10 Typical "Histogram" of measured data with random errors

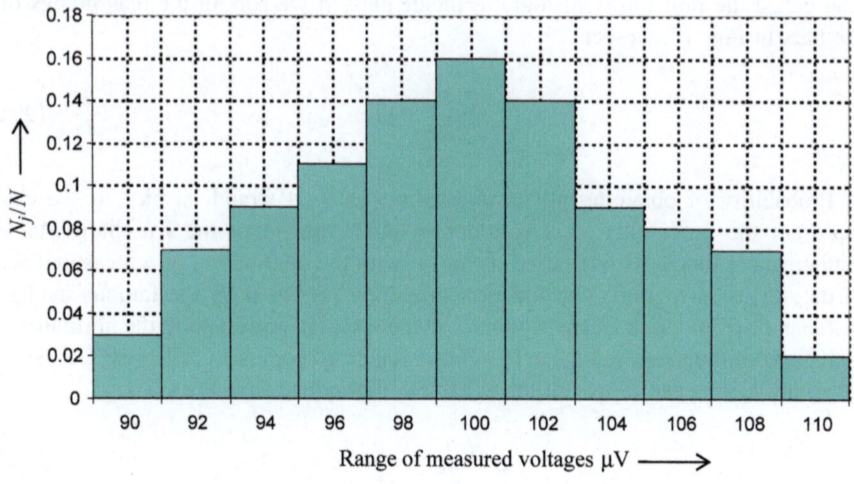

Fig. 11 Normalized "Histogram" of measured data (data with random errors)

measurement (taken to address random errors) then the two sets must have same number of readings *N*. We can not compare two histograms generated from two sets of measurements having different numbers of measurement, say *M* and *N*.

This problem can be easily overcome if we normalize the histogram with respect to the total number of measurements made as illustrated in Fig. 11. Here N_j is the number of measurements obtained in a particular range and *N* is the total number of measurements. In Fig. 11, we see that the y-axis can also be looked at, as the probability of obtaining a reading in that range. Hence a normalized histogram is

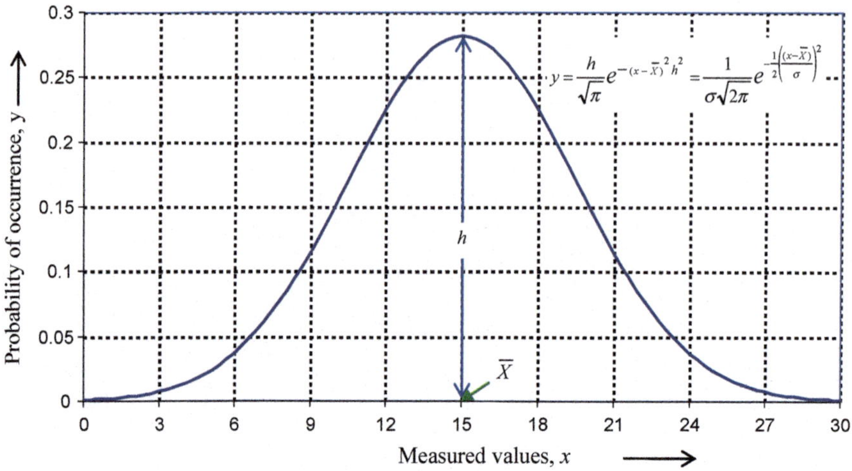

Fig. 12 Gaussian function

also called the probability histogram. Incidentally if we add all the magnitudes of the bars in Fig. 12, we get:

$$\sum_{j=1}^{N}\frac{N_j}{N}=1. \tag{29}$$

Probability of obtaining all the N readings is 1. It would be nice, if we can represent the probability as a continuous curve instead of the bar-chart. Many mathematical functions were tried to 'fit in' with the distribution of a measured set of data or its histogram. The function originally proposed by the famous mathematician Carl Friedrich Gauss is found to be optimal in representing the normalized histogram of the measured data. The Gauss function (popularly called the Gaussian distribution) that can represent the distribution graph of Fig. 11 is:

$$y=\frac{h}{\sqrt{\pi}}e^{-(x-\bar{X})^2h^2} \tag{30}$$

It should be noted here that Gauss did not give the function as given in Eq. (30) and the original Gaussian function $\left(y=he^{-x^2h^2}\right)$ was modified as given in Eq. (30) so that it can be applied to represent the distribution within a measured data with random errors. Figure 12 shows a typical Gaussian function and it is easy to deduce that the average value, since the function peaks at the average value. Now if we compute the standard deviation with the Gaussian function representing the distribution of data, we get:

$$\sigma = \frac{1}{h\sqrt{2}} \text{ or } h = \frac{1}{\sigma\sqrt{2}} \tag{31}$$

Substituting the relationship between h and σ from Eq. (31) in Eq. (30), we get:

$$y = \frac{1}{\sigma\sqrt{2\pi}} e^{-\frac{1}{2}\left(\frac{x-\bar{X}}{\sigma}\right)^2} \tag{32}$$

Thus once we calculate the average value and the standard deviation from the measured data set, we can represent that dada set using Eq. (32). As we reasoned out for specification of accuracy in terms of percentage error, the standard deviation is also expressed as a percentage for the same reasons, as:

$$\sigma\% = \frac{\sigma}{\bar{X}} 100 \tag{33}$$

For example the tolerance of a resistor is always expressed in $\sigma\%$. Typical specification of a resistor will be (100 Ω, ½ W, MFR, 1% tolerance). One should not confuse the standard deviation (expressed as a percentage) with error. While error indicates how close the measurement is to the true value, standard deviation simply indicates how repeatable the measurements were and does not in any way indicate how close the measured value(s) is (are) to the true vale. Since the Gauss function is used wherever natural distribution (popularly called the Gaussian distribution) occurs, it is also called as '*The Standard* (or Normal) *distribution*' (function). 68.2% of all measured data will be covered in the range $\bar{V} \pm \sigma$, 95.4% within the range $\bar{V} \pm 2\sigma$ and 99.6% of all measured data will be covered within the range $\bar{V} \pm 3\sigma$.

Propagation of Random Error

We may use two or more measured values that are afflicted by random errors obtained from sensors and compute another quantity. The average value and standard deviation of the computed quantity depends on the average value and standard deviation of the measured quantities as well as the computational process. Let $A_1, A_2, A_3, \ldots A_N$ with an average value of $\bar{A}$ and standard deviation σ_A. Similarly let $B_1, B_2, B_3, \ldots B_M$ be another set of measured quantity with average value of $\bar{B}$ and standard deviation σ_B. If we compute $C = A + B$, then the average value of C and standard deviation in C can be obtained as given here. We can obtain M N different values of C as:

$$C_{i,j}\big|_{i=1\ to\ N, j=1\ to\ M} = A_{i,}\big|_{i=1\ to\ N} + B_j\big|_{j=1\ to\ M} \tag{34}$$

The average value $\bar{C}$ of C is

$$\bar{C} = \frac{1}{MN} \sum_{\substack{i=1.\\ j=1}}^{\substack{M\\ N}} C_{i,j} = \frac{1}{MN} \sum_{\substack{i=1.\\ j=1}}^{\substack{M\\ N}} (A_i + B_j) \quad \Rightarrow \quad \frac{M}{MN} \sum_{i=1.}^{N} (A_i) + \frac{N}{MN} \sum_{j=1}^{M} (B_j) = \bar{A} + \bar{B} \tag{35}$$

From Eq. (35) we see that "Average of a sum set is equal to the sum of the averages of the individual sets". Standard deviation σ_B is

$$\begin{aligned} \sigma_C &= \sqrt{\frac{1}{MN} \sum_{\substack{i=1.\\ j=1}}^{\substack{M\\ N}} (C_{i,j} - \bar{C})^2} = \sqrt{\frac{1}{MN} \sum_{\substack{i=1.\\ j=1}}^{\substack{M\\ N}} (A_i + B_j - \bar{A} - \bar{B})^2} \\ &= \sqrt{\frac{M}{MN} \sum_{i=1.}^{N} (A_i - \bar{A})^2 + \frac{N}{MN} \sum_{i=1.}^{N} (B_i - \bar{B})^2 - 2\frac{1}{MN} \sum_{i=1.}^{N} (A_i - \bar{A})(B_i - \bar{B})} \\ &= \sqrt{\sigma_A^2 + \sigma_B^2} \end{aligned} \tag{36}$$

If we square Eq. (36) we get:

$$\sigma_C^2 = \sigma_A^2 + \sigma_B^2 \tag{37}$$

The term σ_C^2 is called the variance of C, thus we have "The variance of a sum set is sum of the variances of the individual sets".

Once again it would be worthwhile if we can derive a formula for obtaining the standard deviation or variance for a general case. To do that, let us take that we compute C as a function of measured quantities A, B, X, Y… with variances $\sigma_A^2, \sigma_B^2, \sigma_X^2, \sigma_Y^2, \ldots$ as expressed in Eq. (38)

$$C = f(A, B, X, Y \ldots) \tag{38}$$

Then an incremental deviation ΔC in C can be expressed as:

$$\Delta C = \left(\frac{\partial f}{\partial A} \Delta A\right) + \left(\frac{\partial f}{\partial B} \Delta B\right) + \left(\frac{\partial f}{\partial X} \Delta X\right) + \left(\frac{\partial f}{\partial Y} \Delta Y\right) + \cdots \tag{39}$$

where $\Delta A, \Delta B, \Delta X, \Delta Y \ldots$ are incremental deviations in A, B, X, Y… and $\frac{\partial f}{\partial A}$ is the partial derivative of the function f (that determines C) with respect to the measured variable A, $\frac{\partial f}{\partial B}$ is the partial derivative of the function f with respect to the measured variable B and so on. The variance σ_C^2 in C can be derived as:

$$\frac{1}{N_C}\sum_{N_C}\Delta C^2 = \left(\left(\frac{\partial f}{\partial A}\right)^2 \frac{1}{N_A}\sum_{N_A}\Delta A^2\right) + \left(\left(\frac{\partial f}{\partial B}\right)^2 \frac{1}{N_B}\sum_{N_B}\Delta B^2\right) + \left(\left(\frac{\partial f}{\partial X}\right)^2 \frac{1}{N_X}\sum_{N_X}\Delta X^2\right) + \cdots$$

Resulting in:

$$\sigma_C^2 = \left(\frac{\partial f}{\partial A}\right)^2 \sigma_A^2 + \left(\frac{\partial f}{\partial B}\right)^2 \sigma_B^2 + \left(\frac{\partial f}{\partial Y}\right)^2 \sigma_Y^2 + \left(\frac{\partial f}{\partial X}\right)^2 \sigma_X^2 \cdots \quad (40)$$

It should be noted that Eq. (40) is valid when the variables *A, B, X, Y*, etc. are independent, i.e., the covariance will be zero.

3.5 Gain Error, Offset and Offset Drift of a Sensor

The output of the manometer of Fig. 1 was derived as $(P_1 - P_2) = \rho_U g\, h_U$ making several assumptions. One of those assumptions is that the U-tube's inner wall is homogenous and has uniform surface characteristic throughout the length of the U-tube. In practice this may not be so. Such deviations will then alter the input output characteristic of a sensor. Exercise 11 illustrates this aspect.

Exercise 11 A 100 °C full scale mercury thermometer has a scale made of 100 sub divisions. The output reading *y* (divisions) in terms of the input temperature θ (°C) can be easily deduced as $y = \theta$ (each division represents 1 °C). This characteristic ($y = \theta$) is an ideal input output characteristic and is shown in graphical form in Fig. 13. The actual input output characteristic of a particular thermometer is also plotted in the same figure. In Fig. 13, we see that the input output characteristic obtained for the practical thermometer is: $y = 0.98\,\theta + 4$ (deviates from the

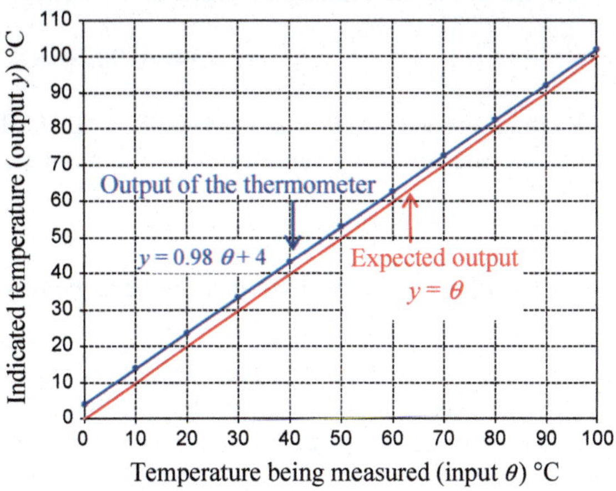

Fig. 13 Input output characteristics of a typical thermometer

expected characteristic of $y = \theta$). We clearly see that the gain (output /input = y/θ) of the sensor instead of being 1 is now 0.98 and the thermometer reads 4 °C instead of 0 °C at the start of the scale. This kind of behavior can be expected out of most sensors. The output of a sensor being a finite value and not equal to zero when the input is zero is called an "*offset*" of the sensor. Generally the offset is given in actual units, but may also be expressed as a percentage of the full scale value. The gain being 0.98 and not 1 is termed as the error in the gain, popularly known as the "*gain error*", once again expressed as a percentage as given below:

$$\%\text{gain error} = \frac{\text{Actaul gain} - \text{Expected gain}}{\text{Expected gain}} \times 100\% \tag{41}$$

The thermometer whose characteristic is indicated in Fig. 13 possesses an offset of 4 °C (or +4% of FS) and a gain error of −2% (−0.02). It should be noted that after the input output characteristic is obtained, the gain error and the offset can be determined. The gain error and offset are, once again, important characteristics of a sensor. Using the knowledge on gain error and offset, compensation can be applied to the readings given by a sensor to find the correct values of the input being sensed by the sensor (obtain readings that are free of offset and gain error). In certain type of sensors, the offset value may not remain constant with respect to time. In such a situation, the sensor not only possesses '*offset*' but also possesses '*offset drift*'. Offset drift is another important characteristic of a sensor defined as:

Offset drift The time rate of change of the offset of a sensor.

For example, a sensor whose full scale output is 100 mV may possess an offset of 2.5 mV and an offset drift of 10 μV/s. Like the systematic error, offset is calculable and hence can be quantified as illustrated in Exercise 12.

Exercise 12 The practical thermometer illustrated in Exercise 11 with its input output characteristic as $y = 0.98\ \theta + 4$ reads 53 °C. Determine the actual value of the temperature of the specimen that is being measured with this thermometer.

We substitute the reading obtained in the equation as:

$$y = 84 = 0.98\theta + 4.$$

Rearranging we get: $\theta = (53 - 4)/0.98 = 50$ °C. Thus the actual (correct value) of the temperature being sensed by this thermometer is 50 °C.

3.6 *Linear and Nonlinear Characteristics of a Sensor*

Some times the input output characteristic of a sensor may contain a "nonlinear" term. Such sensors are said to possess *error due to nonlinearity*.

A sensor is said to possess linear characteristic if the input output relationship obeys:

(i) Cumulative law and
(ii) Associative law.

Cumulative law: Let Y_1 be the output of a sensor for an input X_1. Similarly Y_2 be the output of a sensor for an input X_2. The sensor input output relationship is said to obey the "*cumulative law*" if the sensor out put is $(Y_1 + Y_2)$ for an input of $(X_1 + X_2)$.

Associative law: If the output of the above sensor is K_1Y_1 for an input of K_1X_1, where K_1 is a constant, then the sensor is said to have an input output characteristic that obeys the "*associative law*".

Thus if an input of $(K_1X_1 + K_2X_2)$ to the above sensor results in an out put of $(K_1Y_1 + K_2Y_2)$ then the sensor obeys both cumulative and associative laws and hence possesses a linear input output characteristic. If we plot the input output characteristic of a sensor as an *x-y* graph we will get a straight line for a sensor possessing linear input output characteristic. Figure 14 illustrates typical linear and nonlinear characteristics of sensors. Depending on the extant of deviation from the expected linear response, a sensor can be termed as completely nonlinear or can be taken as nearly linear with a small nonlinearity. In the latter type, the nonlinearity is expressed as an "*error due to nonlinearity*".

For example, the input output characteristic of a sensor possessing small amount of nonlinearity as given in Eq. (42) is shown in Fig. 15.

$$y = x - 0.2x^2, \tag{42}$$

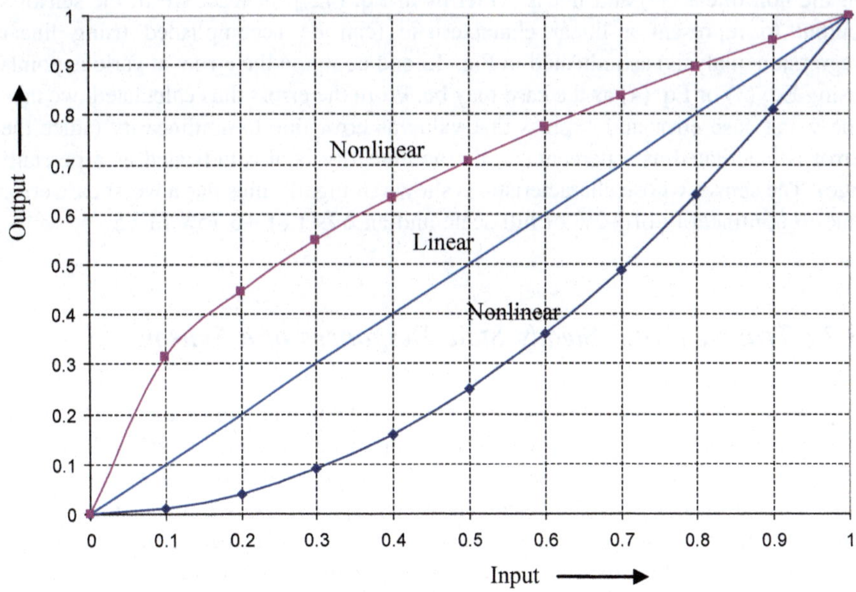

Fig. 14 Linear and nonlinear characteristics of sensors

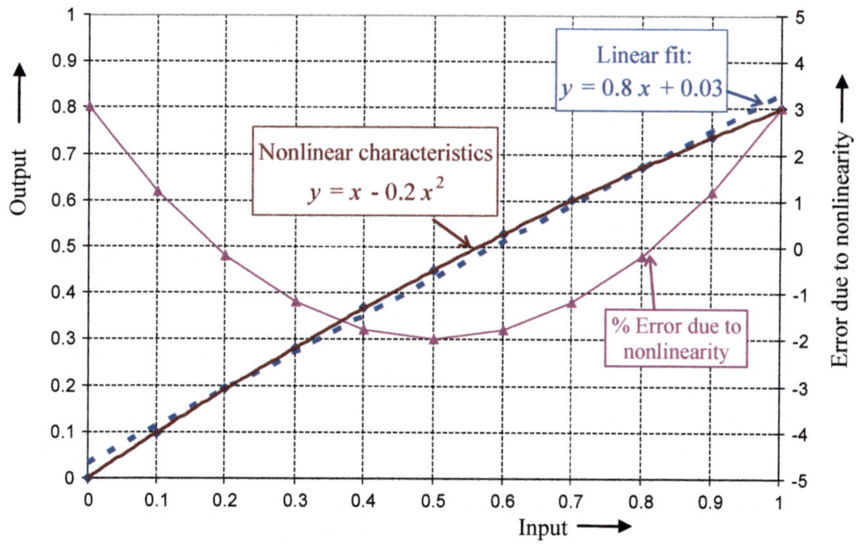

Fig. 15 Error due to nonlinearity in a sensor

It should be noted that we do not indicate the nonlinearity of the sensor whose input output relationship is as given in Eq. 42 as −20% [the ratio of the coefficients of the non linear (x^2) and linear (x) terms in Eq. (42)]. Instead, we fit the sensor's output to represent a linear characteristic (can be accomplished using linear regression analysis) as indicated in Fig. 15 and compute the errors at various points using Eq. (3) or Eq. (4) as the case may be. From the errors thus calculated, we take the worst case error and express that value as error due to nonlinearity (since the error is calculated as a percentage, the nonlinearity is also indicated as a percentage). The sensor whose characteristic is shown in Fig. 15 thus has a worst case error due to nonlinearity of ±2% of full scale and an offset of +3.75% of FS.

3.7 *Transient and Steady State Responses of a Sensor*

A popular sensor everyone is aware of is the thermometer employed for measuring the (body) temperature of a patient. The thermometer is normally kept in the mouth of a patient and the reading is taken only after a specified period of waiting. In the past when a "mercury thermometer", shown in Fig. 16a, is used, the waiting period could be minimum of thirty seconds! Today the mercury thermometers are not very popular mainly due to the fact that mercury pollutes the environment. Though the waiting period of modern electronic thermometer (shown in Fig. 16b is small (few seconds) compared to the mercury thermometer of the yester years, the waiting

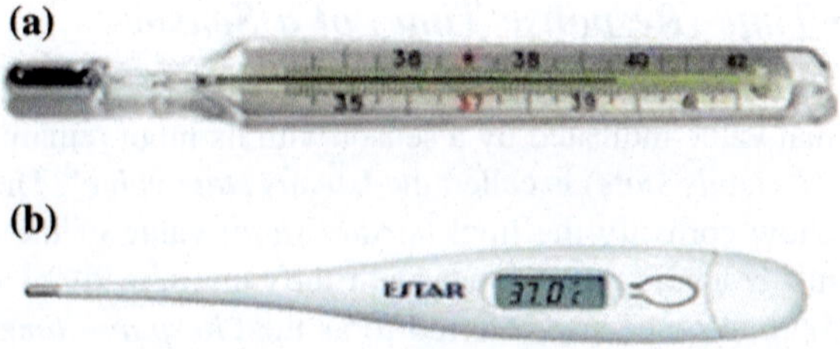

Fig. 16 **a** Mercury thermometer. **b** Digital thermometer

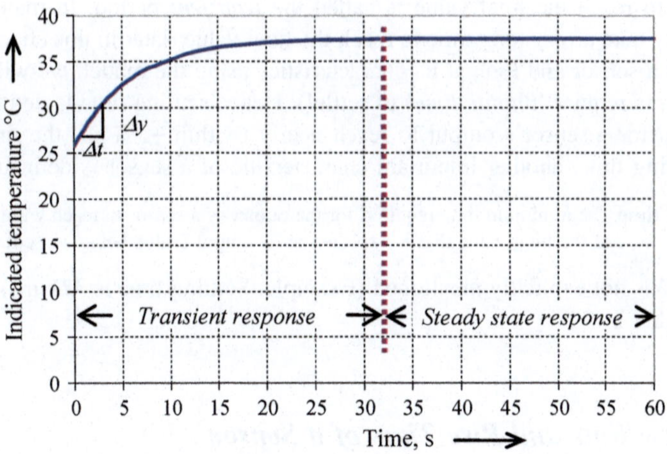

Fig. 17 Time response of a mercury thermometer

period is still not zero. The waiting period is necessitated as the *sensing element* of a thermometer needs sufficient time to attain (in this case through conduction) a temperature that is same as that of the patient to whom the thermometer has been administered and indicate the temperature of the patient correctly. Figure 17 shows the temperature variation of a typical mercury thermometer after it comes into contact with the patient's body. The ambient (room) temperature is approximately 25 °C. In the case of a mercury thermometer, the blob of mercury in the tip of the thermometer has to assimilate the patient's temperature, go through volumetric expansion and thus increase the height mercury occupies in the capillary part of the thermometer. This process takes time as shown in Fig. 17. The initial portion of the *time response* of the output shown in Fig. 17, where the change is taking place, is known as the '*transient response*' of the sensor and the final portion wherein the final value is reached is called the '*steady state response*'.

3.8 Settling Time (Response Time) of a Sensor

In Fig. 10, the final value indicated by a sensor with its input remaining unchanged (i.e., the input is at *steady state*) is called the '*steady state value*'. The time required for a sensor to show correctly the final (*steady state*) value of the quantity being sensed (output just reaching steady state with the input already at steady state) is termed as the "*Settling time*" also referred to as the "*Response time*". The settling time not only provides the minimum time required (waiting period after the input has attained steady state) for a sensor to read but also indicates the time for which the sensor is in the transient state and hence this period where the output is changing to reach the final value is called the *transient* period. In many cases a sensor may take a very long time to reach the final value (later in this chapter, when we model a sensor and look at its characteristics using the model, we will see that most sensors require '*infinite time*' to settle!). Hence settling time is defined as the time taken for a sensor's output to reach nearly (within $\pm z$ % of) the final value. Thus settling time, another important characteristic of a sensor is defined as:

Settling time The minimum time required for the output of a sensor to reach within ±z % of the final value that is expected at its output (value being sensed).

If $\pm z$ % is not explicitly mentioned (Example: Settling time is 321 ms) then it is assumed that $z = 1$.

3.9 Slew Rate and Rise Time of a Sensor

Another important characteristic that can be obtained from Fig. 10 is the maximum rate at which the output can ascend, called the "*slew rate*". The slew rate is indicated by the slope of the output in the transient period. The slope and hence the slew rate is defined as:

$$\text{Slew rate} = \text{slope of output} = \left(\frac{\textit{change in the output y}}{\textit{change in time}} = \frac{\Delta y}{\Delta t} \right). \tag{43}$$

The slope (slew rate) of the thermometer whose input output characteristic is shown in Fig. 10 is also marked therein. The slew rate, another important characteristic of a sensor is defined as:

Slew rate The maximum rate at which the output of a sensor can change (rise or fall).

Another, but not so important characteristics that can be of use sometimes is the rise time as defined below.

Rise time The time taken for the sensor's output to rise (or fall) from 10% of the steady state value of the output to 90% of the steady state value of the output.

3.10 'Static' and 'Dynamic' Characteristics of a Sensor

The input being sensed by a thermometer used for measuring the temperature of a patient (described above) varies very slowly with respect to time. Such an input that does not vary or varies very slowly with respect to time is called "*static*" or *dc* input (signal). The input output characteristic of a sensor obtained for different levels of static inputs is the *static input output characteristic* of a sensor. The *static input output characteristic* is also called, in short form, the "*static characteristic*" of a sensor. Quite often a sensor may have to respond to inputs that are varying with time. Time varying inputs are called *dynamic* inputs and are quite often referred to as dynamic signals. The input output characteristic of a sensor obtained for various values of time varying (dynamic) inputs is called the "*dynamic characteristic*" of the sensor. The time varying inputs (dynamic signals) can be of two distinct types, namely,

(i) Periodic signals or
(ii) Aperiodic (random) signals.

We can easily deduce, from the forgone discussions on settling time of a sensor, that if the input varies within the settling time of a sensor, then the output will not settle to the correct value and hence will not sense the input properly. To obtain an output from a sensor that truly represents the input, the settling time must be less than the time within which the input varies beyond the resolution level of that sensor. Such a condition is hard to test and difficult to comply with. If the dynamic input given to a sensor is of some standard form, then it is easy to test using the standard signal and determine whether a sensor is capable of providing a correct output for that standard dynamic input. The capability of a sensor for other non-standard inputs can be determined easily if those nonstandard inputs are expressed as a combination of standard signals. Two standard signals that fulfill the above criterion are:

(i) The sine/cosine signal (sine/cosine signals are also called sinusoids or phasors) and
(ii) The unit impulse (also popularly known as the Dirac delta function).

3.11 Frequency Response Characteristics of a Sensor

Figure 18 illustrates the sine/cosine signals which are mathematically represented as:

Cosine signal = $\sqrt{2}V_i\text{Cos}(\omega t+\varphi)$ or Sine signal = $\sqrt{2}V_i\text{Sin}(\omega t+\theta)$.

Here V_i is the root mean square (rms) value, $\sqrt{2}V_i$ is the peak value, ω is the angular frequency (in radians per second) of the signal. θ and ϕ are the phase angles of the sinusoids. The frequency f (cycles per second or Hz) is given by: $f = \frac{\omega}{2\pi}$ and

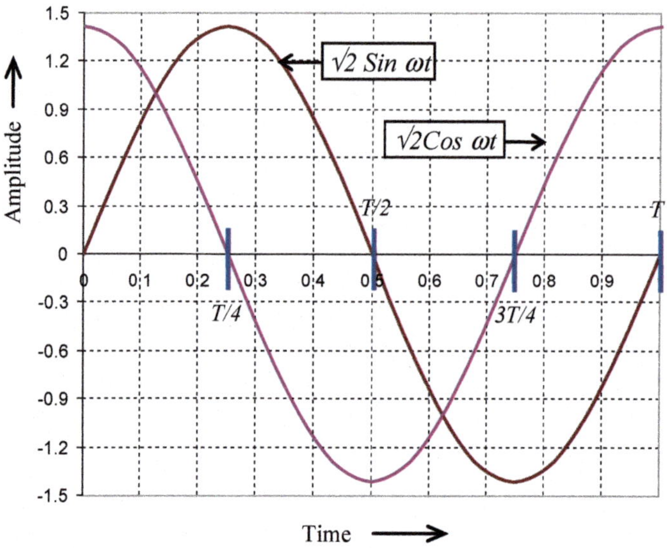

Fig. 18 Sine/Cosine (dynamic) signals

$f = \frac{1}{T}$, where *T* is the time period of the sinusoid. The reason a sinusoid is chosen as the reference waveform is because differentiation or integration when applied to sinusoids results in another sinusoid with changes only in the amplitude and or phase of the signal. Differentiation and integration are the two mathematical processes that inter relate many physical quantities. For example, velocity can be obtained by differentiating the displacement signal or by integrating the acceleration signal. While in the past $\sqrt{2}\text{Sin}\,(\omega\, t)$ was chosen as the reference sinusoid, today $\sqrt{2}\text{Cos}(\omega\, t)$ is chosen as the reference phasor, $\{\text{Re}(\sqrt{2}e^{j\omega t})\text{ or }1\angle 0\}$. A cosine wave is chosen as the reference waveform not only for signal representation and analysis related to sensors but also in many other fields.

Once the performance of a sensor is obtained for sine or cosine signals of varying frequencies, the response to a particular type of dynamic input can be easily ascertained if we are able to represent that dynamic input (signal) as a combination of different cosine waves. For example if a particular signal, say $x(t)$ is resolved into a set of sinusoids (sine/cosine waves with different frequencies, magnitudes and phases) and we test and find that the sensor is capable of responding correctly to all the sinusoids contained in $x(t)$ then we can certainly say that the sensor will respond correctly to $x(t)$.

The process of ascertaining whether a sensor is suitable for a particular signal would become much easier if the manufacturer of a sensor measures the response (input output) of the sensor for sinusoids with different frequencies (normalized with unit magnitude) and provides the result as a characteristic of that sensor. Such a characteristic is called the *frequency response characteristic* of the sensor. The

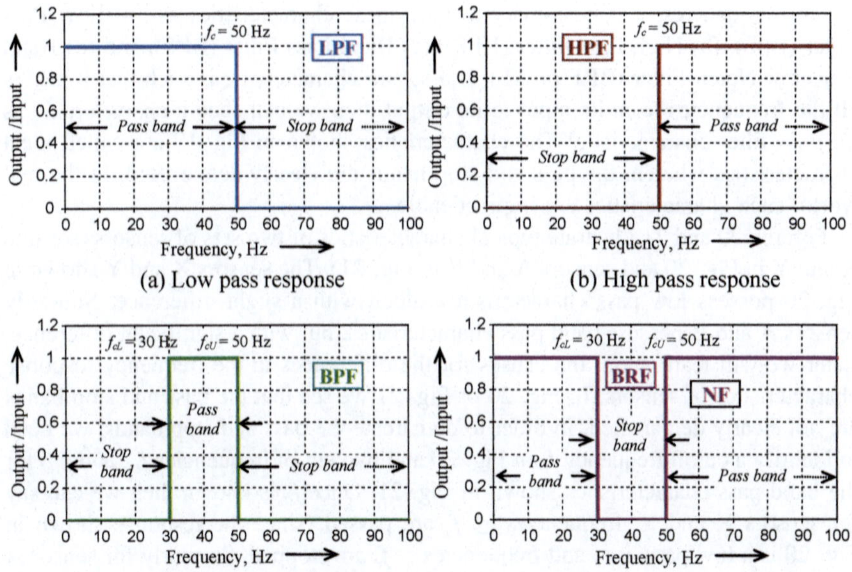

(a) Low pass response (b) High pass response

(c) Band pass response (d) Band reject (Notch) response

Fig. 19 Possible frequency response (ideal) characteristics of a sensor

frequency response characteristic is also known as the *frequency response* or *spectral response or frequency spectrum* or simply *spectrum.* Figure 19 illustrates four different types of frequency responses, an *ideal* sensor can possibly posses. The characteristic shown in Fig. 19a is popularly called the *low pass* characteristic as the sensor passes to the output all sinusoids possessing frequencies less than or equal to a cutoff frequency f_c (In Fig. 19a $f_c = 50$ Hz) to the output and rejects all sinusoids with frequencies $> f_c$. The band of frequencies that are passed to the output is called the *pass band* (0–50 Hz in Fig. 19a) and the range of frequencies for which the input is not passed to the output is called the *stop band.* (stop band for Fig. 19a is 50 Hz to ∞) The pass band is also called the *bandwidth* of the sensor.

The characteristic shown in Fig. 19b is obviously the *high pass* type as the sensor possessing this type of characteristic will pass to the output, from its input, all sinusoidal signals having frequencies equal to or higher than f_c and rejects sinusoids with frequencies $< f_c$. Figure 19c depicts a situation wherein a sensor selectively passes frequencies in a band demarcated by two cutoff frequencies, namely the lower cutoff frequency f_{cL} and the upper cutoff frequency f_{cU} and rejects signals possessing frequencies outside the band $(f_{cU} - f_{cL})$ is aptly called the *band pass* characteristic. $(f_{cU} - f_{cL})$ is the bandwidth (pass band) of the band pass type sensor.

On the other hand a sensor possessing a characteristic as indicated in Fig. 19d selectively rejects signals possessing frequencies in a band (f_{cL} to f_{cU}) and passes all other frequencies and hence is known as the *band reject* type. This type of characteristic is also called the *notch* type. Since all these characteristics filters out certain frequencies, it is customary to term these characteristics as: (a) Low Pass Filter (LPF), (b) High Pass Filter (HPF), (c) Band Pass Filter (BPF), and (d) Band Reject or Notch Filter (BRF or NF). If a sensor characteristic is such that it passes all the frequencies from its input to its output then we call such a characteristic as *all pass* filter characteristic! The characteristics shown in Fig. 19 are called ideal characteristics wherein the pass and stop bands are clearly demarcated. In the real world, such characteristics are impracticable.

Figures 20 and 21 illustrate typical characteristics of two sets of sensors (sensors X and Y in Fig. 20 and sensors A and B in Fig. 21). The sensors X and Y shown in Fig. 20 possess low pass characteristics, albeit with a slight difference. Similarly sensors A and B possess band pass characteristics, but with a significant difference. Later we will learn about the causes for the differences in the frequency response characteristics of sensors. In Fig. 20 or Fig. 21 we see that the pass and stop bands are not clearly demarcated. In order to demarcate the pass and stop bands, we need to identify a cutoff frequency f_c in Fig. 20 and two cut of frequencies f_{cL} and f_{cU} for the band pass characteristics shown in Fig. 21. Once f_c is known then we can say for sensors X and Y: frequencies $\leq f_c$ are passed (since the response shown in Fig. 20 is a low pass one) and frequencies $> f_c$ are stopped. Similarly for sensor A (also for B), we can say that cutoff frequencies f_{cLA} and f_{cUA} (f_{cLB} and f_{cUB} for sensor B) determine the pass and stop bands. A fair criterion to determine the pass band (and thus select f_c) would be to look at the output power and say that the frequencies for which the output power is more than or equal to half the expected (or input or reference) power are deemed to be passed to the output and frequencies for which the output power is less than half the expected value are assumed to be

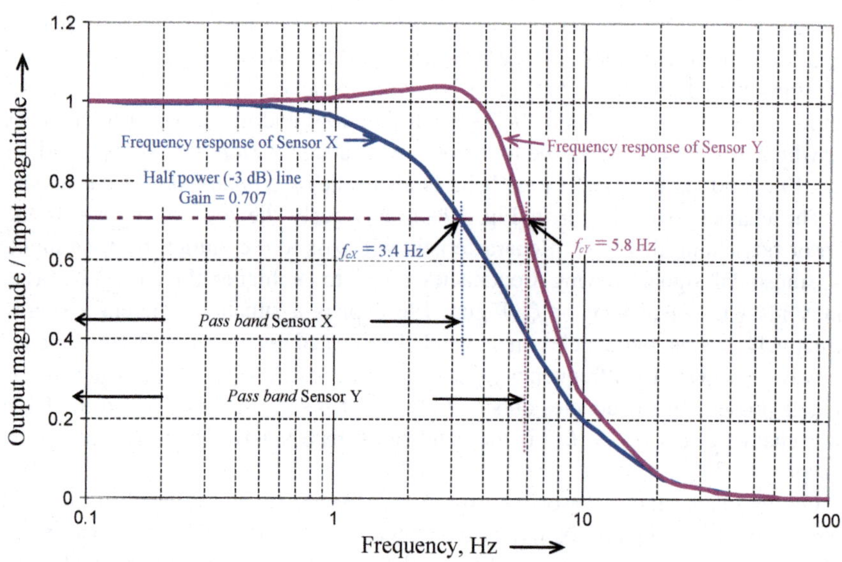

Fig. 20 Frequency response (LPF) characteristics of practical sensors (sensors X and Y)

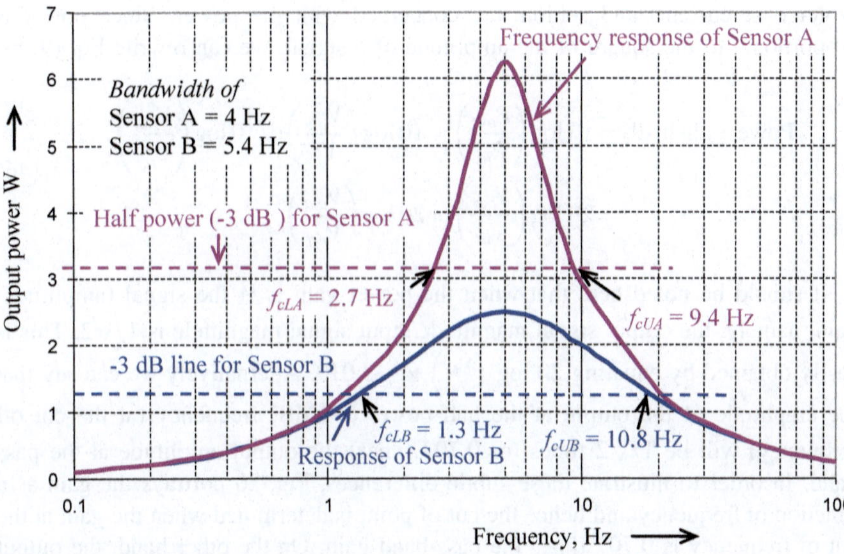

Fig. 21 Frequency response (BPF) characteristics of practical sensors (sensors A and B)

stopped. This norm is known as the half power criterion and the resulting bandwidth is called the '*half power bandwidth*'. The frequency (or frequencies) at which the output power is half is called the '*half power frequency*'. Today this criterion is universally accepted (except when measurement systems are involved; such as sensors!) and if we say bandwidth, it invariably refers to half power bandwidth. It is customary to represent the power gain, the ratio of output power to input power or the ratio of output power to a reference power (P_{out}/P_{in} or P_{out}/P_{ref}) using a unit called *decibel* (short notation used is dB). The decibel unit for power gain is defined as:

$$\text{Power gain in dB} = 10\log\left(\frac{P_{out}}{P_{in}}\right) \text{ or } 10\log\left(\frac{P_{out}}{P_{ref}}\right). \quad (44)$$

If we apply the criterion that the output power is half the expected power then the power gain at the cutoff frequency becomes

$$\text{Power gain(at half power)} = 10\log\left(\frac{1}{2}\right) = -3.010\ \text{dB}. \quad (45)$$

Since the power gain at the cutoff (half power) frequency f_c is −3 dB, f_c is popularly called the −3 dB frequency or simply 3 dB corner frequency. The resulting bandwidth is also aptly called the −3 dB bandwidth or simply '3 dB bandwidth'. In most situations, we look for the amplitude of the output signal (voltage or current) and seldom are concerned with the power. Since power is proportional to the square of the amplitude of a signal, we can rewrite Eq. (9) as:

$$\begin{aligned}\text{Power gain in dB} &= 10\log\left(\frac{P_{out}}{P_{in}}\right) = 10\log\left(\frac{V_{out}^2}{V_{in}^2}\right) \text{or } 10\log\left(\frac{V_{out}^2}{V_{ref}^2}\right). \\ &= 20\log\left(\frac{V_{out}}{V_{in}}\right) \text{or } 20\log\left(\frac{V_{out}}{V_{ref}}\right)\end{aligned} \tag{46}$$

It should be noted here that when the power gain is ½ the signal (amplitude) gain, namely, the output signal magnitude/input signal magnitude is $1/\sqrt{2}$. This is easily obtained by equating 20 log $\left(\frac{V_{out}}{V_{in}}\right)$ to −3.010. Alternatively we can say that the amplitude of the output at the half-power (−3 dB) frequency (at the cut off frequency) will be $1/\sqrt{2}$ times (or 0.707 times) the output amplitude at the pass band. In order to illustrate these subtle differences, Fig. 20 portrays the gain as a function of frequency and hence the cut of point is determined when the gain at the cut of frequency is 0.707 times the pass-band gain. On the other hand, the output power itself is plotted in Fig. 21 and hence the cut-off points are identified by the *half-power* criteria.

When we deal with measuring systems, the half power bandwidth is seldom useful because at the half power point the magnitude is 0.707 times the expected output; hence

$$\%\ \text{error} = \left(\frac{0.707 - 1}{1}\right) \times 100 = -29.3\%.\ (-3\,\text{dB}) \tag{47}$$

If we us the −3 dB criteria, then at the cut off frequency the error in the output is −29.3%! Generally such large errors in a measurement can not be accepted. Thus, for measuring systems such as a sensor, the half power bandwidth is not suitable and the bandwidth for a sensor is specifically determined utilizing the error that can be tolerated in the measurements made with that sensor. For example a −0.5 dB criterion to select the cut off frequency and hence the pass/stop band will result in a worst case error of −5.6% inside the pass band. Similarly a −0.1 dB bandwidth will provide measurements with −1.1% of error in the pass band. Thus when it comes to sensors used for measurement, one has to be careful in looking at the bandwidth of the sensor and not get carried away and use the usual −3 dB bandwidth. Let us look at some practical examples to understand how the bandwidth of a sensor plays a significant role in obtaining a meaningful reading.

For example, the temperature (at steady state) of an oven controlled by an ON-OFF type feedback control system is shown in Fig. 22. We can see from Fig. 22 that the variation in temperature repeats every twenty seconds and is an example of a periodic (but non sinusoidal) dynamic signal.

Figure 23 indicates the volume of blood flow in the peripheral arteries of the finger of a patient (called the *plethysmograph*) and is a typical example of a *quasi*

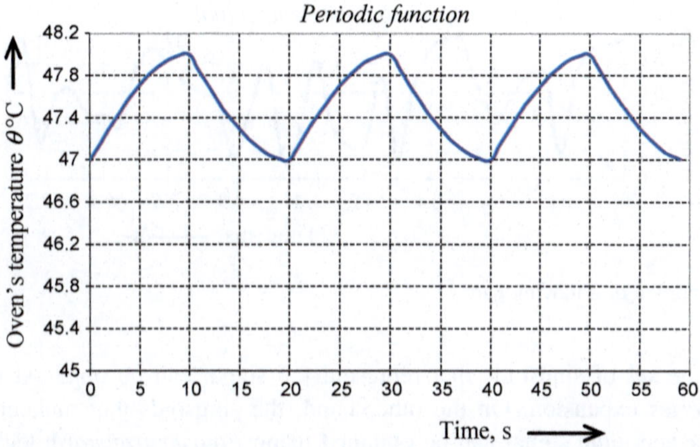

Fig. 22 Temperature of a controlled oven

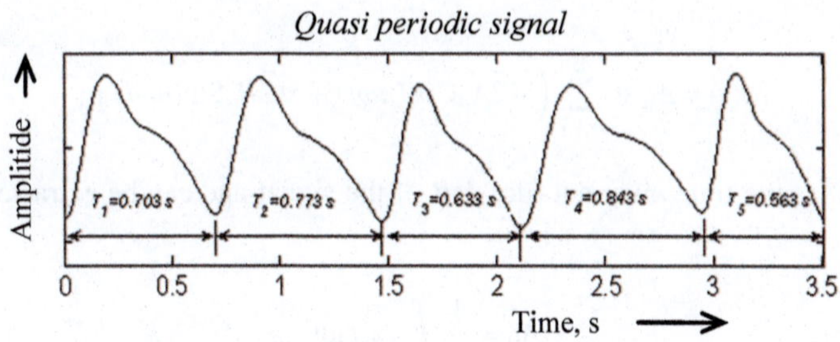

Fig. 23 Blood flow in artery

periodic (neither fully periodic nor fully random) signal. We see that in the quasi periodic waveform shown in Fig. 23, the shape of arterial blood volume change in successive periods look nearly identical, but the period and magnitude (peak to peak values) vary from beat to beat (very characteristic of all physiological systems/signals).

Figure 24 illustrates the waveform that characterizes the vibration of a moving part and is a typical example of an aperiodic (random) signal. Noise signals (noise voltage across a resistor) also come under the category of random signals. Noise signals can be further subdivided into Gaussian noise, white noise, flicker noise etc.

It is seen from these examples that the dynamic input to a sensor can be periodic, quasi periodic or aperiodic (random). In order to assess whether a sensor can provide an output that faithfully represents these types of input (periodic, quasi periodic or aperiodic non sinusoidal signals), these inputs have to be resolved as a set of sinusoids. If all the sinusoids that represent the input are covered within the pass band of a sensor then we are sure that the sensor will function properly for that input.

Jean Baptiste Joseph Fourier, the French mathematician had shown that any signal can be represented as a combination of several sinusoids. If the signal is

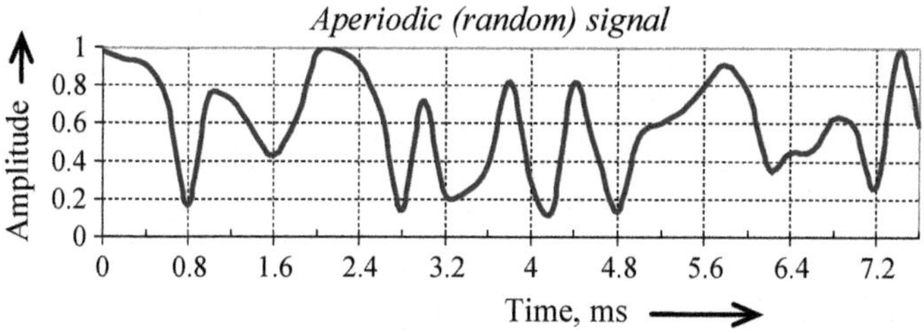

Fig. 24 Vibration of a moving part

periodic, the set of sinusoids that represent the signal can be obtained using the *Fourier series* expansion. On the other hand, the sinusoids that makeup a quasi periodic or aperiodic signal can be obtained using *Fourier transform* technique.

Fourier Series: A periodic signal $y(t)$ whose period is T (hence frequency $f = \frac{1}{T}$ and $\omega = 2\pi f$) possessing finite number of discontinuities within a period can be expressed as:

$$x(t) = A_0 + \sum_{n=1}^{\infty} \left(\sqrt{2} A_n \mathrm{Cos}(n\omega t) + \sqrt{2} B_n \mathrm{Sin}(n\omega t) \right) \tag{48}$$

Here A_0 is the time invariant (dc) part of the signal and can be extracted from $y(t)$ as:

$$A_0 = \frac{1}{T} \int_0^T x(t) \mathrm{dt} \tag{49}$$

The Fourier coefficients A_n and B_n can be determined using the following identities.

$$A_n = \frac{\sqrt{2}}{T} \int_0^T x(t) \mathrm{Cos}(n\omega t)\, \mathrm{dt} \text{ and } \mathrm{B}_n = \frac{\sqrt{2}}{T} \int_0^T x(t)\, \mathrm{Sin}(n\omega t) \mathrm{dt}. \tag{50}$$

Here $n = 1, 2, 3, \ldots \infty$. By combining the sine and cosine terms of Eq. (48), the Fourier series can also be written in a concise form as:

$$x(t) = A_0 + \sum_{n=1}^{\infty} \left(\sqrt{2} C_n \mathrm{Cos}(n\omega t + \phi_n) \right) \tag{51}$$

Equation (51) can also be written using the phasor notation $\underline{C}_n = C_n \angle \phi_n$ as

$$x(t) = A_0 + \sum_{n=1}^{\infty} C_n \angle \phi_n \tag{52}$$

where

$$C_n = \sqrt{(A_n^2 + B_n^2)} \text{ and } \phi_n = -\tan^{-1}\left(\frac{B_n}{A_n}\right) \tag{53}$$

The phasors $\underline{C}_n$ can also be directly computed as:

$$C_n \angle \phi_n = \frac{\sqrt{2}}{T}\int_0^T x(t)e^{jn\omega t}\mathrm{dt}. \tag{54}$$

Here j is the imaginary number $(j = \sqrt{-1})$.

The amplitude (r.m.s. value) of the sinusoid at the frequency f (angular frequency, ω), namely, C_1 is popularly called the amplitude of the fundamental component and ϕ_1 is the phase of the fundamental component. C_2 is the amplitude of the second harmonic component (frequency $2f$ and angular frequency, 2ω) and ϕ_2 is its phase. C_3 is called the third harmonic's amplitude (frequency $3f$ and angular frequency, 3ω) and so on. Since $y(t)$ is periodic, the frequency spectrum will contain sinusoids at the fundamental frequency (the frequency f at which the phenomena or the measurements repeat) and its harmonics (integral multiples of f). If we plot the magnitudes of the fundamental and its harmonics as a function of frequency, we will get the frequency domain representation (magnitude frequency response or magnitude frequency spectrum or simply *magnitude spectrum*) $Y(\omega)$ of $y(t)$. The plot of the phase angles of the output sinusoids as a function of frequency will give the phase response of the sensor (also called the *phase spectrum*). A plot of C_n^2 as a function of frequency will result in the power (frequency) response or simply *power spectrum.*

For the periodic temperature variation (waveform) shown in Fig. 22 the frequency domain representation (*magnitude spectrum*) obtained using the Fourier series expansion is:

$$\begin{aligned}\theta(\omega) = 47.5 &+ 0.337\sqrt{2}\mathrm{Cos}(\frac{\pi}{10}t + 1.15\pi) + 0.028\sqrt{2}\mathrm{Cos}(\frac{3\pi}{10}t + 1.045\pi)\\ &+ 0.011\sqrt{2}\mathrm{Cos}(\frac{5\pi}{10}t + 1.025\pi) + 0.007\sqrt{2}\mathrm{Cos}(\frac{7\pi}{10}t + 1.013\pi)\\ &+ 0.06\sqrt{2}\mathrm{Cos}(\frac{9\pi}{10}t + 1.004\pi) + \cdots\end{aligned} \tag{55}$$

The magnitude spectrum of $\theta(\omega)$ given in Eq. (55) is plotted in Fig. 25. The value of C_0, the dc value (=47.5 °C for the function shown in Fig. 22) is not shown in Fig. 25 as C_0 would dominate and the amplitudes of all other frequencies will not be seen clearly. The waveforms of individual harmonic components (one cycle of the variation shown in Fig. 22) extracted using Eq. (52) are drawn in Fig. 26. The reconstructed waveform using up to 9th harmonic and the original variation are also indicated in Fig. 26. It is seen that while the reconstructed signal in Fig. 26 using the

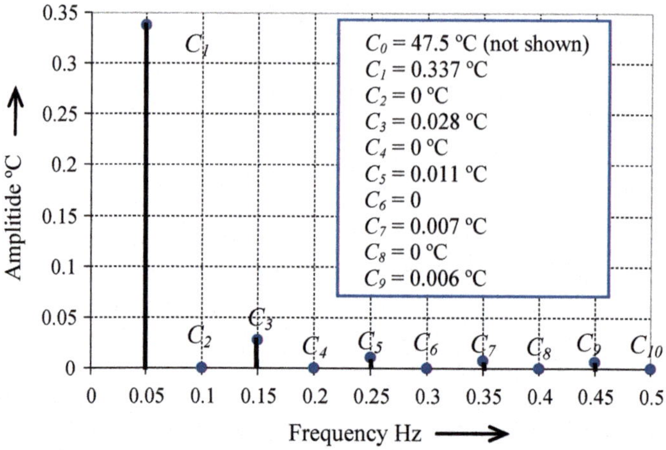

Fig. 25 Magnitude spectrum of the signal shown in Fig. 22

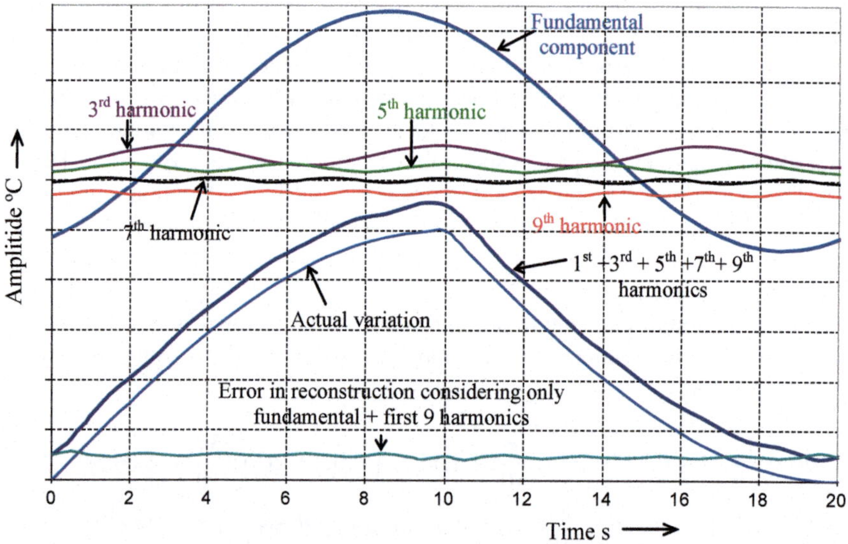

Fig. 26 Harmonics of the signal shown in Fig. 15

first nine harmonics "looks like" the actual variation in temperature of Fig. 22 but is not as smooth as it should be. This condition arises because we stopped at the 9th harmonic. If we include higher harmonics, then we would get the actual variation. Time and again one has to compromise in selecting the number of harmonics to be computed. An informed decision can be arrived at by extracting the error between the original and the reconstructed (utilizing limited number of frequency components) signals. The error that would result if we stopped with nine harmonics alone is also indicated in Fig. 26. Please note that in order to clearly visualize the different waveforms, the waveforms shown in Fig. 26 are arbitrarily shifted in the y direction. In reality all sinusoidal waveforms will cross zero at half their respective periods. It is easily seen that a temperature sensor must have a bandwidth of 1 Hz or more for it to faithfully reproduce the temperature variations shown in Fig. 22.

Note: Thumb rule: If a dynamic parameter that is being sensed (signals) possesses abrupt changes then we will require large number of harmonics to be considered in the Fourier expansion so that the error in the reconstruction of the signal using the derived harmonics is small. Thus to sense signals with abrupt changes in their magnitude, we will require a sensor with a large bandwidth.

As was already brought out, the spectrum of quasi periodic or aperiodic (random) signals can be obtained through Fourier transform. Fourier transformed signal $X(\omega)$ of a time domain signal $x(t)$, provided $x(t)$ is an integrable function, is obtained using the integral given in Eq. (21).

$$X(\omega) = \frac{1}{\sqrt{2\pi}} \int_{-\infty}^{+\infty} x(t) e^{-j\omega t} dt \tag{56}$$

The function $X(\omega)$ in general will be of complex form, resulting in $X(\omega)$ possessing a magnitude and phase at every value of ω. Figure 27 shows the frequency response (magnitude, $X(\omega)$) characteristics of the quasi periodic *plethysmograph* shown in Fig. 23. A comparison of Figs. 25 and 27 shows that while the frequency spectrum (magnitude spectrum) of a periodic signal has only components at discrete (harmonic) frequencies, the frequency spectrum of a quasi periodic signal has significant magnitudes at almost all frequencies but with peaks at certain frequency bands. From the frequency response of the plethysmograph shown in Fig. 27, we can identify four peaks. The peak at 0.175 Hz is due to the influence of breathing

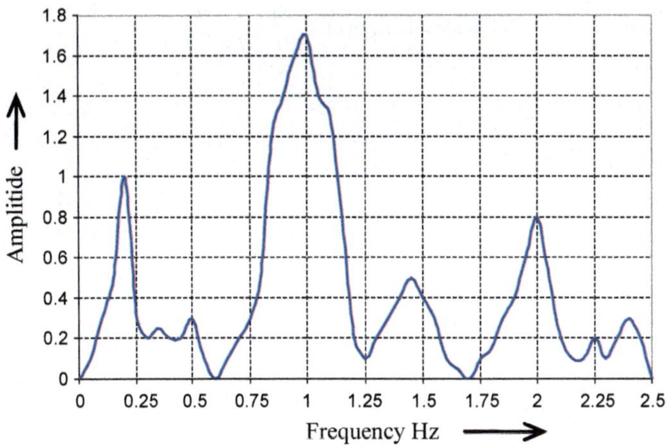

Fig. 27 Magnitude (frequency) spectrum of the signal in Fig. 23

rhythm on the blood volume change (resulting from the exchange of gases that occur in the lungs) in the arteries. The peak near 1 Hz is due to the heartbeat and the rest are small perturbations in the rhythms of the signals from these physical phenomena. Thus to sense the PPG of Fig. 23, we need a sensor whose bandwidth is more than 2.5 Hz. The spectrum of the vibration signal shown in Fig. 24 will also be similar to the one shown in Fig. 27: A continuous line indicating magnitudes at different frequencies, but with peaks and valleys at much higher frequencies (few hundred hertz to few mega hertz).

3.12 Impulse Response Characteristics of a Sensor

Like the sinusoid, another standard input that can help us understand the behaviour of a sensor (for that matter any system) for dynamic inputs is the impulse function. The impulse function was originally formulated by Fourier, but Fourier used a complicated double integral to describe the function and hence its application was not "straight forward". On the contrary the impulse function proposed by Paul Dirac (though mathematically the same as the one formulated by Fourier) is not only elegant but also simple enough to be applied in several fields of science and engineering, including sensors. Figure 28 indicates a pulse $p(t)$ whose width is τ and amplitude is $1/\tau$. Now if we apply the limit as given in Eq. (22) to the pulse $p(t)$ we get the unit impulse $\delta(t)$ (Dirac delta) function as:

$$\delta(t) = \operatorname*{Lim}_{\tau \to 0} p(t) \tag{57}$$

The impulse $\delta(t)$ is also shown in Fig. 28. By evaluating the limit in Eq. (57), we can write:

$$\delta(t) = \operatorname*{Lim}_{\tau \to 0} p(t) = \begin{array}{cc} \infty & t = 0 \\ 0 & t \neq 0 \end{array}. \tag{58}$$

Fig. 28 Pulse $p(t)$ and impulse $\delta(t)$

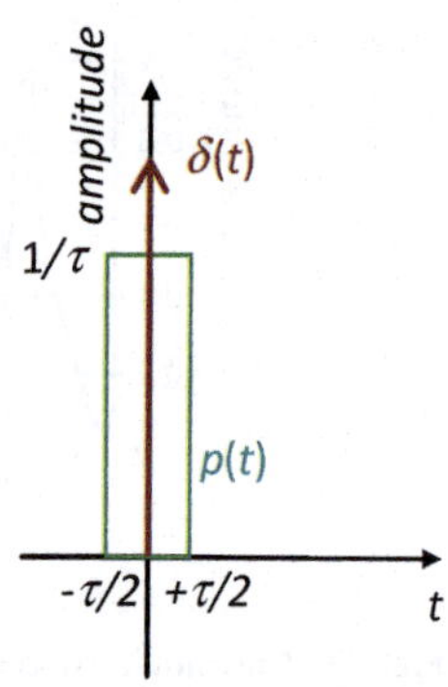

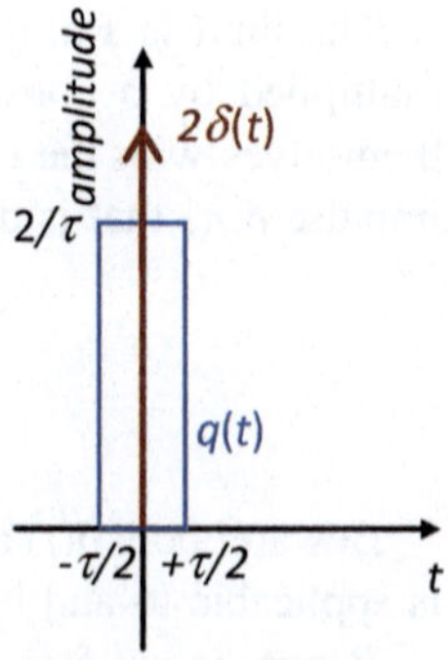

Fig. 29 Pulse $q(t)$ and impulse $2\delta(t)$

Equation (58) is not unique and does not represent only $\delta(t)$! Let us take another pulse $q(t)$ whose width is τ and amplitude is $2/\tau$ as shown in Fig. 29. If we apply the limit as indicated in Eq. (22), we once again get the same result as in Eq. (58) as:

$$\lim_{\tau \to 0} q(t) = \begin{matrix} \infty & t=0 \\ 0 & t \neq 0 \end{matrix}. \tag{59}$$

Definitely the two pulses $p(t)$ and $q(t)$ are very different and hence they cannot converge to the same signal $\delta(t)$. Now if we find the area covered by the pulses $p(t)$ and $q(t)$, we get:

$$\begin{matrix} \int_{-\infty}^{+\infty} p(t) = 1 \\ \text{and} \\ \int_{-\infty}^{+\infty} q(t) = 2 \end{matrix}, \tag{60}$$

If we now combine Eqs. (58) and (60), we can unambiguously define $\delta(t)$ as:

$$\delta(t) = \lim_{\tau \to 0} p(t) = \begin{matrix} \infty & t=0 \\ 0 & t \neq 0 \end{matrix} \quad \text{and} \quad \int_{-\infty}^{+\infty} \delta(t)dt = 1, \tag{61}$$

Thus $\delta(t) = \begin{matrix} \infty & t=0 \\ 0 & t \neq 0 \end{matrix}$ (To be read as $\delta(t)$ is infinity when $t = 0$ and 0 otherwise.) *does not* fully define $\delta(t)$ and the integral $\int_{-\infty}^{+\infty} \delta(t)dt = 1$ must accompany the definition of $\delta(t)$. The impulse $\delta(t)$ as defined in Eq. (26) is popularly known as the Dirac delta function and is also called the "*Unit impulse*" since the area covered by the impulse is unity. We can also say that $\delta(t)$ is an impulse whose *strength* is unity. The limiting value of $q(t)$ can be obtained as:

$$\lim_{\tau \to 0} q(t) = \begin{matrix} \infty & t=0 \\ 0 & t \neq 0 \end{matrix} \text{and} \int_{-\infty}^{+\infty} q(t)dt = 2 \Rightarrow 2\delta(t), \tag{62}$$

The limit in Eq. (62) gives us an impulse whose strength is 2 or unit impulse multiplied by a constant of value 2 or simply 2 $\delta(t)$. One should not confuse themselves with the continuous time, unit impulse $\delta(t)$, and the discrete time unit impulse $\delta(n)$ that is defined as:

$$\begin{aligned}\delta(n) &= 1 \text{ for } n=0;\\ &= 0 \text{ otherwise}\end{aligned} \tag{63}$$

Beware! both $\delta(t)$ and $\delta(n)$ are called unit impulses! While $\delta(t)$ is universal, $\delta(n)$ is applicable to and hence can be used only in analyzing discrete time systems.

Properties of $\delta(t)$ that are useful in analyzing the performance of a sensor (or any system).

If we multiply a signal $x(t)$ with a unit impulse $\delta(t)$ we get:

$$x(t)\delta(t) \Rightarrow \begin{matrix} \infty & t=0 \\ 0 & t\neq 0 \end{matrix}$$

$$\text{and} \int_{-\infty}^{+\infty} x(t)\delta(t)dt = x(0)$$

If we If we multiply a signal $x(t)$ with a delayed unit impulse $\delta(t-\tau)$ we get:

$$x(t)\delta(t-\tau) \Rightarrow \begin{matrix} \infty & t=\tau \\ 0 & t\neq 0 \end{matrix}$$

$$\text{and} \int_{-\infty}^{+\infty} x(t)\delta(t-\tau)dt = x(\tau)$$

Thus by multiplying a signal $x(t)$ with delayed impulse $\delta(t-\tau)$ and by varying τ, from zero in steps of T_s (sampling time), we get the samples of $x(t)$ at various instants of time as: $x(0)$, $x(T_s)$, $x(2T_s)$, $x(0), x(T_s), x(2T_s), x(3T_s) \ldots$ · This property is called the sampling property of the unit impulse.

Figure 30 shows the block schematic of a typical sensor possessing linear input output characteristic. For an input $x(t)$, let the output of the sensor be $y(t)$. The output, say $h(t)$, obtained for a unit impulse $\delta(t)$ as input is called the impulse response of the sensor. If we give an impulse $\delta(t-\tau)$ (unit impulse delayed by time τ) as the input to the sensor, then we will get the output from the sensor as $h(t-\tau)$, impulse response delayed by τ as shown in successive steps in Fig. 30. We now give an input $x(\tau)\delta(t-\tau)$ then the sensor's output (we assume linear input output characteristics for the sensor) will be $x(\tau)h(t-\tau)$. We now integrate the input over a range (-∞ to +∞) with respect to the variable τ, then the output will be:

$$\int_{-\infty}^{+\infty} x(\tau)\delta(t-\tau)d\tau \Rightarrow Sensor \Rightarrow \int_{-\infty}^{+\infty} x(\tau)h(t-\tau)d\tau \tag{64}$$

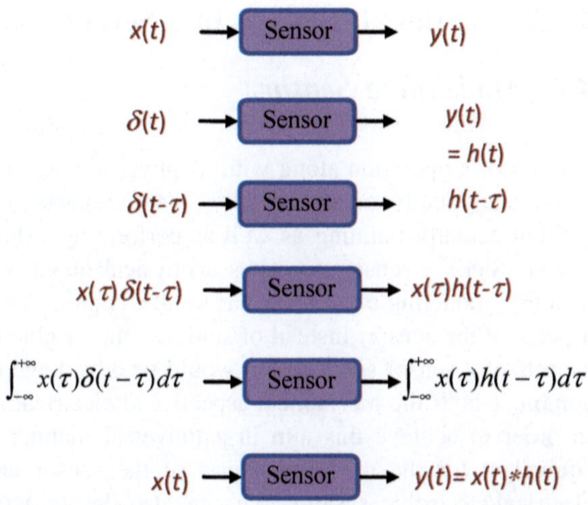

Fig. 30 Output $y(t)$ utilizing impulse response $h(t)$

as indicated in Fig. 23. The integral $\int_{-\infty}^{+\infty} x(\tau)\delta(t-\tau)d\tau$ gives all possible samples of $x(t)$ and hence $\int_{-\infty}^{+\infty} x(\tau)\delta(t-\tau)d\tau = x(t)$. The output of the sensor indicated by the integral $\int_{-\infty}^{+\infty} x(\tau)h(t-\tau)d\tau$ is popularly known as the convolution integral and is denoted as:

$$y(t) = x(t) * h(t) = \int_{-\infty}^{+\infty} x(\tau)h(t-\tau)d\tau. \tag{65}$$

The symbol '*' represents the process of convolution. Thus once we have the impulse response $h(t)$ of a sensor, then for any particular input $x(t)$, the corresponding output $y(t)$ can be obtained by simply convolving the input signal $x(t)$ with the impulse response $h(t)$ of the sensor. Though evaluating a convolution integral in time domain as given in Eq. (65) is fraught with problems, evaluation of convolution using Laplace transform technique is quite simple as we will see in the sequel.

In spite of this nice property of impulse, namely, one can obtain the output of a sensor for any given input by simply convolving the input with the impulse response; the impulse response characteristics of sensors are not popular and manufacturers of sensors do not normally provide such a characteristic. However, in situations where reliability of a sensor is of paramount importance (example aerospace applications), impulse response analysis is essential.

Time and again (especially during the design stage) it may be necessary to analyze the response of a sensor for various operating conditions and environments. A popular technique employed for such an analysis is the transfer function method of analysis.

4 Model Based Analysis of a Sensor's Characteristic

4.1 Modeling a Sensor

If a sensor's operation along with its physical and electrical properties is modeled, either analytically or numerically, many aspects of the sensor can be studied without actually building as well as performing exhaustive testing of a prototype sensor. Since a sensor translates a physical quantity into a measurable electrical quantity, modeling of a sensor has to take cognizance of the physical and electrical aspects of the sensor. Instead of studying the mechanical aspects and the electrical aspects of a sensor separately it would be advantageous if a composite model that combines both the mechanical aspects and electrical aspects together is arrived at. In order to achieve this aim in a universal manner, we can derive an electrical equivalent for the mechanical part of the sensor and analyze the sensor as an electrical/electronic system. We can also derive a mechanical equivalent of the electrical aspects of the sensor and analyze the sensor's performance in the mechanical domain. Since circuit analysis techniques are well researched, it is a universal practice to convert the mechanical aspects of a sensor to an equivalent electrical system and analyze the sensor as an electrical/electronic system.

A typical sensor may possess a mass and elasticity, can involve frictional and or viscous damping on the mechanical side and can be represented by a general mechanical system indicated in Fig. 31 as a block diagram. The differential equation that describes the mechanical system of Fig. 31 is:

$$F(t) = M\frac{d^2x_M(t)}{dt^2} + D\frac{dx_M(t)}{dt} + K_s x_M(t). \tag{66}$$

Here $F(t)$ is the force applied to the sensor, $x_M(t)$ is the displacement of the mass M of the sensor, D is the coefficient of damping (frictional and or viscous damping) of the damper and K_s is the spring constant.

Now we look at the series electrical circuit made of inductor L_{Ms}, resistor R_{Ds} and capacitor C_{ss} shown in Fig. 32a. The total circuit voltage e_s of this series circuit is:

$$e_s = e_{L_{Ms}} + e_{R_{Ds}} + e_{C_{ss}} \tag{67}$$

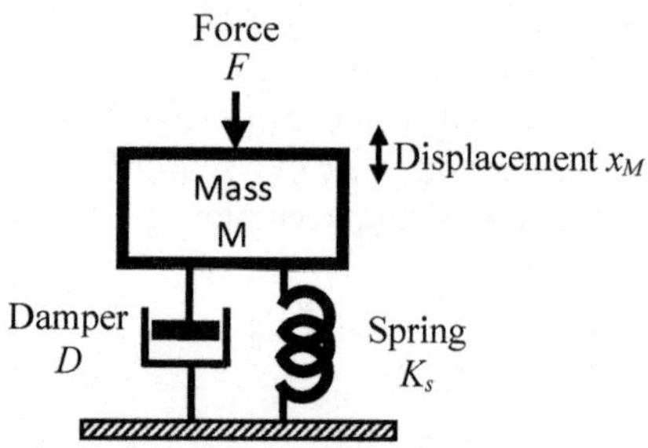

Fig. 31 Block schematic of the mechanical side of a sensor

Here $e_{L_{Ms}} = L_{Ms}\frac{di_s}{dt}$ is the voltage across the inductor L_{Ms}, $e_{R_{Ds}} = R_{Ds}i_s$ is the voltage across the resistor R_{Ds}, $e_{C_{ss}} = \frac{1}{C_{ss}}\int i_s dt$ is the voltage across the capacitor C_{ss} and i_s is the current in the circuit. The integral-differential equation that relates the voltage e_s and the current i_s of Fig. 32a is:

$$e = L_{Ms}\frac{di_s}{dt} + R_{Ds}i_s + \frac{1}{C_{ss}}\int i_s dt. \tag{68}$$

If we substitute $i_s = \frac{dq_s}{dt}$ in Eq. (68), where q_s is the electrical charge, we get:

$$e_s = L_{Ms}\frac{d^2q_s}{dt^2} + R_{Ds}\frac{dq_s}{dt} + \frac{1}{C_{ss}}q_s. \tag{69}$$

A look at Eqs. (66) and (69) tells us that on the analytical side these equations are similar differential equations except for the differences in the variables and coefficients. Thus we can say with certainty that the series *L-R-C* circuit of Fig. 32a is an electrical equivalent (in the mathematical or analytical sense) of the mechanical system of Fig. 31. We can also infer using Eqs. (66) and (69) the following equivalence given in Table 1. Thus when we model a sensor, the

Fig. 32 Block schematic of the mechanical side of a sensor

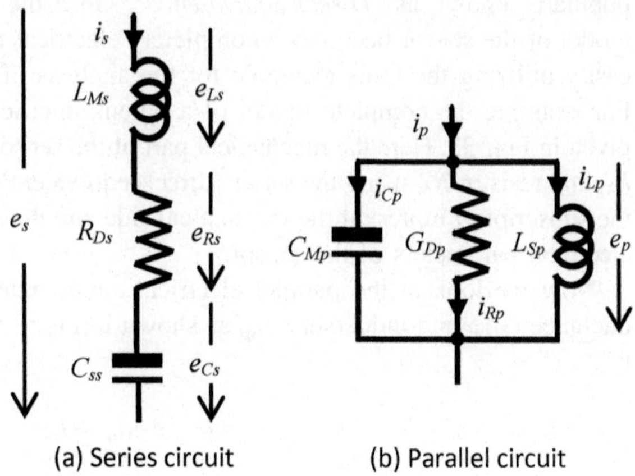

Table 1 Electrical equivalence (direct) of mechanical variables

Sl. No.	Mechanical			Equivalent electrical (direct)		
	Variable	Symbol	Unit	Variable	Symbol	Unit
1	Displacement	x_M	m	Charge	q_s	C
2	Force	F	N	e.m.f	e_s	V
3	Velocity	v_M	m/s	Current	i_s	A
4	Mass	M	kg	Inductance	L_{Ms}	H
5	Damping coefficient	D	Ns/m	Resistance	R_{Ds}	Ω
6	Spring constant	K_s	N/m	1/Capacitance	$1/C_{ss}$	F^{-1}

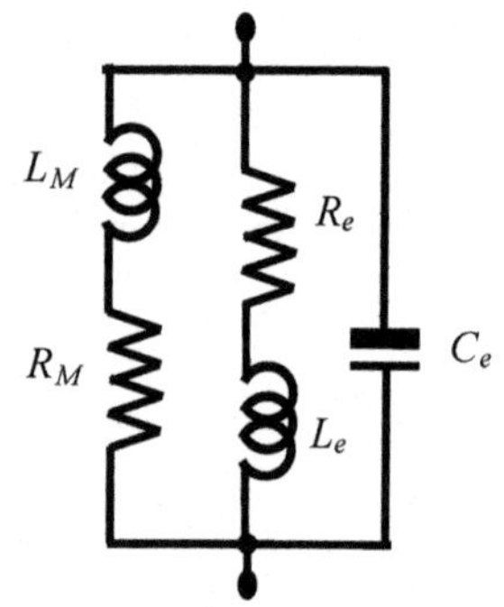

Fig. 33 Equivalent circuit model of a potentiometric sensor

mechanical (physical) characteristics can be converted into an equivalent electrical series circuit consisting of

(i) An inductance L_{Ms} to represent the mass,
(ii) Resistance R_{Ds} to represent the damping characteristics and
(iii) Capacitance C_{ss} to represent the elasticity of the mechanical part of the sensor.

The series *L*-*R*-*C* representation of the mechanical characteristics of a sensor is popularly known as "*Direct equivalence*". Once the equivalence is obtained, the model of the sensor becomes a completely electrical network and can be analyzed easily utilizing the tools available for the analysis of electrical/electronic circuits. For example the complete model of a potentiometric type displacement sensor is given in Fig. 33. Here the mechanical part of the sensor is replaced with an inductor L_M and resistor R_M using the series (direct) equivalent circuit. The components with the subscript *M* represent the mechanical side and the subscript *e* denotes the actual electrical parameters of the sensor.

Now we look at the parallel electrical circuit made of capacitance C_{Mp}, conductance G_{Dp}, and inductance L_{sp} as shown in Fig. 32b. The total circuit current i_p is:

$$i_p = i_{C_{Mp}} + i_{G_{Dp}} + i_{L_{sp}} \tag{70}$$

The current $i_{C_{Mp}}$ through the capacitor C_{Mp} can be derived as $i_{C_{Mp}} = C_{Mp}\frac{de_p}{dt}$, where e_p is the voltage across the capacitor C_{Mp} and is also the voltage drop across the entire parallel circuit. The current $i_{G_{Dp}}$ through the conductance G_{Dp} is: $i_{G_{Dp}} = G_{Dp}e_p$. Similarly the current $i_{L_{sp}}$ through the inductor L_{sp} will be $i_{Lp} = \frac{1}{L_{sP}}\int e_p dt$. The integral-differential equation that relates the circuit voltage e_p and the circuit total current i_p of Fig. 32b is:

$$i_p = C_{Mp}\frac{de_p}{dt} + G_{Dp}e_p + \frac{1}{L_{sp}}\int e_p dt. \tag{71}$$

Table 2 Electrical equivalence (indirect or inverse equivalent) of mechanical variables

Sl. No.	Mechanical			Equivalent electrical (indirect)		
	Variable	Symbol	Unit	Variable	Symbol	Unit
1	Displacement	x_M	m	Magnetic flux	ϕ_p	Wb
2	Force	F	N	Current	i_p	A
3	Velocity	v_M	m/s	e.m.f	e_p	V
4	Mass	M	kg	Capacitance	C_{Mp}	F
5	Damping coefficient	D	Ns/m	Conductance	G_{Dp}	S
6	Spring constant	K_s	N/m	1/Inductance	$1/L_{sp}$	H^{-1}

If we substitute $e_p = \frac{d\phi_p}{dt}$ in Eq. (24), where ϕ_p is magnetic flux, we get:

$$i_p = C_{Mp}\frac{d^2\phi_p}{dt^2} + G_{Dp}\frac{d\phi_p}{dt} + \frac{1}{L_{sp}}\phi_p. \tag{72}$$

Comparison of Eqs. (66) and (72) will lead us to conclude that these equations are also similar differential equations that bring out another set of equivalent parameters as shown in Table 2. The parallel electric circuit equivalent of Fig. 32b is termed as *indirect or inverse equivalent model.* Whether it is the mechanical system described by Eq. (66) or the equivalent electrical systems described by Eq. (69) or Eq. (72), the output of a system described by ordinary differential equation for a given input can be obtained by solving the underlying differential equation. An elegant way of solving an ordinary differential equation is to employ the Laplace transform technique.

4.2 The Laplace Transform

The Laplace transform $X(s)$ of a signal $x(t)$ is defined as:

$$X(s) = \mathcal{L}\{x(t)\} = \int_0^\infty x(t)e^{-st}dt \tag{73}$$

Here s is the Laplace variable [$s = (\sigma + j\omega)$] and is called the "*complex frequency*". Laplace transform exists for a given $f(t)$ if and only if $f(t)$ is integrable in the range $(0, \infty)$. It should be noted here that the Laplace transform is described as given in Eq. (73) in the range $(0, \infty)$ and is known as the single sided or unilateral Laplace transform. In a more general form, the Laplace integral can be obtained in the range $(-\infty, +\infty)$. If the Laplace transform is defined in the range $(-\infty, +\infty)$ then the transform is popularly called the double sided or bilateral Laplace transform. While the Fourier transform, transforms $x(t)$ into the frequency domain, the Laplace transform transforms $x(t)$ into the "moments" (complex frequency domain) of $x(t)$.

Table 3 lists the Laplace transform of some commonly encountered functions. Many difficult to ascertain operations in the time domain become simple mathematical operations in the Laplace domain. For example the convolution of two signals $x_1(t)$ and $x_2(t)$ given in Eq. (76), difficult to compute in the time domain becomes simple multiplication of the Laplace transformed signals $X_1(s)$ and $X_2(s)$ of signals $x_1(t)$ and $x_2(t)$ as given in Eq. (77).

$$x_1 * x_2 = \int_{-\infty}^{\infty} x_1(\tau)x_1(t-\tau)d\tau \tag{74}$$

$$x_1 * x_2 = \int_{-\infty}^{\infty} x_1(\tau)x_1(t-\tau)d\tau \quad \Rightarrow^{LT} \quad X_1(s)X_2(s) \tag{75}$$

Table 3 also lists two very useful theorems, called the initial value theorem and the final value theorem wherein we exploit the Laplace transform and obtain the value of the output at time $t = 0$ and $t = \infty$ respectively.

Table 3 Laplace transform of functions

Sl. No.	$x(t)$	$X(s)$	Range of convergence
1	Unit impulse (Dirac delta), $\delta(t)$	1	$(-\infty, +\infty)$
2	Delayed unit impulse, $\delta(t-\tau)$	$e^{-\tau s}$	$(-\infty, +\infty)$
3	Unit step $[u(t) = 1 \text{ for } t \geq 0 \text{ and} = 0 \text{ for } t < 0]$	$\frac{1}{s}$	Re(s) > 0
4	Delayed Unit step $u(t-\tau)$	$\frac{e^{-\tau s}}{s}$	Re(s) > 0
5	Ramp $[r(t) = t\,u(t)]$	$\frac{1}{s^2}$	Re(s) > 0
6	$t^n\, u(t)$	$\frac{n!}{s^{n+1}}$	Re(s) > 0 and n > −1
7	Exponential decay $e^{-\alpha t}u(t)$	$\frac{1}{s+\alpha}$	Re(s) > $-\alpha$
8	Exponential build-up $(1-e^{-\alpha t})u(t)$	$\frac{\alpha}{s(s+\alpha)}$	Re(s) > 0
9	Sin(ωt) $u(t)$	$\frac{\omega}{(s^2+\omega^2)}$	Re(s) > 0
10	Cos(ωt) $u(t)$	$\frac{s}{(s^2+\omega^2)}$	Re(s) > 0
11	Differentiation: $\frac{dx(t)}{dt}$, initial condition $x(0^-)$	$sF(s) - x(0^-)$	
12	$\frac{d^2x(t)}{dt^2}$ with $x(0^-), x'(0^-)$ and $\frac{dx(0^-)}{dt}$	$s^2F(s) - sx(0^-) - x'(0^-)$	
13	Integration: $\int_0^t x(\tau)d\tau$	$\frac{1}{s}F(s)$	
14	Initial value theorem $x(0) = \lim_{t \to 0} x(t)$	$x(0) = \lim_{s \to \infty} sF(s)$	
15	Final value theorem $x(\infty) = \lim_{t \to \infty} x(t)$	$x(0) = \lim_{s \to 0} sF(s)$	

Once we have the Laplace transform, it is possible to obtain the time domain signal by performing the inverse Laplace transform. In a concise form the inverse Laplace transform can be computed using Eq. (76) or Eq. (77)

$$x(t) = \mathcal{L}^{-1}\{X(s)\} = \frac{1}{2\pi j}\int_0^\infty X(s)e^{st}ds \tag{76}$$

If we have the Laplace transform as a polynomial in s then it is possible to apply the inverse Laplace transform and obtain the result in the time domain. A unique function $X(s)$ with the Laplace variable s will result only when the function $x(t)$ is continuous. It is possible that we may get the same function in s for different functions in the time domain if these functions contain discontinuities. For example we will get $1/s$ as the Laplace transform for unit step function $u(t)$ ($u(t)$ is defined as $u(t) = 1$ for $t \geq 0$ and $= 0$ for $t < 0$) as well as for another function $g(t)$, where

$$g(t) = \begin{matrix} 1 & \text{for } 0 < t < 5 \\ -7 & at\, t = 5 \\ 1 & \text{for } t > 5 \end{matrix}.$$

Note: the integrals in Eqs. (73) and (76) over the range (0, ∞) are 'improper integrals'. The proper mathematical representation for computation of, say, inverse Laplace transform should be:

$$x(t) = \mathcal{L}^{-1}\{X(s)\} = \frac{1}{2\pi j}\lim_{T\to\infty}\int_{k-jT}^{k+jT} X(s)e^{st}ds \tag{77}$$

Easiest way of obtaining inverse Laplace transform is to employ partial fraction, resolve $X(s)$ into smaller entities and then obtain the inverse using "pattern matching" with the help of Table 3.

4.3 *Sensor's Performance Analysis Using Laplace Transform*

The discussions and methods of analysis given here are applicable not only for sensors but also to any system, governed by a linear input output relationship. We now look at the mass spring accelerometer, whose input output relationship is governed by the ordinary differential equation given in Eq. (66).

If we apply Laplace transform to Eq. (66), we get (Assuming all initial conditions to be zero):

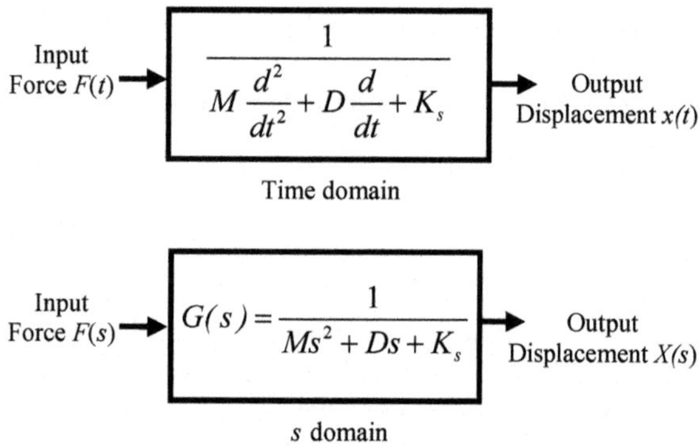

Fig. 34 Concept of the transfer function of a sensor

$$F(s) = Ms^2X(s) + DsX(s) + K_sX(s) \tag{78}$$

Here $X(s)$ is the Laplace transform of the output [displacement $x(t)$] and $F(s)$ is the Laplace transform of the input (force $F(t)$) of the mass-damper-spring (accelerometer) sensor. The process of applying the Laplace transform is illustrated in Fig. 34. Rearranging Eq. (78), we get:

$$X(s) = F(s)\frac{1}{Ms^2 + Ds + K_s} \tag{79}$$

The ratio of the Laplace transformed output $X(s)$ and input $F(s)$ gives us the *transfer function* $G(s)$ as shown in Eq. (77):

$$G(s) = \frac{X(s)}{F(s)} = \frac{1}{Ms^2 + Ds + K_s} \tag{80}$$

Dividing numerator and denominator by M and multiplying the result by $\frac{K_S}{K_S}$, we get:

$$G(s) = \frac{\frac{K_S}{M}}{\left(s^2 + \frac{D}{M}s + \frac{K_S}{M}\right)} \frac{1}{K_S} \tag{81}$$

Substituting $\omega_n = \sqrt{\frac{K_S}{M}}$ and $\alpha = \frac{D}{2\sqrt{MK_S}}$ we get:

$$G(s) = \left(\frac{1}{K_S}\right)\left(\frac{\omega_n^2}{s^2 + 2\xi\omega_n s + \omega_n^2}\right). \tag{82}$$

ω_n in Eq. (82) is popularly known as the natural frequency of the mass spring accelerometer sensor (or any system that is governed by a second order transfer function) and ξ is called the damping constant of the sensor or system.

Note: Obtaining a transfer function to represent the behavior of a sensor is similar to obtaining an analytical equation for a practical input output relationship. Thus the transfer function of a sensor can be derived utilizing any of the two distinct techniques listed below.

(i) Use the analytical input output relationship (obtained from a comprehensive electrical equivalent circuit model of different parts with which the sensor is made) and derive a transfer function by applying Laplace transform. Or
(ii) Use the input output characteristic and derive the transfer function directly (curve fitting).

Equation (82) describes the normalized input output relationship of a sensor (system) whose input output relationship is described by a second order differential equation in the Laplace domain (*s*-domain) and is applicable for the mechanical system shown in Fig. 24 and hence is equally applicable to the equivalent series and parallel electrical circuits (systems) shown in Fig. 25. All these systems can be viewed as a system described by the block schematic shown in Fig. 27.

In general *G*(*s*) can be a function consisting of a numerator polynomial and a denominator polynomial of the Laplace variable *s* as given in Eq. (83).

$$G(s) = \frac{N(s)}{D(s)} = \frac{N_n S^n + N_{n-1} s^{n-1} + \cdots + N_0}{D_n S^n + D_{n-1} s^{n-1} + \cdots + D_0} \tag{83}$$

The coefficients of the Laplace variable *s* in Eq. (83), namely the N_ns and D_ns are constants. It should be noted here that N_0 and D_0 are coefficients of s^0. The highest power of *s* in the transfer function determines the *order* of the system (of course the order of the transfer function itself). The system described in Eq. (80) is of order *n* or the system is an *n*th order system. If the input output relationship is governed by a simple constant [only s^0 terms N_0 and D_0 exist both in the numerator and denominator of *G*(*s*)] then the order of the system is zero! We call these systems as zero order systems. In a zero order system, the output is a true representation of the input (for both static and dynamic inputs) and the frequency response has an "*all pass*" characteristic (passes all frequencies or the sensor is capable of responding to all frequencies and hence can sense perfectly inputs possessing any type of waveform). It would make every ones life simpler if all sensors are of zero order!

The function *G*(*s*) shown in Eq. (83) can be rearranged as:

$$G(s) = \frac{N(s)}{D(s)} = \frac{K(s+z_n)(s+z_{n-1})(s+z_{n-2})\cdots(s+z_1)}{(s+p_n)(s+p_{n-1})(s+p_{n-2})\cdots(s+p_1)} \tag{84}$$

It may so happen that some of the terms (*zs* and *ps*) in Eq. (84) may turn out to be of negative or complex values. It is easily seen that the value of *G*(*s*) is zero whenever *s* is equal to $-z_n$ or $-z_{n-1}$ or $-z_{n-2}$ or ... $-z_1$. Hence the *z*s are called the "zeroes" of the transfer function *G*(*s*). Similarly whenever

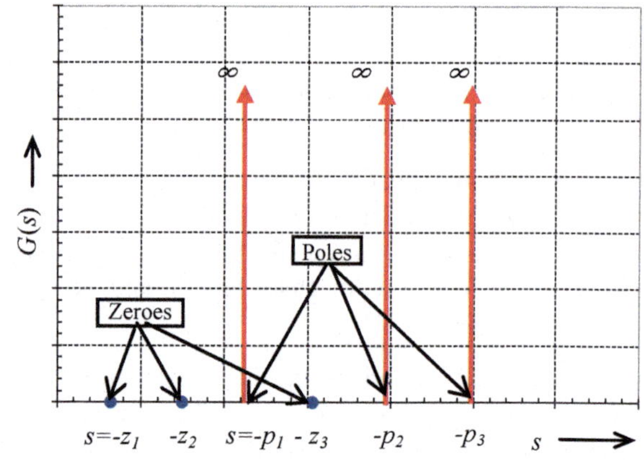

Fig. 35 Poles and zeroes of $G(s)$

$s = -p_n$ or $-p_{n-1}$ or $\ldots -p_1$, the value of $G(s)$ goes to infinity. Figure 35 plots the values of $G(s)$ at these special points and it is seen that whenever $s = -p_n$ or $-p_{n-1}$ or $\ldots -p_1$, a "*pole*" of height infinity is placed at these points to represent $G(s)$ in Fig. 35. Thus the points $-p_n, -p_{n-1} \ldots -p_1$ are aptly called the poles of $G(s)$. Once we have the transfer function $G(s)$ of a sensor, several types of analysis can be obtained easily using the values of poles and zeroes of the transfer function. An added advantage in describing the characteristic of a sensor using the transfer function is the by simply replacing the Laplace variable s with the complex frequency $j\omega$ we can obtain the frequency response of the sensor as given in the example below.

Example 1 The transfer function of a pressure sensor is given as $G(s) = \frac{9}{s^2 + 4s + 9}$. Determine its frequency response.

Solution: We get the frequency response by substituting $s = j\omega$ in $G(s)$ as:

$$Y(\omega) = G(j\omega) = \frac{9}{-\omega^2 + j4\omega + 9} \text{ resulting in}$$

$$\text{Magnitude of } Y(\omega) = |Y(\omega)| = \frac{9}{\sqrt{(9 - \omega^2)^2 + 16\omega^2}} \text{ and}$$

$$\text{Phase of } Y(\omega) = \Phi(\omega) = -\tan^{-1}\left(\frac{4\omega}{(9 - \omega^2)}\right).$$

The magnitude spectrum and phase spectrums of the pressure sensor are shown in Fig. 36.

Computation of the response of a sensor for a given input.

Once we obtain the transfer function, the output of the sensor for any input, can be obtained by simply multiplying the transfer function by the Laplace transform of the input and derive the inverse Laplace transform of the resulting polynomial in

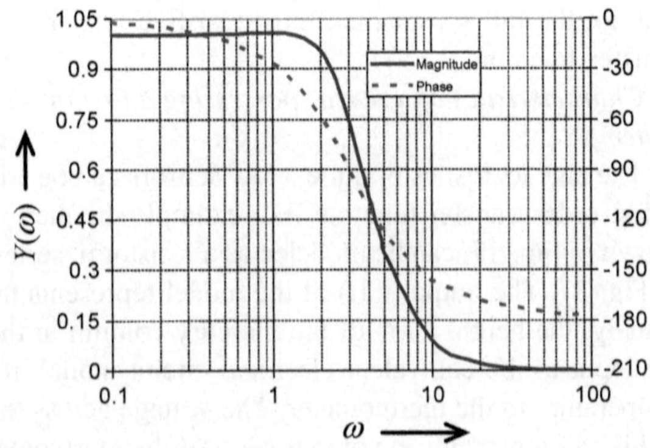

Fig. 36 Magnitude and phase plots of the output from transfer function

s. Thus the accelerometer whose characteristics is described by its transfer function of Eq. (82), its output $x(t)$ can be derived for any input force $F(t)$ as:

$$x(t) = \mathcal{L}^{-1}\{G(s)F(s)\} \tag{85}$$

4.4 *Analysis of the Behavior of a Sensor Using Transfer Function*

Once the transfer function of a sensor (or a system) is derived, the behavior of the sensor can be easily obtained through analysis.

Characteristics of a sensor possessing a transfer function of order zero (*Zero order system*).

The transfer function of a sensor falling under the category of zero order system will be:

$$G(s) = \frac{Y(s)}{X(s)} = K, \tag{86}$$

Here $Y(s)$ is the Laplace transform of the output of the sensor $y(t)$ and $X(s)$ is the Laplace transform of its input $x(t)$. Then we have

$$y(t) = K\,x(t), \tag{87}$$

where $y(t)$ is the output of the sensor with $x(t)$ as its input. From Eq. (87) we can easily infer that the sensor will respond to dc as well as ac of all frequencies. Thus a sensor of zero order transfer function possesses an all pass characteristic with an infinite bandwidth and infinite slew rate, and hence is an ideal sensor. Sensors possessing zero order transfer function are rare. If we model a sensor element alone and not the full sensor then most resistive sensor elements will have a zero order transfer function.

Characteristics of a sensor possessing a first order transfer function (*First order system*).

The characteristics of quite a few sensors can be adequately represented utilizing a first order transfer function. For example the thermometer, whose characteristic, shown in Fig. 17, can be modeled as a resistor in series with a capacitor as indicated in Fig. 37. The output $v_o(t)$ of the model represents the output of the thermometer, namely, the height $h_\theta(t)$ of the mercury column in the capillary. The voltage $v_i(t)$, the input to the equivalent electrical circuit model, represents the input (patient's) temperature to the thermometer. The voltage across the capacitor $v_o(t)$ in the circuit of Fig. 37 represents the output $h_\theta(t)$ of the thermometer (and hence the reading of the temperature θ_P). $v_o(t)$ can be derived as:

$$v_o(t) = \frac{1}{C_T}\int i(t)dt, \tag{88}$$

where $i(t)$ is the circuit current. If we apply Laplace transform to Eq. (88) then we get:

$$V_o(s) = \frac{1}{sC_T} I(s) \tag{89}$$

Applying Kirchoff's voltage law (KVL) to the circuit in Fig. 30, we get:

$$v_i(t) = R_T i(t) + \frac{1}{C_T}\int i(t)dt \tag{90}$$

Applying Laplace transform to Eq. (90), we get:

$$V_i(s) = I(s)R_T + \frac{1}{sC_T} I(s). \tag{91}$$

Rearranging Eq. (91), we obtain

$$I(s) = \frac{sC_T}{sR_T C_T + 1} V_i(s). \tag{92}$$

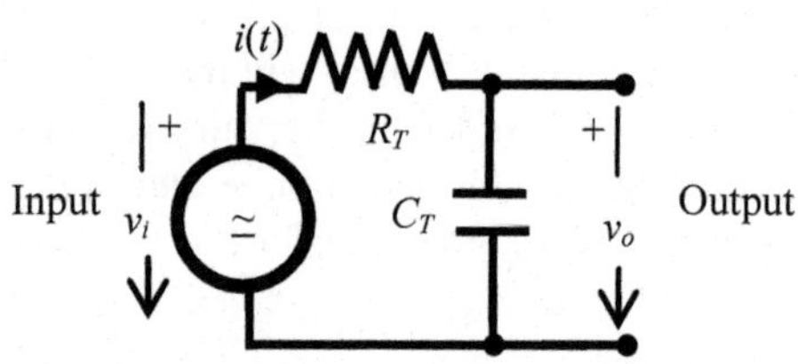

Fig. 37 Electrical equivalent of mercury thermometer

Substituting $I(s)$ from Eq. (92) in Eq. (89) we get:

$$V_o(s) = \frac{1}{sR_TC_T + 1} V_i(s) = \frac{1}{1 + s\tau} V_i(s) \tag{93}$$

Here $\tau = R_TC_T$, is known as the time constant of a series *R-C* circuit. Rearranging Eq. (93), we obtain the transfer function of the thermometer as

$$G(s) = \frac{h_\theta(s)}{\theta_P(s)} = \frac{V_o(s)}{V_i(s)} = \frac{1}{1 + sR_TC_T} = \frac{1}{1 + s\tau} = \frac{1/\tau}{(s + 1/\tau)} \tag{94}$$

The transfer function in Eq. (94) is of order one (maximum power of the Laplace variable s is 1) and has a single pole at $-\frac{1}{\tau}$ and a zero at $s \to \infty$. Thus the thermometer is a "*first order system*". When we place the thermometer into a patient's mouth, we apply an input suddenly at the that time (say time $t = 0$) to the thermometer. Mathematically we can represent this input as:

$$\theta(t) = \begin{matrix} 0 & t < 0 \\ \theta_P & t \geq 0 \end{matrix} \quad \Rightarrow \theta(t) = \theta_P u(t), \tag{95}$$

where θ_P is the temperature of the patient. For such an input, the output of the thermometer can be obtained from its model described by Eq. (94) as:

$$h_\theta(s) = \frac{1/\tau}{(s + 1/\tau)} \theta(s); \text{ where } \theta(s) = \mathcal{L}\{\theta_P u(t)\}. \tag{96}$$

Resulting in

$$h_\theta(s) = \frac{1/\tau}{(s + 1/\tau)} \frac{\theta_P}{s}. \tag{97}$$

The time response of the thermometer can now be obtained by applying inverse Laplace transform to Eq. (97) as:

$$h_\theta(t) = \mathcal{L}^{-1}\left\{\frac{1/\tau}{(s + 1/\tau)} \frac{\theta_P}{s}\right\} = \mathcal{L}^{-1}\left\{\frac{A}{s} + \frac{B}{(s + 1/\tau)}\right\}. \tag{98}$$

From Eq. (98), we can derive:

$$\theta_P(1/\tau) = A(s + 1/\tau) + Bs. \tag{99}$$

Putting s = 0 in Eq. (99), we get $A = \theta_P$. Again substituting s = $-1/\tau$, we get $B = \theta_P$, resulting in:

$$h_\theta(t) = \mathcal{L}^{-1}\left\{\frac{\theta_P}{s} + \frac{\theta_P}{(s+1/\tau)}\right\}. \tag{100}$$

Using Table 3, we can get the inverse Laplace transform of Eq. (100) as:

$$h_\theta(t) = \theta_P\left(u(t) - u(t)e^{-\frac{t}{\tau}}\right) = \theta_P\left(1 - e^{-\frac{t}{\tau}}\right)u(t) \tag{101}$$

Plot of the output of the thermometer as given by Eq. (101) is shown in Fig. 38, assuming the patient's temperature $\theta_P = 37.4\,°\mathrm{C}$ (same as in Fig. 10) and time constant τ to be 6 s. A comparison of Figs. 17 and 38 indicates that the responses are nearly the same. The difference is that while the response in Fig. 17 starts from the ambient temperature, the response in Fig. 38 starts from 0. This situation has arisen because, we assumed zero initial condition.

The reader is urged to obtain a new transfer function including the initial condition, namely, the temperature of the thermometer is equal to the room temperature (by assuming a charge in and hence a voltage across the capacitor at time $t = 0$) and plot the output of the thermometer. A practical mercury thermometer will have a notch at the entrance of the capillary so that the mercury column does not fall back as soon as the thermometer is taken out of the patient's mouth facilitating easy and leisurely reading of the indicated value. After a reading is taken, one has to rigorously shake the mercury thermometer so that the mercury from the capillary is drained back into the reservoir and the thermometer is again ready for a new measurement. We can introduce this aspect also into the model of the thermometer given in Fig. 37 by inserting an ideal diode in parallel with a switch between the voltage source and the resistor!

Characteristics of a sensor possessing a second order transfer function (*Second order system*).

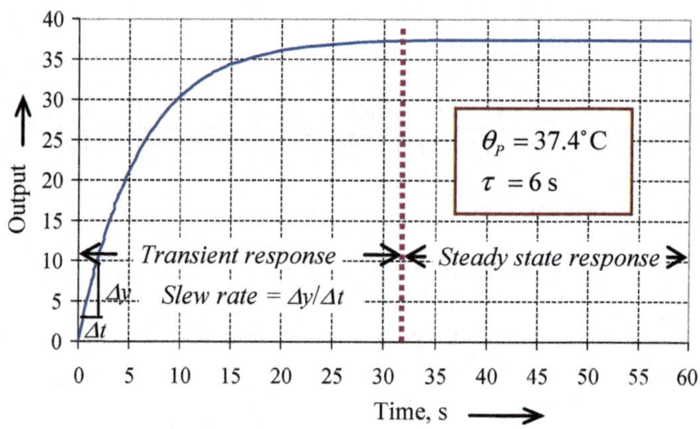

Fig. 38 Time response of a mercury thermometer obtained from its model

We have already seen that normalized transfer function of a mass-spring accelerometer sensor is:

$$G(s) = \frac{X_M(s)}{F(s)} = \left(\frac{1}{K_S}\right)\left(\frac{\omega_n^2}{s^2 + 2\xi\omega_n s + \omega_n^2}\right) \tag{102}$$

Once again if we apply an input force $F(s)$, as soon as the input is applied, the sensor will go through the transient period and then reach steady state. To understand the transient and steady state behaviors of a second order system (sensor possessing a second order transfer function) we excite the sensor with a unit step input and analyze its behavior. For this condition Eq. (102) gets modified as:

$$G(s) = \frac{X_M(s)}{U(s)} = \left(\frac{1}{K_S}\right)\left(\frac{\omega_n^2}{s^2 + 2\xi\omega_n s + \omega_n^2}\right) \tag{103}$$

Substituting $U(s) = \frac{1}{s}$ in Eq. (103) and rearranging, we get:

$$X_M(s) = \left(\frac{1}{K_S}\right)\left(\frac{\omega_n^2}{s^2 + 2\xi\omega_n s + \omega_n^2}\right)\frac{1}{s} \tag{104}$$

Depending on the value of the damping constant ξ, the nature of the roots α and β of the quadratic equation $s^2 + 2\xi\omega_n s + \omega_n^2 = 0$ will vary and hence the solution to Eq. (104) will vary. The roots of Eq. (105) can be derived as:

$$\alpha = \left(-\xi + \sqrt{\xi^2 - 1}\right)\omega_n \text{ and } \beta = \left(-\xi - \sqrt{\xi^2 - 1}\right)\omega_n \tag{105}$$

(i) If $\xi > 1$, then roots α and β will be real and distinct as given in Eq. (105). In this situation, we can simplify Eq. (104) using partial fractions as:

$$X_M(s) = \left(\frac{A}{s}\right) + \left(\frac{B}{s-\alpha}\right) + \left(\frac{C}{s-\beta}\right). \tag{106}$$

Using Eqs. (104) and (105) we can derive:

$$A(s-\alpha)(s-\beta) + Bs(s-\beta) + Cs(s-\alpha) = \frac{\omega_n^2}{K_s} \tag{107}$$

Substituting $s = 0$, α and β in turn in Eq. (107) we get:

$$A = \frac{\omega_n^2}{\alpha\beta K_s}, B = \frac{\omega_n^2}{\alpha(\alpha-\beta)K_s} \text{ and } C = \frac{\omega_n^2}{\beta(\beta-\alpha)K_s}. \tag{108}$$

Substituting the values of A, B and C from Eq. (108) in Eq. (106) and applying inverse Laplace transform results in:

$$x_M(t) = \frac{1}{K_s}\left(1 - \frac{\beta e^{-\alpha t} - \alpha e^{-\beta t}}{(\beta-\alpha)}\right)u(t)$$
$$x_M(t) = \frac{1}{K_s}\left(1 + \frac{\left(-\xi-\sqrt{\xi^2-1}\right)e^{-\omega_n\left(-\xi+\sqrt{\xi^2-1}\right)t} - \left(-\xi+\sqrt{\xi^2-1}\right)e^{-\omega_n\left(-\xi-\sqrt{\xi^2-1}\right)t}}{2\sqrt{\xi^2-1}}\right)u(t) \tag{109}$$

The condition $\xi > 1$ is popularly called the ***over damped condition***. The time response of an over damped second order system for a unit step input is shown in Fig. 39.

(ii) If $\xi = 1$, then roots α and β will be real and equal with $\alpha = \beta = -\omega_n$ resulting in

$$X_M(s) = \left(\frac{1}{K_S}\right)\frac{\omega_n^2}{(s+\omega_n)^2}\frac{1}{s} \tag{110}$$

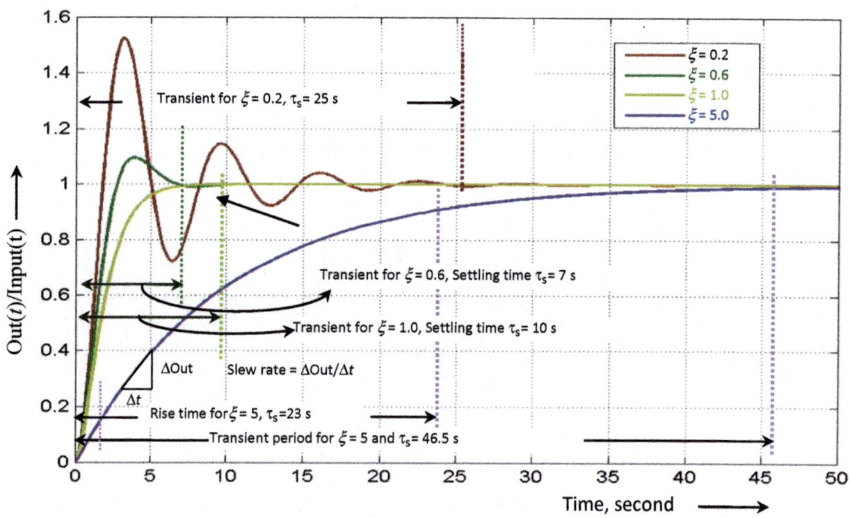

Fig. 39 Step response of a second order transducer

In this situation, we can simplify Eq. (110) using partial fractions as

$$X_M(s) = \left(\frac{A}{s}\right) + \left(\frac{B}{s+\omega_n}\right) + \left(\frac{C}{(s+\omega_n)^2}\right) \tag{111}$$

Using Eqs. (110) and (111) we can derive:

$$A(s+\omega_n)^2 + Bs(s+\omega_n) + Cs = \frac{\omega_n^2}{K_s} \tag{112}$$

Substituting $s = 0$ and $s = -\omega_n$ in turn in Eq. (112) we get:

$$A = \frac{1}{K_s}, \quad \text{and} \quad C = \frac{-\omega_n}{K_s}, \tag{113}$$

Differentiating Eq. (112) and substituting $s = -\omega_n$, we get;

$$B = \frac{-1}{K_s}, \tag{114}$$

Substituting the values of A, B and C in Eq. (111) and applying inverse Laplace transform, we obtain:

$$x_M(t) = \frac{1}{K_s}[1 - (1 + t\omega_n)\, e^{-\omega_n t}]\, u(t) \tag{115}$$

When $\xi = 1$, the sensor is called a '*critically damped*' sensor. The response of a critically damped second order type sensor is also plotted in Fig. 39.

(iii) If $\xi < 1$, then roots α and β will be complex conjugates as:

$$\alpha = \left(-\xi + j\sqrt{1-\xi^2}\right)\omega_n \text{ and } \beta = \left(-\xi - j\sqrt{1-\xi^2}\right)\omega_n$$

We can handle this situation in a slightly different way. We split the Laplace transform given in Eq. (104) namely, $X_M(s) = \left(\frac{1}{K_S}\right)\left(\frac{\omega_n^2}{s^2+2\xi\omega_n s+\omega_n^2}\right)\frac{1}{s}$ as:

$$X_M(s) = \left(\frac{1}{K_S}\right)\left[\frac{A}{s} - \left(\frac{B}{s^2+2\xi\omega_n s+\omega_n^2}\right)\right] \tag{116}$$

Solving we get $A = 1, B = s + 2\xi\omega_n$, resulting in:

$$X_M(s) = \left(\frac{1}{K_S}\right)\left[\frac{1}{s} - \left(\frac{\xi\omega_n}{(s+\xi\omega_n)^2 + \omega_n^2(1-\xi^2)}\right) - \left(\frac{s+\xi\omega_n}{(s+\xi\omega_n)^2 + \omega_n^2(1-\xi^2)}\right)\right] \tag{117}$$

Rearranging Eq. (117) gives:

$$X_M(s) = \left(\frac{1}{K_S}\right)\left[\frac{1}{s} - \left(\frac{\xi}{\sqrt{1-\xi^2}}\frac{\omega_n\sqrt{1-\xi^2}}{(s+\xi\omega_n)^2 + \omega_n^2(1-\xi^2)}\right) - \left(\frac{(s+\xi\omega_n)}{(s+\xi\omega_n)^2 + \omega_n^2(1-\xi^2)}\right)\right] \tag{118}$$

Applying inverse Laplace transform, we get:

$$x_M(t) = \left(\frac{1}{K_S}\right)\left[1 - \left(\frac{\xi}{\sqrt{1-\xi^2}}e^{-\xi\omega_n t}\mathrm{Sin}\omega_n\sqrt{1-\xi^2 t}\right) - \left(e^{-\xi\omega_n t}\mathrm{Cos}\omega_n\sqrt{1-\xi^2 t}\right)\right]u(t). \tag{119}$$

Equation (119) can be further simplified as:

$$x_M(t) = \left(\frac{1}{K_S}\right)\left[1 - \frac{e^{-\xi\omega_n t}}{\sqrt{1-\xi^2}}\mathrm{Cos}(\omega_d t + \theta)\right]u(t). \tag{119}$$

where

$$\theta = -\tan^{-1}\left(\frac{\xi}{\sqrt{1-\xi^2}}\right) \text{ and } \omega_d = \omega_n\sqrt{1-\xi^2}. \tag{120}$$

ω_d is called the '*damped natural frequency*' of a under damped sensor (system). Figure 39 also illustrates the step response of a sensor possessing an under damped second order transfer function. It is seen from Fig. 39 that the output of an under damped sensor over shoots the expected value, goes through certain number of oscillations and then settles down to steady state condition. The amplitude of the first overshoot, the subsequent decay rate on the amplitudes of successive overshoots are dictated by how small is the damping constant. These and other characteristics of a sensor possessing a slightly under damped transfer function ($0.6 < \xi < 1$) can be derived using Eqs. (119) and (120). The amplitude of the first overshoot in percentage (peak overshoot expressed as a percentage of the steady date value) is:

$$\% \text{ overshoot}|_{\max} = e^{-\frac{\pi\xi}{\sqrt{1-\xi^2}}} \times 100 \approx \left(1 - \frac{0.6}{\xi}\right) \times 100; \text{ for } 0.6 < \xi < 1 \quad (121)$$

and occurs at a time (when the first peak occurs) $\tau_p = \frac{\pi}{\omega_n\sqrt{1-\xi^2}} = \frac{\pi}{\omega_d}$. The *'Rise time'* of an under damped second order sensor can be derived as:

$$\text{Rise time} = \frac{1.8}{\omega_n}. \quad (122)$$

The settling time τ_s of an under damped sensor, for the sensor's output to settle within ±1% of the steady sate value can be derived as:

$$\text{Settling time } \tau_s|_{\pm 1\%} = \frac{4.6}{\xi\omega_n} \quad (123)$$

Decay rate (logarithmic) of an under damped sensor can be derived as:

$$\text{Decay Rate } \delta = \frac{1}{m}\ln\left(\frac{\hat{x}_k}{\hat{x}_{k+m}}\right) = \frac{1}{m}\ln\frac{2\pi\xi}{\sqrt{1-\xi^2}} \quad (124)$$

where $\hat{x}_k$ is the peak of the kth overshoot and $\hat{x}_{k+m}$ is the peak value of the $(k + m)$ th overshoot.

The settling times for different damping constants and slew rate are also marked in Fig. 39.

Note: The behavior of a sensor predicted through the transfer function approach matching correctly the sensor's actual behavior depends on how well the transfer function matches the elements that constitute a sensor. For example we may model a sensor that is made of an coil (inductor) as a series combination of the winding resistance R_c of the coil and its inductance L_c as shown in Fig. 40a. This model will produce an excellent match between the actual sensor behavior and the one calculated from the transfer function of the senor derived using this model if the inductor is an air cored one and is excited by a source possessing a frequency in the kilohertz range. If the inductor type sensor element is to be excited by a source possessing high frequency (f in hundreds of kilohertz or megahertz range) then the effect of inter turn and inter terminal capacitances become significant. The circuit model shown in Fig. 40a will not be adequate for this case and the circuit model shown in Fig. 40b is more appropriate.

Similarly if the inductor is constructed using a ferromagnetic core then a resistance R_i that represents the losses in the ferromagnetic core placed in parallel with the inductor L_c in the circuit model as shown in Fig. 41a will provide a better match between the actual response of the sensor and its model at low frequencies. On the other hand, if the same sensor is excited by a high frequency signal, then the equivalent circuit given in Fig. 41b is more appropriate.

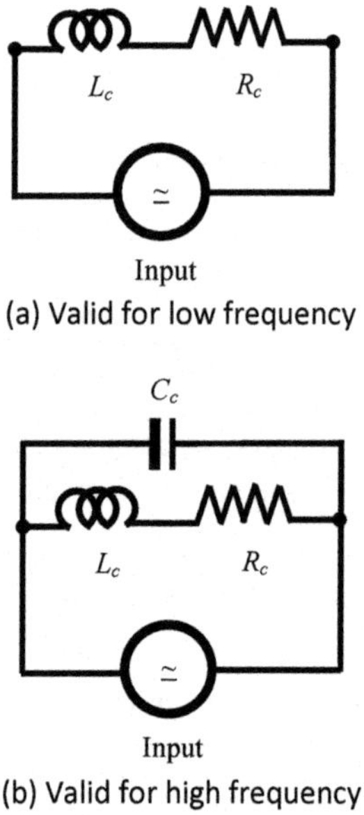

Fig. 40 Equivalent electrical circuit of an inductive sensor

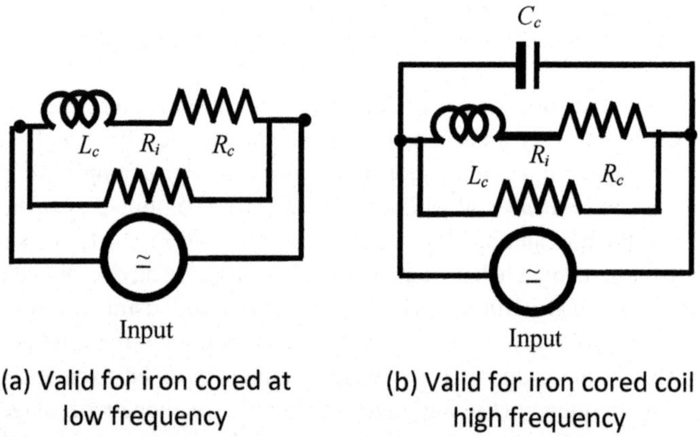

Fig. 41 Equivalent electrical circuit of an iron cored inductive sensor

4.5 Operating Environment Based Characteristics of a Sensor

Apart from the characteristics listed above a sensor is also characterized for the environment in which the sensor will work properly. The environmental conditions to be checked when using a sensor are:

(i) *The operating temperature range*: A sensor's behavior, in most cases, is also dependent on the temperature of the environment in which it is operated. Most manufacturers test this aspect and will specify a range of temperature over which the sensor is expected to work satisfactorily. Thus the range of temperatures of the environment in which a sensor is expected to work satisfactorily must be verified before selecting a sensor for a particular application.

(ii) Over and above the temperature the humidity of the environment in which the sensor is operating may also affect the operation of the sensor. If excessive humidity is expected in the environment in which the sensor is to be operated then a sensor which is capable of operating at that humidity level must be chosen. Thus manufacturers of sensor also test the sensor at different humidity levels and provide the result of the test. A sample specification is given below for a typical out door temperature sensor.
"Humidity: <95% and non condensing"

(iii) Apart from the temperature, the environment in which a sensor is operating may be polluted by debris, dust, water (humidity) and vibration. Thus it is also essential to verify whether a particular sensor is suited for the operating environment in which it is expected to work. The suitability of a sensor for environmental condition is tested and certified by the manufacturer. The test results are indicated through a standard code called the Ingress Protection code (IP Code) as listed in Table 4. IP code is also enshrined in national (AS 1939) and international standards (IEC 529, EN60529)

Table 4 Nomenclature for ingress protection code: IP klmXY

Single digit numerals			Alphabets			
k	*l*	*m*	*X* (defines the object for *k*)		*Y*	
Describes the protection against contact with object or dust (vide Table 5)	Details the protection against water ingression (vide Table 6)	Impact energy withstand capability (vide Table 7)	A	Hand	H	Can withstand high voltage environment
			B	Finger	M	Can withstand movement in water
			C	Tool	S	Can not withstand movement in water
			D	wire	W	Can operate in special weather conditions (to be specified)

Table 5 IP code—classification for numeral k (protection against contact with objects)

No	Type of protection guaranteed by the manufacturer	
0	No protection	The sensor has no protection against contact of an object, big or small
1	 Objects < 50 mm can enter	Sensor can withstand contact of soft parts >50 mm (brief contact with a hand can be tolerated—but no protection against deliberate contact)
2	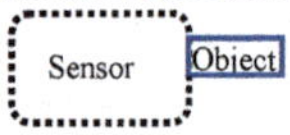 Objects < 12.5 mm can enter	Sensor can withstand contact of soft parts >12.5 mm (such as fingers)
3	 Objects < 2.5 mm can enter	Sensor can withstand contact of parts >2.5 mm (such as tools, thick wires)
4	 Objects < 1 mm can enter	Sensor can withstand contact with parts >1 mm (thin wires, screws ….)
5	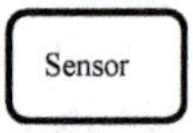 Dust proof	Sensor can work in a moderately dusty environment
6	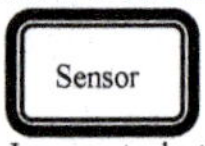 Immune to dust	Sensor can work in a any type of dusty environment

In the past, the IP code contained three numerals and may also contain two additional alphabets as listed in Table 4. The third numeral in the past identified the suitability of a sensor in an atmosphere wherein the sensor may be exposed to vibration and or shock. Today the IP code is restricted to only two numerals followed by one or two alphabets. The protection information conveyed by the first two numerals that are still in vogue is listed in Tables 5 and 6. The first numeral, as illustrated in Table 5, indicates the ability of a sensor to survive accidental contact with a foreign body or dust. The second numeral, as illustrated in Table 6, provides information on whether the sensor can survive/function normally in different atmospheric condition such as mist or rain or inundated in water.

The letter m in the earlier version of the IP code provided information on the ability of a sensor to withstand accidental impact. Table 7 lists the detail regarding the values m can take in a typical IP protection code IP $klmXY$ and the resultant impact withstanding capacity. It should be noted that in 2002, after the introduction of the international standard IEC 62262: 2002 the impact resistant capability was separated from the Ingress Protection code and new code, namely, the IK code was introduced. Table 8 lists the details of the IK code. For example if the IP

Table 6 IP code—classification *l* (protection against ingress of water)

No	Protection against ingress of water		Test conditions
0		Sensor will work only in dry atmosphere. Can not handle ingress of liquid in any form	Duration: nil Criteria: nil
1	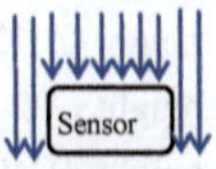 Immune to dripping water (vertical)	Sensor will work even if exposed to vertically dripping water environment	Duration: 10 min Criteria: 1 mm/min rainfall
2	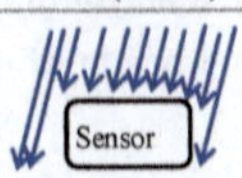 Immune to water spray up to 15°	Sensor will work if exposed to sprayed water impinging within ±15° to vertical position	Duration: 10 min Criteria: 3 mm/min rainfall
3	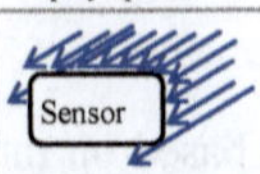 Immune to water spray up to 60°	Sensor will work if exposed to spray of water within ±60° to vertical position	Duration: 5 min Criteria: spray at 0.7 L/min at 100 kPa
4	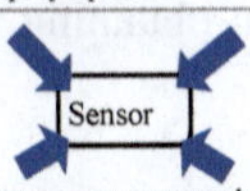 Immune to water splash from any direction	Sensor can withstand water splashing from any direction	Duration: 5 min Criteria: splashed at 10 L/min at 100 kPa
5	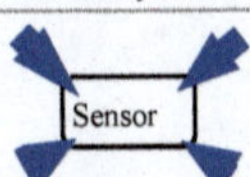 Immune to water jets from any direction	Sensor can withstand water jets from nozzle (6.3 mm) in any direction	Duration: 3 min Criteria: jet at 10 L/min at 100 kPa
6	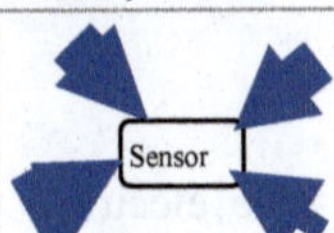 Immune to water jets from any direction	Sensor can withstand water jets from nozzle (12.5 mm) from any direction	Duration: 3 min Criteria: 12.5 L/min with a pressure of 30 kPa at a distance of 3 m
7	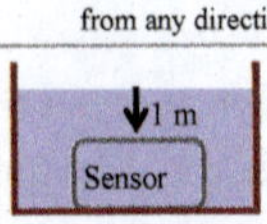 Can be immersed in water up to 1 m	Sensor will work even if submerged in water continuously up to a depth of 1 m	Duration: 30 min Criteria: submerged at a depth of 1 m
8	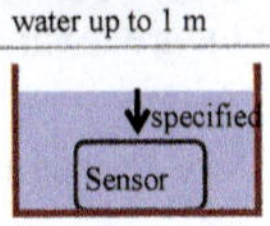 Can be immersed in water up specified depth	Sensor will work even if submerged in water continuously. Depth to be specified	Duration: continuous (normally 72 h) Criteria: submerged at a depth to be specified

specification of a particular sensor says the sensor conforms to IP50, then the sensor can be employed in a dusty environment but can not be employed in situations where excess moisture in the atmosphere is expected. On the other hand, a sensor specified to conform to IP67 can be used in any atmosphere (wet, dry and dry with dust) but the water pressure acting on it must be less than 10 kPa.

Table 7 IP XX*m* impact withstanding capability (no longer in use)

Value of m	Withstanding impact energy	Remarks
0	0	Can not withstand impact—fragile
1	0.225 J	Can withstand a 150 g mass dropped from a height of 15 cm
2	0.375 J	Can withstand a 250 g mass dropped from a height of 15 cm
3	0.500 J	Can withstand a 250 g mass dropped from a height of 20 cm
5	2.000 J	Can withstand a 500 g mass dropped from a height of 40 cm
7	6.000 J	Can withstand a 1.500 kg mass dropped from a height of 40 cm
9	20.000 J	Can withstand a 5.00 kg mass dropped from a height of 40 cm

5 Classification of Sensors

Sensors can be classified based on the principle of operation or based on the input or output parameters. Classification of sensors helps understand the advantages and shortcomings of a class of sensors and also provide insights in obtaining the best performance out of a class of sensors.

5.1 Active and Passive Sensors

In a general sense, sensors can be classified as:

(i) *Active sensors* and
(ii) *Passive sensors.*

Active sensors are the sensors which provide a measurable electrical output without the need for an auxiliary (external) power. Tachogenerators and thermocouples are examples of active sensors.

A passive sensor, on the other hand, requires power (and hence energy) from an external voltage or current source for sensing (providing an electrical output proportional to the physical input that is being sensed). Strain gage type pressure sensors and Linear Variable Differential Transformers (LVDT) are examples of passive type sensors.

Most of the sensors in vogue today are passive sensors and very few sensors are of the active type.

Table 8 Impact withstanding capability of a sensor as per IK cpode (EN62262)

IK code	Impact withstand capacity (J)	Remarks	Additional remarks
IK00	Nil	Sensor is fragile	IK01–IK10 compliance is tested normally with a pendulum type hammer. Spring type hammer can also be employed for ascertaining IK01–IK07 compliance. Free fall hammer can be employed only for testing codes IK07–IK10
IK01	0.14	Can withstand a 200 g mass dropped from a height of 7.5 cm	
IK02	0.20	Can withstand a 200 g mass dropped from a height of 10 cm	
IK03	0.35	Can withstand a 200 g mass dropped from a height of 17.5 cm	
IK04	0.50	Can withstand a 200 g mass dropped from a height of 25 cm	
IK05	0.70	Can withstand a 200 g mass dropped from a height of 35 cm	
IK06	1.00	Can withstand a 500 g mass dropped from a height of 20 cm	
IK07	2.00	Can withstand a 500 g mass dropped from a height of 40 cm	
IK08	5.00	Can withstand a 1.7 kg mass dropped from a height of 20 cm	
IK09	10.00	Can withstand a 5.0 kg mass dropped from a height of 20 cm	
IK10	20.00	Can withstand a 5.0 kg mass dropped from a height of 40 cm	

5.2 *Classification Based on a Sensor's Parameter(s)*

We have already learnt that a sensor (or a transducer) takes a physical quantity as input and provides an easily measurable electrical quantity as output. For example a tachogenerator (rotational speed sensor) takes the physical rotation of a shaft as input and provides an electrical voltage (which can be easily measured with a simple voltmeter) proportional to the rotational speed of the shaft as the output. Hence we can classify sensors based on the input or output parameter they are sensing as:

Classification of sensors based on the input parameters
When we group sensors based on the input parameter we get:

- pressure sensors
- displacement sensors
- temperature sensors
- angle sensors (clinometers)
- acceleration sensors (accelerometers)
- flow sensors (flow meters)
- rotational speed sensors (tachometers)

Classification of sensors based on the output parameters
It is also possible to group sensors based on the electrical output as:

- Analogue (output) sensors.
- Digital (output) sensors.

We can also classify sensors based on the parameter associated with the sensing element of the sensors. Examples are:

- Resistive type sensors.
- Inductive type sensors.
- Capacitance type sensors.
- Peizoelectric type sensors.

Since the sensors we are going to learn are mostly made of a sensing element operated upon by an electrical or electronic signal conditioning circuitry (most sensors available today are passive sensors and there are only a few active type sensors), it would be logical to group sensors based on the type of sensor element. In such a case, the electrical/electronic signal conditioning schemes applicable to sensors sensing different physical inputs but using similar sensing elements can be discussed in a coherent manner. In this book we take this approach and group sensors based on the sensing element characteristic and study them together from the electrical or electronic signal conditioning point of view.

6 Exercises

1 A 100 kPa full scale pressure gage has a scale marked with 100 divisions. Determine the resolution of the gage.
2 A transducer has an accuracy specification of 1% of FS and a nonlinearity of 0.5% FS. Calculate the worst case error that can be expected in the reading when a reading of 0.5 FS is obtained from the transducer.
3 A 500 kPa full scale pressure sensor's accuracy specification is given as 1.0% of reading and a nonlinearity of 0.2% full scale. Determine the worst case error

that can be expected in the output if the sensor reads (i) 500 kPa, (ii) 200 kPa, (iii) 50 kPa and (iv) 10 kPa. What do you infer from these results?

4 A sensor possess a characteristics of $y = 100x + 0.2x^2 + 0.1$. If the maximum value x can take is 1.0, determine the offset and the nonlinearity of the sensor.

5 A condenser type microphone employs a fixed circular plate and a clamped corrugated circular plate that can vibrate when sound waves impinge on it. Derive an equivalent circuit of the sensor. From the equivalent circuit determine its transfer function. Assume: Area of the plates is A_p, distance between the plates is d_p, air as the insulator, the mass of the corrugated plate is M_p, its stiffness is S_p and the viscous damping is D_p.

6 A transducer is modeled as a first order system with a transfer function $\frac{1}{(s+1)}$. Determine the magnitude and phase of the output of the transducer when its input is 10 Sin 2t.

7 A sensor is modeled as a second order system with a gain of 2, natural frequency of 1 rad/s and damping constant of 1. Determine its transfer function. If an input of 2u(t) is given to the sensor, plot its output. Determine the sensor's (i) Sensitivity, (ii) settling time (iii) slew rate (iv) bandwidth so that the error in the output within the bandwidth is <0.5%.

8 Repeat Exercise 5, for a similar sensor but having a damping constant of 0.6. Additionally calculate the number of overshoots and the peak value of the first overshoot.

9 A sensor has a forward characteristics (when input is monotonically increased from zero to full scale) defined by the equation $y = m_1x + C$ and exhibits a reverse characteristics (when input is monotonically decreased from full scale to zero) as $y = m_1x + D$ with $C \neq D$. What kind of characteristics the sensor possesses?

10 A displacement sensor has a characteristics $y = 100x$, where y is the output in volts and x is expressed in micrometer. The bandwidth of the transducer is 500 kHz and the slew rate is 0.1 V/μs. The input to the transducer is a 0.1 V amplitude sine wave at a frequency of 100 kHz. Determine its output.

Advanced Interfacing Techniques for the Capacitive Sensors

Tarikul Islam

Abstract Today, capacitive sensors are playing an important role in the measurement of different types of physical and chemical parameters. The capacitive sensor is very old and still it finds many new applications in different fields. Different types of the capacitive sensors and the interface electronics can be used for the instrumentation and the transduction application. The present chapter briefly reviews the important capacitive sensors suitable for the transduction applications. Some advanced electronic circuits for interfacing the perfect and the lossy capacitive sensors are discussed. The electronic circuits combine the advantages of the bridge method and the oscillator method of interfacing.

1 Introduction

The term "sensor" means to perceive something. The sensor represents the most important element in instrumentation, measurement and control applications. It is a device which senses the presence of the chemical or the physical parameters to be measured [1]. It is mainly used to collect different types of information from different sources which may be physical, chemical or biological in nature. Most often, the sensor converts any non-electrical quantity into an electrical signal. The transducer normally means a device, which converts the information from one form to the other form. The output energy of the transducer may be same or different from the input energy. According to the Instrument Society of the United States, the transducer is a device, which produces a measurable output in response to the specified physical or chemical measurand [2, 3]. A transducer is commonly called a passive transducer, if the output energy is almost entirely supplied by the input measurand while, an active transducer is one, which is having an auxiliary power

T. Islam (✉)
Jamia Millia Islamia, Central University, New Delhi, India
e-mail: tislam@jmi.ac.in

B. George et al. (eds.), *Advanced Interfacing Techniques for Sensors*,
Smart Sensors, Measurement and Instrumentation 25,
DOI 10.1007/978-3-319-55369-6_2

source, which supplies most of the power to the output signal [2]. On the other hand, an actuator acts opposite to a sensor, it generally receives the electrical signal and converts it into a nonelectrical signal. For example, a motor can be called an actuator [3].

The physical parameters to be measured by the sensors are generally pressure, force, inertia, acceleration, torque, temperature, humidity, flow, level, viscosity, displacement, speed, etc. The chemical parameters are of generally various toxic gases such as CO, NH_3, NO_2, toxic organic vapors, ionic concentration, chemical ware fare agents or biological species [1]. The sensors which work on the electrical effects can be fabricated utilizing different techniques such as resistive, capacitive, inductive, impedance, amperometric, electrolytic, ISFET, ChemFET, piezoelectric (SAW) etc. [1]. Among various types, the capacitive method is widely used for sensing a large number of physical and chemical parameters [2, 4–6]. For example, in automobile industry, practically all accelerometer sensors are of the capacitive type and most of the humidity sensors employ the capacitive method for humidity detection [2, 7, 8]. Similarly, many toxic vapors such as ethanol, methanol, iso-propyl alcohol, etc. can be detected by the capacitive sensor [9, 10]. The main advantages of the capacitive sensor are easy fabrication, high sensitivity, low power consumption, small size and low cost [2, 4, 5]. However, toxic gases are normally detected by the resistive sensors and very recent, piezoelectric SAW devices get popularity day by day [1]. The present chapter mainly focuses on the capacitive type sensors and the recent advancement in the related electronic circuits. Three important classes of the capacitive sensors are used for the transduction applications [2, 6, 11–13]. These are

(a) Parallel plate capacitors
(b) Cylindrical co-axial capacitors
(c) Cylindrical cross-capacitors

Among them, the parallel plate capacitors are extensively used for the measurement applications [2]. The design of the capacitive sensor can be broadly classified into two categories: grounded and floating. In the grounded capacitive sensors, one of the electrodes of the sensor is connected to the ground terminal, while in the floating type capacitive sensors, neither of the electrodes are grounded [14].

2 Capacitive Sensors for Sensing Applications

2.1 Parallel Plate Capacitive Sensor

A parallel plate capacitor shown in Fig. 1 with the dielectric of permittivity ε_r can be expressed by the following expression [2, 11]

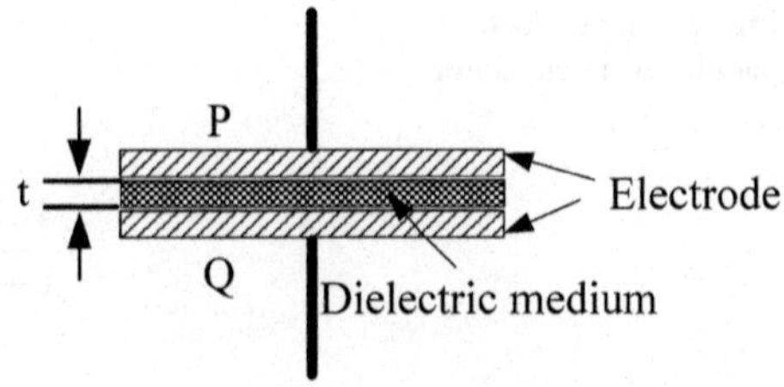

Fig. 1 Parallel plate capacitor

$$c = \varepsilon_0 \epsilon_r \frac{A}{t} \tag{1}$$

where 'A' is the cross-sectional area of the plate electrode, 't' is the gap between the electrodes. The expression (1) is true, when the gap between the electrodes is much smaller than the overlapping area. This is due to the fact that for t ≪ A, the fringing field is negligible. The capacitance (C) of (1) can be varied by (a) the cross-sectional area, (b) the gap between the electrodes and (c) the relative permittivity of the dielectric medium, (d) by inserting grounded plate between the stationary main Electrodes (4). Both the translation and the rotational motions can be measured by varying A and t. When the displacement is large, the gap is kept fixed and the capacitance variation due to the variation of A is preferred. But for smaller displacement, the area is kept fixed and the capacitance variation due to the variation of the gap is preferred. The capacitive sensor, which works on the variation of the gap is more useful for the transduction applications provided the variation of the gap is not large. The capacitance, with the gap variation can be given by [2]

$$c = \varepsilon_0 \epsilon_r \frac{A}{t + \Delta t} = \varepsilon_0 \epsilon_r \frac{A}{t\left(1 + \frac{\Delta t}{t}\right)} \tag{2}$$

The capacitance varies nonlinearly with the variation of t. In order to determine the linear range, expand the expression using Taylor series expansion,

$$c = \varepsilon_0 \epsilon_r \frac{A}{t}\left[1 - \frac{\Delta t}{t} + \left(\frac{\Delta t}{t}\right)^2 - \left(\frac{\Delta t}{t}\right)^3 + \cdots\right] \tag{3}$$

It is shown in the expression that for small change in gap, the change in C is directly proportional to the change in t. For example, in MEMS based sensor, the gap between the electrodes is of few μm only, so the proportionality can be maintained when Δt is ~nm order [3]. The nonlinearity due to the gap variation can be eliminated by the use of the three electrode differential form of the structure as shown in Fig. 2. The sensor consists of three electrodes with air as the dielectric medium and forms two capacitances C_1 and C_2 with respect to the middle one. Both C_1 and C_2 can be varied either by moving the middle electrode keeping the outer

Fig. 2 Three electrodes parallel plate capacitor

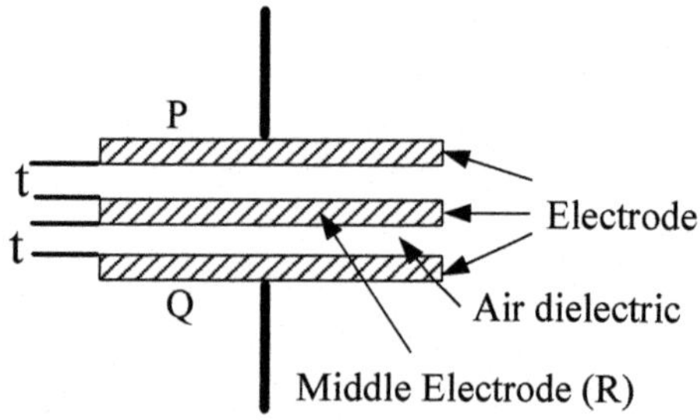

electrodes fixed or by moving the outer electrodes keeping the middle one fixed. At the exact middle position, both the capacitances are equal. But if the middle electrode moves, one of the capacitances will increase, while the other will decrease. The capacitance C_1 and C_2 can be written as

$$C_1 = \varepsilon_0 \epsilon_r \frac{A}{t - \Delta t} \text{ and } C_2 = \varepsilon_0 \epsilon_r \frac{A}{t + \Delta t} \tag{4}$$

The differential form of the capacitance $C_d = (C_1 - C_2)$ for $(\Delta t^2 \ll t^2)$ can be written as

$$C_d = \varepsilon_0 \epsilon_r \frac{A}{t} 2 \frac{\Delta t}{t} \tag{5}$$

If we compare this expression with (2), the three electrode capacitor has linear relationship between C and delta t and the sensitivity is almost double. If one of the capacitances C_1 is excited by an alternating voltage source (V, ω) and the other by −(V, ω), then the current through the capacitor is given by

$$I = j\omega(C_2 - C_2)V = j\omega \cdot \varepsilon_0 \epsilon_r \frac{A}{t} 2 \frac{\Delta t}{t} \cdot V \tag{6}$$

The current is proportional to the difference of two capacitances, which is proportional to the variation of the air gap due to the movement of the middle electrode. This current can be converted into a voltage signal using a current to voltage converter circuit. If the electrodes are fixed, then the three electrode capacitive structure can be used to measure the permittivity of the solid or the liquid dielectrics. For example, if C_1 is a free space air capacitor and C_2 is a capacitor with dielectric ε_r, whose permittivity is to be measured, then the permittivity of the dielectric medium can be given by the following expression [15]

$$\varepsilon_r = \frac{C_2}{C_1} \tag{7}$$

Thus, the permittivity can be determined by taking the ratio of the two capacitances. Sometimes, there is a need to fabricate a multi dielectric capacitive sensor

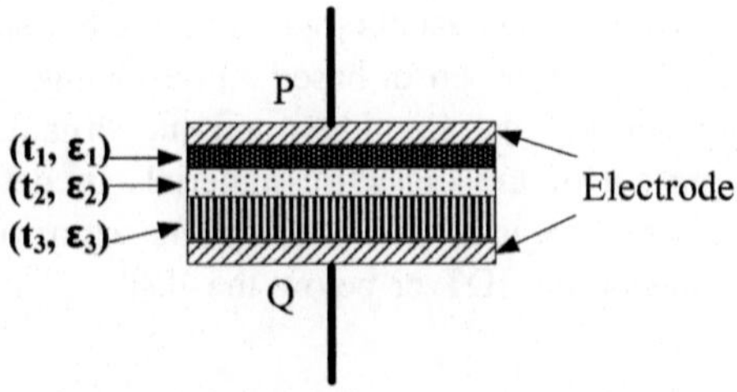

Fig. 3 Multi dielectric parallel plate capacitor

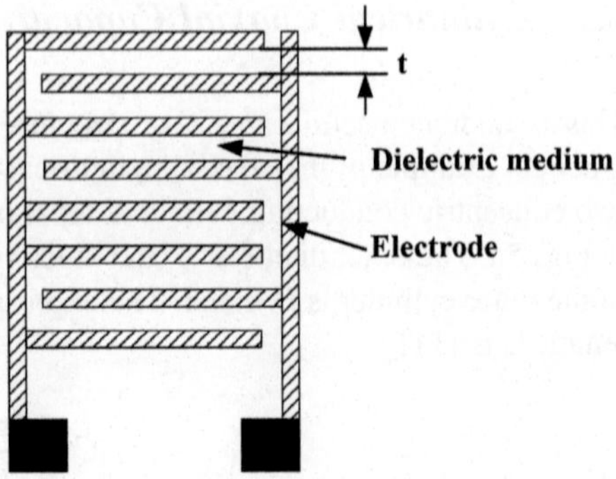

Fig. 4 Interdigitated (multi-electrode) capacitor

for sensing the measurands. For a parallel plate capacitive sensor with multiple dielectric layers shown in Fig. 3, the capacitance value can be represented by [11].

$$\frac{1}{C} = \frac{1}{C_1} + \frac{1}{C_2} + \frac{1}{C_3} = \frac{A}{\frac{t_1}{\varepsilon_1} + \frac{t_2}{\varepsilon_2} + \frac{t_3}{\varepsilon_3}} \tag{8}$$

where t_1, t_2, t_3 are the thickness of the dielectric layers of permittivity ε_1, ε_2, ε_3 respectively. For example, a capacitive humidity sensor fabricated by depositing the moisture adsorbing metal oxide film on the polyimide substrate, the resulting capacitor structure with multiple dielectrics is formed. Such sensor is employed to measure the humidity over a wide dynamic range [16].

Another interdigitated parallel plate (IDT) capacitive structure, which is very popular in the literature for humidity and organic vapors sensing is shown in Fig. 4. This is a multi-electrode capacitor fabricated on the dielectric medium [17, 18]. For an interdigitated electrode capacitor with permittivity ε_r having N number of plates (fingers), the capacitance can be approximately (electrodes are close to each other) given by

$$C = \frac{\varepsilon_0 \varepsilon_r (N-1)}{t} \tag{9}$$

The equation shows that the capacitance of the parallel plate capacitor can be increased to a large value by using the multi plates with identical dielectric. The number of plates on both sides of the capacitor may be same or be different. Some interdigitated capacitors with guard electrode are also reported in the literature [19]. The capacitive sensor based on the change in relative permittivity is most often used for sensing the humidity and the organic vapors or determining the dielectric constant of the dielectric materials or detecting the biological species [20]. For sensing the water and the organic vapor molecules, the sensing film is deposited either on the IDT or below the IDT.

2.2 *Cylindrical Coaxial Capacitive Sensor*

This is another useful capacitive structure for the sensing application. The most practical example of the coaxial cylindrical capacitor, is the coaxial cable. Consider two concentric conducting cylinders separated by the dielectric medium ε as shown in Fig. 5a. The outer diameter of the inner cylinder is d meter and the inner diameter of the outer cylinder is D meter. The expression of the capacitance with the cylinder length L is [11]

$$C = \frac{2\pi \varepsilon L}{ln\frac{D}{d}} \tag{10}$$

Such structure can find application for the measurement of non-conducting liquid level or other parameters [21, 22]. The structure for the measurement of the non-conducting liquid level using the cylindrical sensor is shown in Fig. 5b. Suppose, the length of the cylinder is L meter and the level of the liquid is H meter, then the capacitance for determination of the liquid level can be given by following expression

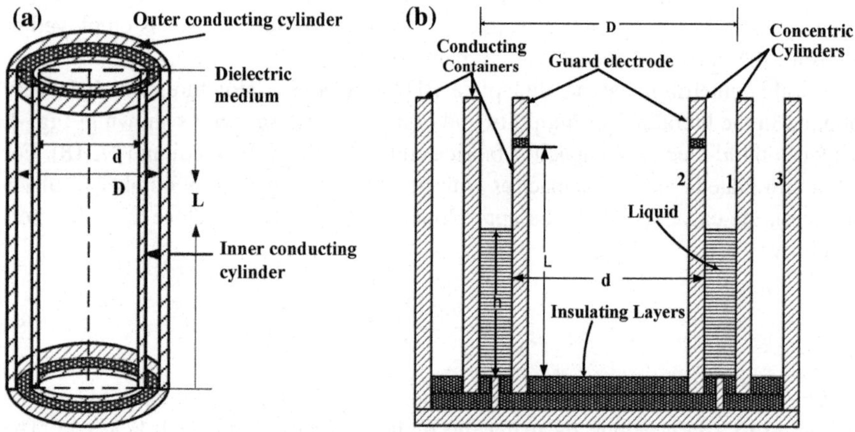

Fig. 5 **a** Cylindrical coaxial capacitor. **b** Coaxial capacitor for liquid level measurement

$$C_{12} = 2\pi \frac{[H\varepsilon_r + (L-H)\varepsilon_{r1}]}{ln\frac{D}{d}} \tag{11}$$

where, ε_r and ε_{r1} are the permittivity of the liquid and air respectively. The (11) can be written as

$$C_{12} = KH(\varepsilon_r - 1) + KL \tag{12}$$

where $K = \frac{2\pi}{ln\frac{D}{d}}$. Thus, the capacitance is directly proportional to the level of the liquid.

2.3 *Cylindrical Cross-Capacitor*

This is another very useful capacitive structure which is used as a primary standard due to its single dimensional accuracy. Consider a cross-section of a conducting cylinder of length L in a dielectric medium ε shown in Fig. 6. The cylinder is divided into four equal parts (A, B, C, D) with very small insulating gaps at points (a, b, c, d) [6].

The cross-capacitances (C_1, C_2) between the opposite parts A-C and B-D can be given by

$$e^{-\frac{C_1\pi l}{\varepsilon 0}} + e^{-\frac{C_2\pi l}{\varepsilon 0}} = 1 \tag{13}$$

For $C_1 = C_2 = C_0$, then $C_0 = \ln(2)L$.

In a well-designed cross-capacitor, $C_1 = C_2$ can be easily achieved. However, if there is a deviation from the symmetry, C_0 will differ from C_1 or C_2 by ppm only.

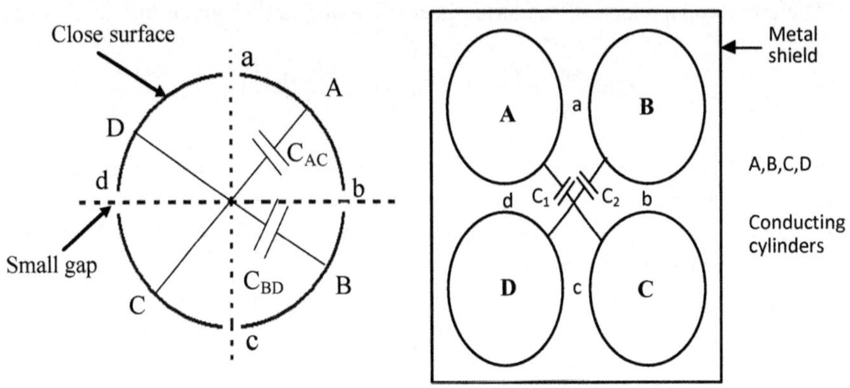

Fig. 6 Cylindrical capacitors with metal shield

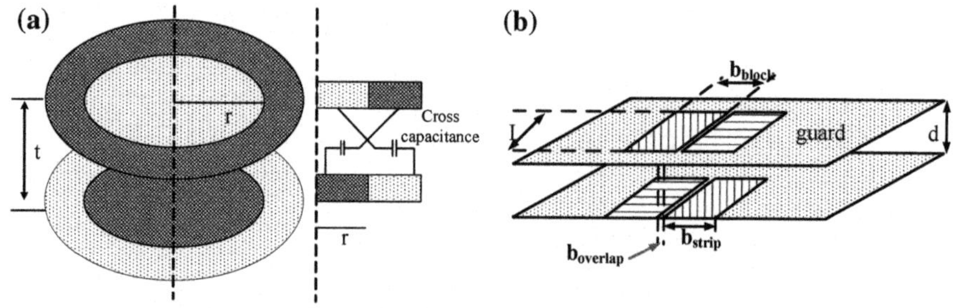

Fig. 7 The cross-capacitor **a** circular **b** strip sheet planar

The most important advantage of the cylindrical cross-capacitance is that the capacitance value can be determined by a single length measurement. Therefore, the accuracy of the capacitor depends on the accuracy of the length measurement only. However, the structure proposed by the Thomson and Lampard is not suitable for sensing application as the materials and the fabrication of the structure are costly. Also the capacitance value is small, unless the length is larger. Rahman and VGK Murty utilize the cross-capacitive structure for the measurement of the pressure [22]. An alternative structure as suggested by the Heerens [12, 13] shown in Fig. 7a is more suitable for the fabrication and may be useful for the sensing applications.

When r ≫ t, the capacitance value C is given by the following expression [12]

$$C = \frac{\varepsilon_0 \varepsilon_{r \ln(2)}}{\pi} 2\pi r \left[1 + 0.00043507\left(\frac{t}{r}\right)\right] - 0.050631437\left(\frac{t}{r}\right)^2 + \cdots \tag{14}$$

The analytical calculation of the circular structure is complex and the numerical solution based on the finite element method does not give desired accuracy. This circular capacitor can be replaced by a lateral strip type capacitor. This strip sheet capacitor shown in Fig. 7b can be treated as the transformation of the circular cross-capacitor having guard electrode with infinite radius r [12].

The cross-capacitance of the strip sheet capacitor can be given by

$$C \cong \frac{\varepsilon_0 \varepsilon_r}{\pi} L\left(\ln(2) + ln\left(\cosh\frac{\pi b_{overlap}}{2d}\right)\right) \tag{15}$$

However, this capacitor is less symmetrical and hence, the sensitivity to the parameter variations may be comparatively more than the circular cross-capacitor. Fabrication of the strip capacitor is easy and may be useful for the precise measurement of some of the measurands.

3 Precautions Necessary for the Use of the Capacitive Sensors

3.1 *Shielding of the Capacitive Sensor*

For the capacitive type transducer for sensing applications, most of the times, it is absolutely necessary to shield the capacitors from the external electrostatic fields. There are different sources, which produce the unwanted electrostatic field [6, 11]. Shielding is provided by enclosing the capacitor inside a grounded conducting metal. The conducting shield may be made of aluminum, copper, brass or the wire mesh [11]. Due to the presence of the shield, the charges accumulated on the shield pass to the ground without affecting the overall capacitance value of the capacitors. However, in one of the research papers, the metal shield of a three-electrode capacitor has been used for the measurement of the speed of a rotating device like motor [23]. A parallel plate capacitor within a metallic shield can be represented by a three-terminal capacitor as shown in Fig. 8a. C_{AB} is the actual desired capacitance, C_{AG}, and C_{BG} are the parasitic capacitances formed at terminals A and B respectively with respect to the grounded metal shield. C_{AG}, and C_{BG} can be minimized by selecting a suitable circuit configuration of an interfacing electronic circuit as reported in [4, 10, 15].

3.2 *Guarding of the Capacitor Electrodes*

To avoid the inaccuracy in the calculation of the parallel plate capacitor due to the fringing fields at the edges of the electrodes, the guarding is necessary [11]. The distribution of such unwanted fields is uncertain and this causes, an inaccuracy in the exact value of the capacitance. Figure 9 shows the arrangement of a guard ring of a parallel plate capacitor. The ring, which surrounds the electrode P of the capacitor is of the metallic plate. The thickness of the guard ring may be of the same plate thickness P. There is a small gap between the guard ring and the plate P. The outer diameter of the guard ring is usually same as the diameter of Q, the other plate of the capacitor.

If the guard ring is the same potential as that of the electrode P, then the fringing field is transferred to the extreme edges of the guard and in this situation, the effective area of the capacitor is now equal to the electrode of the plate P. The guard electrode is essential for other types of the capacitive sensors such as the cylindrical coaxial and the cylindrical-cross capacitor including the structure proposed by Heerens [12]. For the cylindrical capacitor, the guard ring is also in cylindrical shape and of the same diameter as that of the cylindrical electrode to which the guard is attached. It is placed at one end and is coaxial to the cylindrical electrode. In the cylindrical capacitor shown in Fig. 5b, for the liquid level measurement, the guard ring is coaxial and placed at one end of the inner electrode.

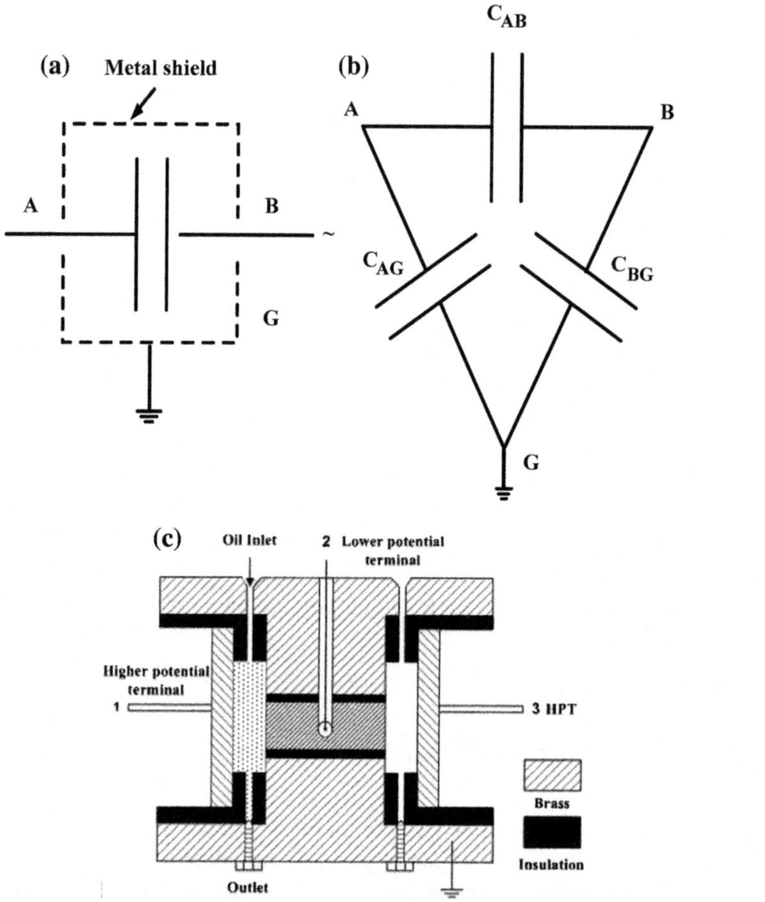

Fig. 8 Capacitor **a** within a grounded metal shield, **b** equivalent circuit, **c** three electrode capacitors within metallic shield

Fig. 9 Parallel plate capacitor with guard ring

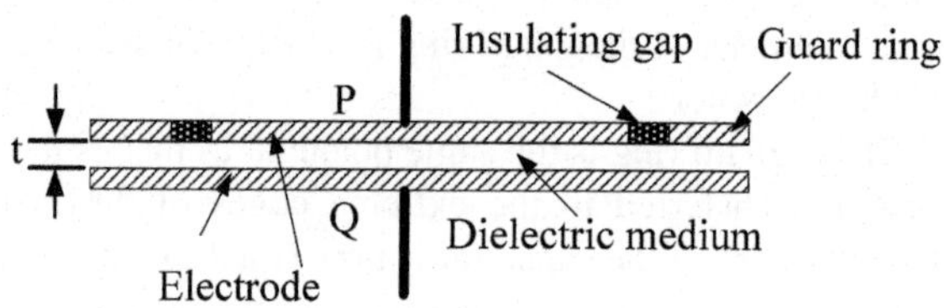

3.3 *Shielded Cable and the Cable Length*

These are other important precautions necessary for the capacitive sensors. These requirements are more stringent for the low value capacitive sensors. The connecting wire is essential every time the sensor is to be interfaced with the detection electronic circuit. Normally, the cable is having two insulated wire, which has an unnoticed distributed wire capacitance along its length. The wire capacitance depends on the type of the insulation of the conducting wires and the length. This capacitance is combined with the desired sensor capacitance. The typical value of the wire capacitance is 30 pF/ft and the resistance is of 0.01 Ω. One solution is to use the shielded cable, which is having the metallic mesh or the thin sheet on the outer periphery of the insulated signal wire. This grounded metallic shield prevents the external electrostatic field to couple with the wire. However, the metal shield is not effective to prevent the electromagnetic noise without magnetic shielding. Magnetic noise picks up can be reduced by using a twisted signal wire (minimize loop area). A commercially shielded cable is having twisted wires wrapped within a metal shield. However, the internal distributed capacitance between the signal conductors can be reduced by insulating the wires with high quality insulation like Teflon and by reducing the length of the wire as far as possible [2]. Also, it is important to avoid the formation of the ground loop, when the interface circuit is connected to more than one ground point.

4 Lossy Capacitive Sensor

When a capacitor is excited by a constant DC voltage source of V volts, the charging current flows through the capacitor for a short duration of time till the capacitor is charged to the full voltage. The current through the capacitor stops flowing, when the voltage, across the charged capacitor reaches the input voltage level. A small conduction current continuously flows through the capacitor due to the presence of very high resistance of the dielectric medium. But, it is observed in practice, in most of the dielectrics, the initial charging current is much higher than the conduction current and the current decays exponentially taking longer duration to stop. This suggests, the other phenomena like the adsorptive nature of the dielectrics. If a capacitor with an absorptive dielectric is initially charged, and then discharged, it charges itself and the potential across its plates rises again. This absorptive behavior is due to the viscous movement of the ions and the molecules in the dielectrics.

Initially, when a capacitor is connected across a voltage source, the ions and the molecules move very fast for a short duration of time giving rise to large initial charging current, but later on due to slow motion, the current takes longer times. For an absorptive capacitor, the time response of accumulation of the charges is shown in Fig. 10a [10]. The time response of the charges Q consists of three important parts: (i) the initial high charging current (ii) slowly rising absorptive current and the steady state conduction current. An absorptive capacitor can be approximately represented by an equivalent circuit shown in Fig. 10b. C is the geometric capacitance due to the permittivity of the dielectric medium, R_b is the resistance of the dielectric contributing the conduction current and (C_a, R_a) are

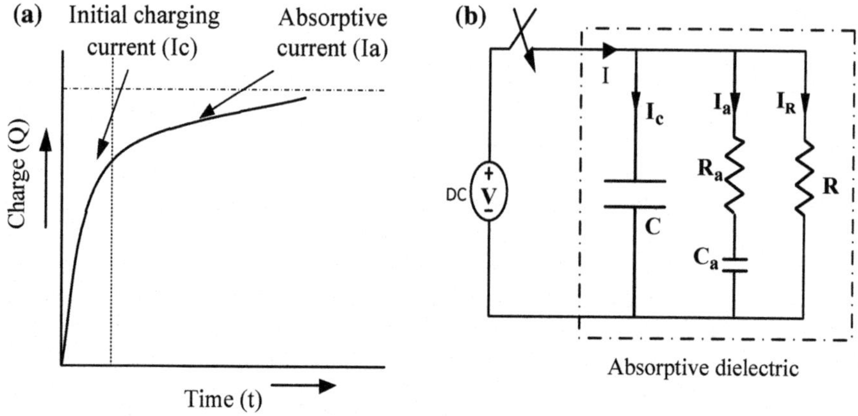

Fig. 10 **a** Time response of charge accumulation in the absorptive capacitor, **b** its equivalent circuit

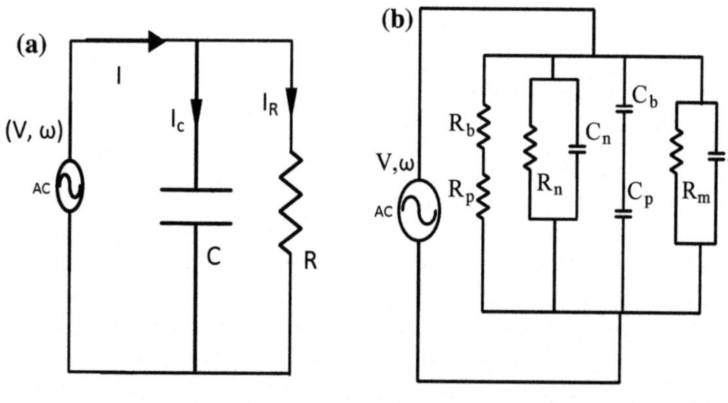

Fig. 11 **a** AC equivalent circuit of a lossy capacitor, **b** equivalent circuit of an Al_2O_3 moisture sensor

the resistance and the capacitance contributing the absorbing current. In an actual capacitor, there may be several such branches of (C_a, R_a) in the equivalent circuit.

When the capacitor is connected across an alternating voltage source V (rms), the absorption effect causes the power loss. The current I (rms) through the dielectric leads the voltage by angle Φ. This angle is equal to 90° for the perfect capacitive sensor but for the lossy capacitor, it is less than 90°. The loss angle δ is equal to (90 – θ)°. The dielectric loss is given by VIcosΦ or IVsinδ. An imperfect capacitor at any signal frequency can be equivalently represented by a perfect capacitor in parallel with a resistance showing the power loss (Fig. 11a). The equivalent resistance in general varies with the frequency. For example, a humidity sensor or an organic vapor sensor works as a lossy capacitor. The capacitance and the resistance of the sensor vary with the signal frequency [11, 18]. The power loss in the imperfect capacitor can be given by $P = V^2\omega C\tan\delta$.

For example, consider a porous aluminum oxide capacitive humidity sensor. The porous oxide is formed by anodization of a pure aluminum sheet in the presence of acidic electrolytic solution (sulphuric or oxalic acid). Anodization is one of the well-established methods to prepare the nanostructure of the metal oxide, which is used for the measurement of the humidity over a wide dynamic range from %RH to trace moisture in ppm level. The approximate equivalent circuit of the metal oxide humidity sensor is shown in Fig. 11b [24]. In Fig. 11b, C_n, R_n are the capacitance and the resistance of the alumina nano-walls, C_p, R_p are the capacitance and the resistance of the void (pores), C_m, R_m denote the multi-dielectric (barrier layer, chemisorbed layer, water condensed in the voids) capacitance and resistance respectively.

5 Advancement in the Design of Electronics Interface for the Capacitive Sensors

5.1 *Introduction*

An interface electronics or read out is a detection electronic circuit, which is designed and realized according to the type and the output of a primary sensing element. Sensors and the detection electronics are the most important parts of a measurement system. Performance of a measurement system is generally specified by accuracy, repeatability, sensitivity, resolution, linearity, response and recovery times, hysteresis, drifts, cross-sensitivity [1]. The functional units of a sensor are based on measurement system in general, and can be represented by the following block diagram (Fig. 12).

The primary sensor/the transducer senses the presence of the desired physical or chemical parameter to be measured. The output of the sensor may or may not be in a suitable form for further processing. The sensor may be either the resistive or the inductive or the capacitive type, sometimes, it provides the output in the form of the voltage or the current signal. Most often, the desirable output of the electronic circuit is to be in the form of voltage, current or frequency (time period). Since, the real world phenomena are continuous in time, the output of the sensor is analog. However, the digital output of the sensor is desirable for easy interfacing, noise

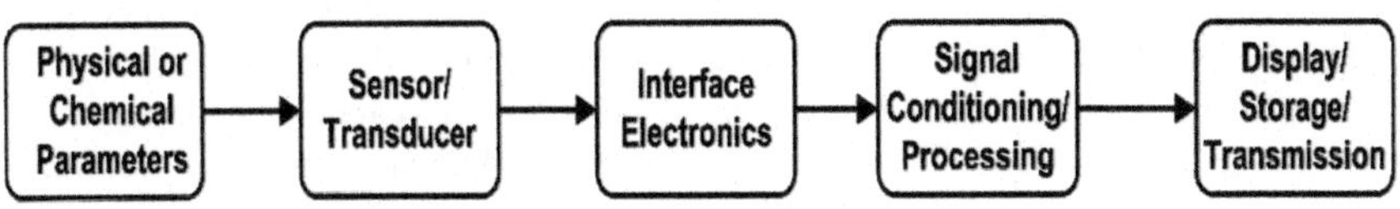

Fig. 12 A general block diagram of a measurement system

immunity, storing and communication with a digital system. The parameters of the sensor can be measured by a LCR meter, but for a low cost dedicated electronic test system, the interface electronic circuit is needed. The interface circuit provides an easy manipulation and the conditioning of the electrical signal such as amplification, filtration, minimization of loading effects, (impedance matching) etc. When, the sensor is remotely placed to monitor the measurand, it is necessary to transmit data. Finally, the data should be represented in the form, which can be easily displayed (analog or digital), recorded and stored. The signal conditioning unit is another important unit in the measurement system. When a sensor is to operate in a particular environment, there are some undesired inputs which invariably affect the desired output signal causing significant measurement error. One of the inputs may be called interfering input, which is often present in the environment such as 50 Hz electromagnetic fields or electrostatic fields generated by the nearby electrical apparatus. The ambient temperature or the humidity may be treated as other interfering input. Sometimes, a situation becomes worse, when the interfering input modifies the input–output relation of the desired signal. Therefore, care must be taken to avoid the errors to ensure the correct measurement accuracy.

Different interface electronic circuits, based on resonant oscillators, AC bridge, charge–discharge, capacitance to frequency (or time period) or capacitance to phase angle have been reported in the literature to interface the different capacitive sensors [25–28]. The suitability of the techniques for the measurement applications depends on the requirement of accuracy, resolution, minimum baseline drift, immunity to the stray capacitance, effects of humidity and ambient temperature and elimination of the shunt conductance of the sensor [25]. The resonant method is very old and effective for measuring both capacitance and conductance of the sensor. In this, an unknown capacitor is a part of a resonant circuit, which is having known value of the inductance. The resonance condition is obtained by varying the input signal frequency and is detected either by measuring the voltage or the current. The capacitance and the resistance of the sensor are measured from the expression derived at the resonant condition. This method is rarely used nowadays because of manual adjustment of the frequency of the source voltage for obtaining the resonant condition and the difficulties of minimizing the stray effects [25].

In the charge-discharge method, the sensor capacitance is charged with a fixed voltage V using an analog switch and then discharged through a current detector using a second switch. The charge-discharge steps are repeated for several cycles (f) with the controlled clock pulse. The average discharge current ($=V\ f\ C_x$) is then converted into a DC voltage. For better accuracy and the minimization of the conductance of the capacitor, sometime differential configuration having another identical charge–discharge network with reference capacitance C_r is employed. However, it also suffers from poor immunity to the stray capacitance effect [25].

The AC bridge method is still considered as the most accurate and stable method for interfacing a capacitive sensor, which is having small capacitance change [25, 26]. The bridge consists of four arms formed by unknown impedance, reference impedance, and two out of phase identical voltage sources. If the two impedances are not in balance, an imbalance current flows through the bridge which can be converted into a voltage signal using a current to voltage converter (I-V). This is immune to various sources of the errors, including the most prominent stray capacitance. A phase detector circuit at the output of the I-V converter reduces various sources of the noise, thus giving highly sensitive and accurate results. The accuracy can be further improved by incorporating an auto balancing arrangement to minimize the fluctuation of the excitation voltage. By auto balancing of the bridge, the capacitance value lower than pF can be measured [29, 30]. The main drawback of the AC bridge is the limited measurement range of the capacitance value [25].

In the oscillation method, the unknown capacitance C_x is the part of an LC or RC oscillator circuit [25, 26]. The oscillation frequency (or time period) depends on the value of the C_x, therefore, by measuring the frequency, the capacitance value can be determined. Nowadays, the relaxation oscillator based interface electronics are getting popularity day by day for both capacitive and resistive sensors [31–36]. This method is suitable for interfacing the sensor, whose capacitance value varies over a wide range. However, the oscillation method suffers from poor immunity to the stray capacitance, the frequency stability and the effects of the shunting conductance. There are some commercially available evaluation boards ICs which can be used to directly interface the capacitive sensors like analog device AD7746, Texas Instrument FDC1004EVM [37]. These evaluation boards are capable of measuring very small capacitance with Femto farad resolution, but these boards do not provide much flexibility to the user for different type of the capacitive sensors and are costly.

5.2 *Interface Electronics Circuit for the Capacitive Sensors*

A simple opamp based interface read out for a capacitive sensor is shown in Fig. 13a. Often a large value resistance is connected across the feedback

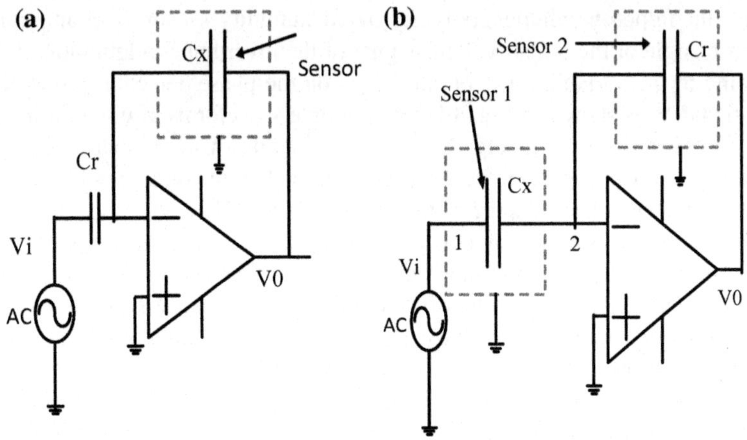

Fig. 13 The simple interface circuit using opamp as inverter **a** single sensor **b** double sensors

capacitance to provide a path for the DC current due to the opamp offset voltage and the DC bias current. Assuming ideal opamp, the voltage output can be given by [2]

$$V_0 = -V_i\frac{C_r}{C_x},\ or\ V_0 = -V_i\frac{t}{A}\frac{1}{\in_0\in_r}C_r \tag{16}$$

The output voltage is directly proportional to the gap (t) between the electrodes of the senor. The output voltage can be directly calibrated into the motion of the moving plate. If C_x and C_r are the two identical air capacitors with a common electrode (terminal 2), the inverting opmap configuration shown in Fig. 13b can be used to measure the relative permittivity (ε_r) of the dielectric medium [15, 38]. Such three-electrode capacitive transducer is shown in Fig. 8c. For example, if C_r is the capacitance with a dielectric medium and C_x is the free space air capacitor, then Eq. (16) can be written as

$$\begin{aligned} &V_0 = -V_i\in_r \\ &\text{since } C_r = \frac{\in_0\in_r}{t}A \text{ and } C_x = \frac{\in_0}{t}A \end{aligned} \tag{17}$$

Thus, the output voltage of the interface is directly proportional to the permittivity of the dielectric medium to be measured. The circuit may be suitable for the precise measurement of a capacitance with the immunity of the stray capacitance. If the capacitive sensor is within a conducting grounded shield as shown in Fig. 8a, then it can be represented by an equivalent circuit as shown in Fig. 8b. C_{2G} and C_{1G}, are the parasitic capacitances. Output voltage V_0 of the interface circuit will be almost independent of the stray capacitances. This is because, the stray capacitance C_{1G} appears across the input voltage and the node 2 is at the virtual ground potential, therefore C_{2G} will be negligible. The same inverting opmap based interface circuit can also be utilized to interface a lossy capacitive sensor like porous silicon/porous alumina/polymer based humidity sensor. The amplitude or the phase angle of the output voltage signal of the inverting configuration will vary according to the variables. The circuit based on the phase detection principle with input signal V_i is shown in Fig. 14. The possible waveforms at different terminals of the circuit are shown in Fig. 15. The phase angle of the opmap output V1 is shifted due to the change in impedance (Zx), caused by the measurand for example, the humidity. The phase change of the output signal (V1) with respect to the input signal (Vi) can be extracted by converting both the input and the output signals to the square wave outputs using the comparators and then by comparing the outputs by a digital XOR gate. The output of the XOR gate is a pulse wave signal, which can be converted into a DC voltage (V_{DC}) using a demodulator or into the proportional clock pulses [39].

In Fig. 14, $Z_x(=R_x || C_x)$ is a capacitive sensor at the input and $Z_f(= R_f || C_f)$ is the impedance at the feedback path of the opmap respectively. The output V_1 at the output of the inverter is given as $V_1 = -V_iZ_fY_x$, where the admittance

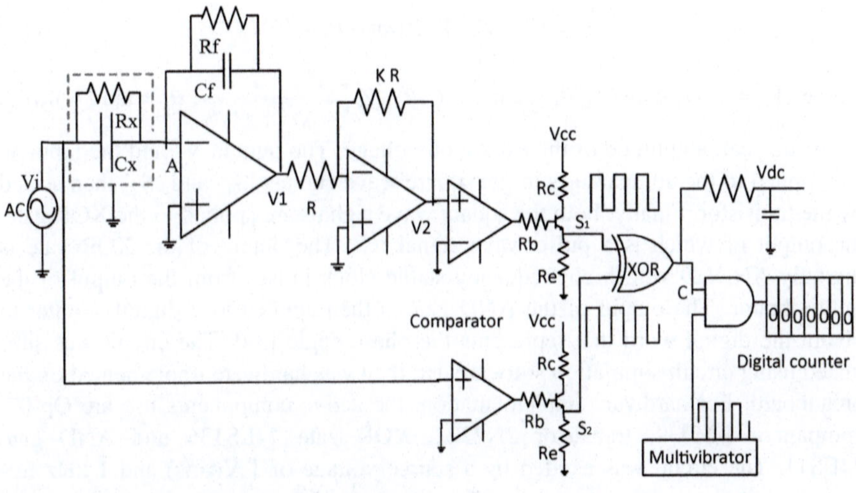

Fig. 14 Opamp configuration to interface a lossy capacitive sensor

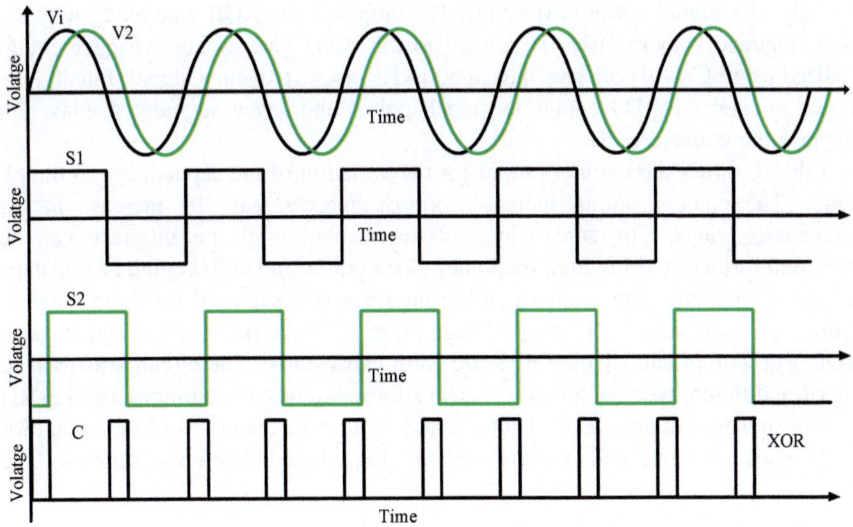

Fig. 15 Timing diagram at different points of the circuit

$Y_x = \frac{1}{Z_x} = G_x + j\omega C_x$ and the impedance $Z_i = \frac{R_f}{1+j\omega C_f R_f}$, $G_x\left(=\frac{1}{R_x}\right)$, ω is the signal frequency $(=2\pi f)$. The output of the second stage inverter, $V_2 = V_i Z_f Y_x$, which can be written in polar form as

$$V_2 = KV_m|Z_f||Y_x|\angle(\omega t + \theta_1 - \theta_2) \quad (18)$$

where $|Y_x| = \sqrt{G_x^2 + \omega^2 C_x^2}$, $\theta_1 = \tan^{-1}\omega C_x R_x$, $|Z_f| = \frac{R_f}{\sqrt{1+\omega^2 C_f^2 R_f^2}}$, $\theta_2 = \tan^{-1}\omega R_f C_f$, V_m is the peak amplitude of the excitation voltage. The output, V_2 and the input V_i are applied to the comparators to get square wave signals 'S_1' and 'S_2' then scaled by the transistor. Finally, both the square wave signals are applied to the XOR gate, the output of which is a pulse wave signal 'C'. The output of the XOR gate is digitally ANDED with high frequency stable clock pulses from the output of the multivibrator. The output of the AND gate is then applied to a digital counter to obtain the digital word that represents the phase angle [39]. The circuit was simulated using circuit simulation software and then was hardware implemented on the breadboard. For hardware implementation, the active components use are Op-07, comparator LM339, transistor 2N2222, XOR gate 74LS136 and AND gate 74LS11. The circuit was excited by a source voltage of 1 V(rms) and 1 kHz frequency. Initial experiments were performed using discrete resistance and ceramic capacitance for Z_x and $C_f = 100$ pF, $R_f = 1$ MΩ for Z_f. Suitably scaled square wave signals using transistors 2N2222 are applied to the inputs of XOR gate with the help of a digital buffer (74LS244). The output of the XOR gate along with the high frequency clock pulses is applied to the AND gate. Schmitt trigger circuit realized using CD40106B generates a 500 kHz clock frequency signal. Finally, the digital counter CD4029 counts the clock pulses and seven segment display unit displays the counted pulses.

Table 1 shows the counter output for the variation of the capacitance in the pF range. The counter output increases almost linearly with the increase in the capacitance value. The sensitivity and the resolution of the interface can be increased further by using high frequency clock pulses and utilizing the fast opamps for the comparator. The accuracy of the interface was analyzed for three different sources of errors such as (i) offset voltage (ii) error due to rising and falling edge of XOR gate output and (iii) the parasitic earth capacitance. The circuit was used to interface different types of humidity sensors for measuring the humidity from % RH to trace moisture in ppm level. Tables 2 and 3 show the responses of the circuit for the porous alumina and porous silicon (PSi) based humidity sensors. The

Table 1 Response of the circuit for the discrete capacitance

Vi = 1 V, f = 1 kHz, R_x = 980 kΩ							
Capacitance (pF)	100	150	200	250	300	350	400
Counter output	2	5	8	10	11	13	15

Table 2 Response of the interface circuit for PA humidity sensor

Humidity (%)	11	25	45	70	83	97
Counter output	20	27	38	52	57	59

Table 3 Response of the interface circuit for PSi humidity sensor

Humidity (%)	11	25	45	70	83	97
Counter output	21	23	25	28	32	34

effectiveness of the interface to the lossy capacitive sensor is clearly visible. However, the interface uses a single sensor may suffer from the baseline drift and the effect of the ambient temperature.

5.3 *Oscillator Based Transformer Ratio Arm Bridge for Interfacing the Capacitive Sensor*

The transformer bridge based AC bridge method is one of the oldest and accurate techniques to measure the small capacitance. The AC bridge method has limited capacitance measurement range. The oscillator based interface is popular for interfacing the wide range capacitive sensor. If both the AC Bridge and the oscillator based methods are combined, it is possible to measure the capacitance of a sensor accurately with a wide measurement range. Therefore, in the present circuit, AC bridge and oscillator has been combined to improve the range and the measurement accuracy of the capacitive sensor [40].

5.3.1 Working of the Interface Electronic

Figure 16 shows the interface circuit. Part 1 enclosed within a block is the transformer ratio arm bridge circuit (TRA) realized using an audio frequency centre tap transformer. The primary of the transformer is excited by the output of a relaxation oscillator and the secondary of the transformer generates two 180° output of phase signals (V_i and $-V_i$). The secondary outputs of the TRA are connected to the reference capacitance C_r and the sensing capacitor C_x respectively. (C_{p1} and C_{p2}) and (C_{r1} and C_{r2}) are the stray capacitances of C_x and C_r. C_{p1} and C_{r1} capacitances that appear across the voltage sources of the transformer secondary and $(C_e = C_{p2} + C_{r2})$ is the parasitic capacitance at the detector terminal D, which is at the virtual ground potential. The stray capacitances have negligible effects on the output voltage V_{01} of the opamp A1 because of minimum voltage source impedance and the virtual ground potential at point D. The unbalance detector current I_D, which is proportional to the difference of the capacitance ($C_r - C_x$) is converted into the voltage signal V_{01} using a current to voltage converter circuit. Part 2 is the relaxation oscillator circuit consisting of an integrator and a Schmitt trigger (hysteresis comparator). The output V_{01} of the TRA bridge with a buffer is applied to one of the inputs of the integrator and the other input of the integrator is the output of the comparator V_C. The output of the integrator is connected to the positive input

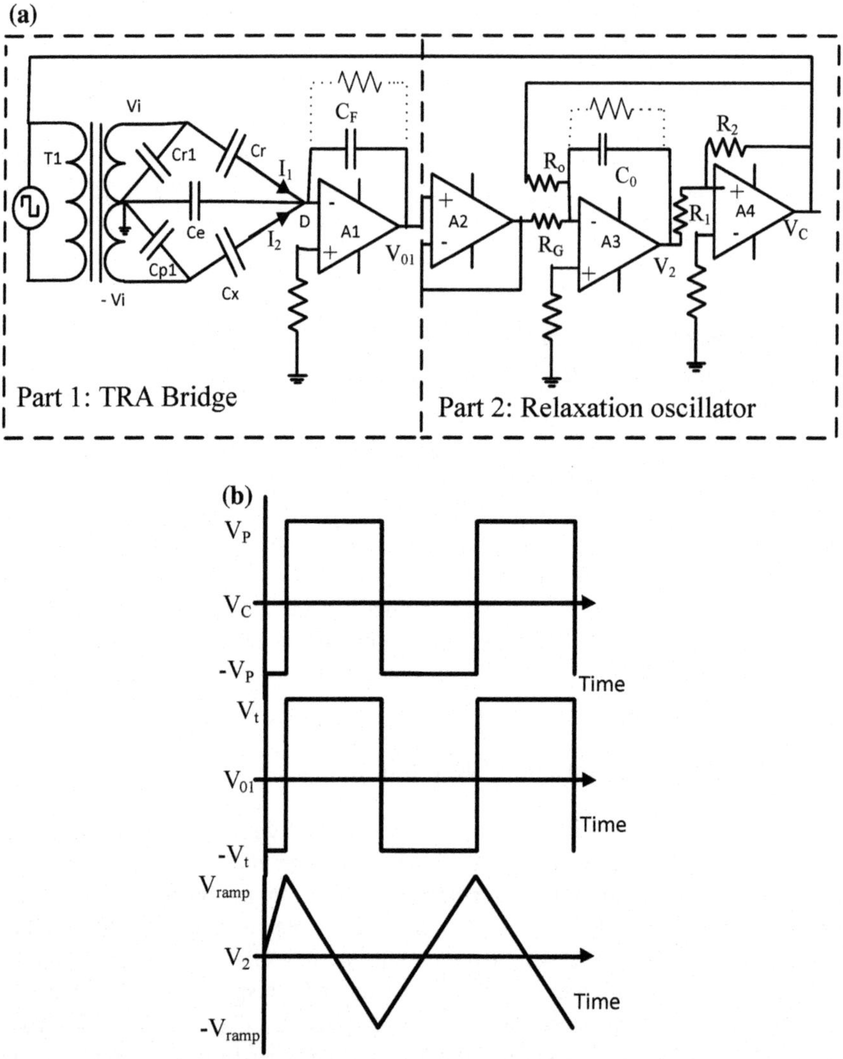

Fig. 16 **a** Schematic diagram of the signal conditioning circuit. **b** Significant waveforms at the different output terminals of the circuit

terminal of the comparator. The V_{01} output of the TRA bridge, at unbalance condition in the time domain can be given by

$$V_{01} = -(j\omega C_r V_i - j\omega C_x V_i)\frac{1}{j\omega C_F} \text{ or, } V_{01} = \left(\frac{C_x - C_r}{C_F}\right) V_i \quad (19)$$

5.3.2 Capacitance to Frequency Converter

When the peak of the comparator voltage output $|+V_p| = |-V_P| = V_C$ is symmetrical, the output voltage of the integrator is a triangular wave with V_2(p-p) is $2\frac{R_1}{R_2}V_c$, which is set by the Schmitt trigger circuit (comparator hysteresis) [19, 26]. During an unbalance of TRA bridge condition, the frequency of the output signal can be given by

$$f = f_0\left(1 + \frac{R_0}{R_G}\eta\delta\right) \tag{20}$$

where, the central frequency at the balance condition $f_0 = \frac{R_2}{4R_0C_0R_1}$, $\delta = \frac{C_x - C_s}{C_f}$, and η is transformation ratio of the transformer ($\eta = V_i/V_C$). When the capacitance of the sensor C_x in presence of measurand changes to $C_x + \Delta C_x$, then the frequency of the output signal can be written as

$$f = f_0\left(1 + \frac{R_0}{R_G}\eta\frac{\Delta C_x}{C_f}\right) \tag{21}$$

Thus, the change in the frequency is directly proportional to the change in the capacitance of the sensor. Simulation work has been performed with the help of the equivalent circuit of the bridge to study the frequency error due to the stray capacitance and the effect of the temperature of the sensor. Errors due to the offset voltage of the opamps, fixed time delay of the comparator, and the components values are also analyzed [40]. The circuit provides satisfactory results.

5.3.3 Hardware Realization of the Circuit and Determination of the Response Characteristics of the Humidity Sensor

The circuit was implemented using a high slew rate and fast response opamp LF-351. The central frequency f_0 is selected to be at 11.4 kHz for the response of a capacitive moisture sensor. The component values are: $C_0 = 0.954$ nF, $R_0 = 0.232$ MΩ, $R_1 = 0.978$ kΩ, $R_2 = 9.89$ kΩ, $R_G = 35$ kΩ and $C_F = 100$ pF. For the TRA bridge, an audio frequency centre tap transformer with transformation ratio $\eta = 1/20$ for the generation of the secondary voltage ($\pm V_i$) of 1.18 V has been employed. Both the primary and the secondary windings are placed on the high permeability core. The interface was initially tested with discrete ceramic capacitances and the variable gang capacitors and finally with the capacitive humidity sensor. C_r is a reference capacitance and C_x is an unknown variable capacitance arm. The output voltage V_{01} has been measured by varying the capacitance C_x with values $C_r < C_x$. Table 4 shows the values of V_{01} for different values of C_x with $C_r = 110$ pF (fixed). It shows that the output voltage proportionally increases with an increase in the difference $(C_x - C_r)$ almost linearly with slop 0.0082 V/pF.

Table 4 Output voltage V_{01} with change in the capacitance C_x

C_r = 110 pF (fixed)									
Cx (pF)	110	134.3	177.4	199.2	218.6	233.7	247.7	261.7	279
V_{01} (V)	0.13	0.32	0.66	0.82	0.96	1.08	1.18	1.28	1.42

Table 5 Frequency (f) with the variation of the capacitance C_x

C_x (pF)	8.38	20.29	37.14	59.1	86.82	112.7	134.3	199.22
f (kHz)	3.94	4.486	5.02	6.14	7.37	8.58	9.60	12.43

Table 6 Reading of the prototype moisture meter

Commercial moisture meter (ppm)	21	39	56	63	101	108
Prototype meter moisture (ppm)	20	39	54	62	100	107

Change in frequency of square wave output of the interface measured using a digital oscilloscope (Agilent Technologies DSO1002A) with the discrete ceramic capacitance at the central oscillator frequency of 3.95 kHz is shown in Table 5.

The interface was successfully utilized to develop a prototype digital hygrometer using a laboratory fabricated trace moisture sensor [40–43]. For the development of the meter for the moisture measurement, the output of the oscillator is interfaced with a PIC microcontroller. The microcontroller was programmed in the timer mode to measure the frequency. The frequency values are calibrated in-terms of the moisture. The nonlinearity of the frequency variation due to the moisture is addressed by the lookup table that is stored in the μC memory. The reading of the prototype meter is compared to a commercial dew point meter. The comparison of the reading of the prototype meter with the electronic circuit is shown in Table 6. The frequency sensitivity and the nonlinearity of the sensor for 0–110 ppm moisture range are found to be 10.94 Hz/ppm and ~1% respectively. This is a linear, sensitive, low baseline drift and simple differential capacitive bridge circuit for converting the incremental capacitance change in the frequency with high resolution (1 ppm moisture) and accuracy (nearly 1%). However, because of the transformer, the circuit is not CMOS compatible and is suitable for the lossless capacitive sensor.

5.4 *A Microcontroller Compatible Oscillator Based Active Bridge Circuit for Interfacing Capacitive Sensors*

5.4.1 Working of the Circuit

An alternate form of the TRA bridge realized using the active devices are shown in Fig. 17a [41]. It combines the properties of the capacitance to frequency convertor

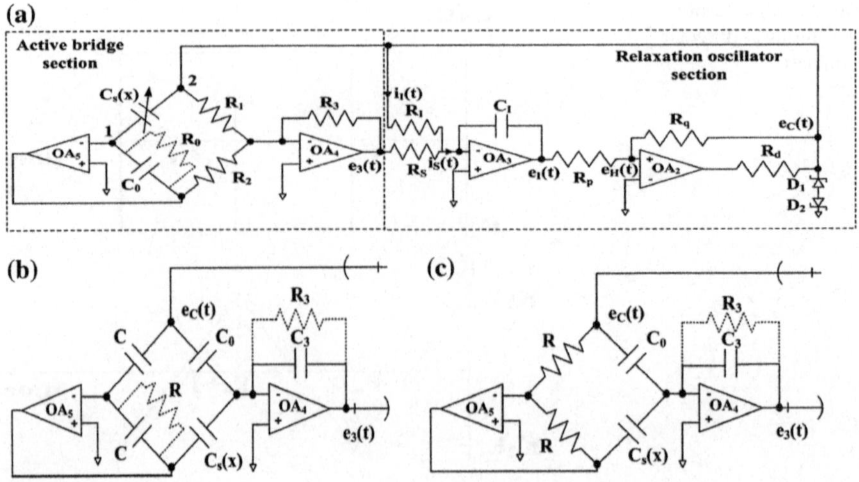

Fig. 17 **a**. The oscillator based active bridge for the capacitive sensor. **b** Full active bridge. **c** Half active bridge

and the capacitive ratio arm bridge technique with an output signal, which is compatible to the microcontroller [25, 32–34, 41]. The circuit offers all the features of the TRA bridge, including the CMOS compatibility. The circuit requires few hardware components and is easy to integrate for an application specific integrated circuit in a standard CMOS technology with configurable sensitivity. Figure 18 is the output waveforms. Element R_1, R_2, C_0 and $C_s(x)$ are the four arms of the bridge, which, along with the opamp OA_5 form an active bridge and, the opamp OA_4 forms a bridge amplifier. The operational amplifiers OA_2 and OA_3 work, as the integrator and the Schmitt trigger circuit of the relaxation oscillator respectively. The bridge is driven by the square wave voltage $e_C(t)$ (suitably scaled by diode clipper circuit) obtained from the output of OA_2. The bridge unbalanced voltage is amplified by an op-amp OA_4 to give the voltage, $e_3(t)$. This amplified voltage $e_3(t)$ and the voltage signal $e_c(t)$ are applied to the summing integrator of the oscillator. The frequency of the output waveform can be given by

$$f = f_o\left[1 + K_1\left\{\frac{C_s(x) - C_o}{C_o}\right\}\frac{R_I}{R_S}\right] \tag{22}$$

where, $f_0 = \frac{1}{K_2 R_I C_I\left(2+\alpha+\frac{1}{\alpha}\right)}$ is the central frequency at the balance condition when $\delta = 0$.

The capacitance of the humidity sensor can be written as

$$C_S(x) = C_o\left[1 + \frac{1}{K_1}\left\{\frac{f - f_o}{f_o}\right\}\frac{R_S}{R_I}\right] \tag{23}$$

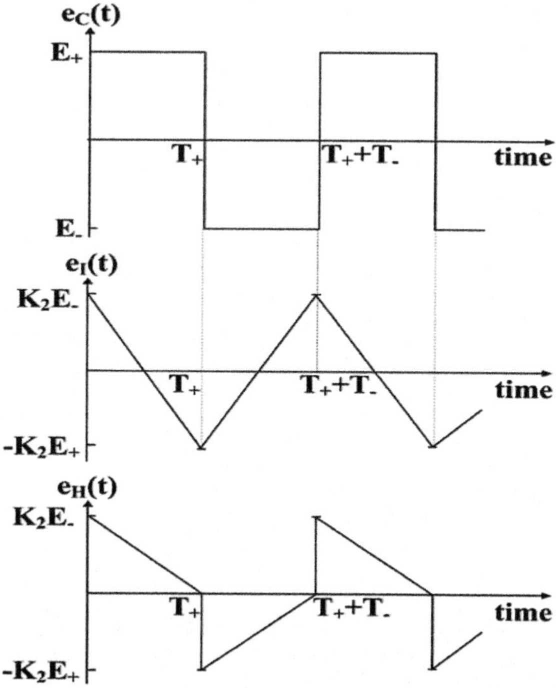

Fig. 18 Significant waveforms at different terminals

If C_x and C_0 are the two capacitors of a three electrode capacitive transducer, then the interface can be used to determine the relative permittivity of the dielectric medium, for example, the permittivity of edible oils or the transformer oil [15, 41]. To measure the permittivity of the dielectric using the interface, the relative permittivity can be given by

$$\in_r(x) = \left[1 + \frac{1}{K_1}\left(\frac{f - f_o}{f_o}\right)\frac{R_S}{R_I}\right] \tag{24}$$

$e_3(t) = e_C(t)\left\{\frac{C_s(x) - C_0}{C_3}\right\}$, for $sRC \gg 1$ and $sR_3C_3 \gg 1$ for the full active bridge and $e_3(t) = e_C(t)\left\{\frac{C_s(x) - C_0}{C_3}\right\}$, for $sR_3C_3 \gg 1$ for the half active bridge circuit.

5.4.2 Experimental Verification of the Interface Circuit

To confirm the performance as described, the circuit was hardware implemented by using the active devices having high slew rate, ultra-low offset voltage. Operational amplifier OP-07 was used for the integrator OA_3 and the amplifier OA_4, respectively and for the remaining op-amps, LF-351s were used. 2.7 V Zener diodes were used for D_1 and D_2. For appropriate op-amp selection, the Gain Bandwidth Product (GBWP) should be a factor of approximately 100 times higher than the maximum oscillation frequency. The LF-351 op amp has a GBWP = 4 MHz (typ.) with Vs $= \pm 15$ V [29] and hence, the excitation frequency or the oscillator frequency should be below 40 kHz to achieve the satisfactory performance. The excitation frequency has been selected around 10 kHz. The oscillator frequency was adjusted to nearly 10 kHz by using the following component values: $R_1 = R_2 = 2.189\,\mathrm{k\Omega}$, $R_3 = 0.976\,\mathrm{k\Omega}$, $R_p = 1.108\,\mathrm{k\Omega}$, $R_q = 2.227\,\mathrm{k\Omega}$, $R_d = 100\,\Omega$, $R_I = 382\,\mathrm{k\Omega}$, $C_s(x) = 500\,\mathrm{pF}$, $C_I = 100\,\mathrm{pF}$ and $C_0 = 500\,\mathrm{pF}$. A high resistance value $R_0 = R_f = 10\,\mathrm{M\Omega}$ was connected across the feedback capacitor that provides a dc current path due to the dc offset voltage and the bias current. Initially, the discrete capacitors were used in place of the capacitive sensor $C_s(x)$. Variation of the frequency with the discrete capacitance values with corresponding relative error is shown in Fig. 19. The maximum relative error of the capacitance value measured by the interface is found to be nearly 0.62%. The circuit is also tested with the capacitive sensors for the measurement of ppm level moisture and the dielectric constant of the edible oils respectively. The change in frequency of the circuit with the variation of the moisture in ppm is shown in Table 7. To determine the dielectric constant of the edible oils, the components values are redesigned to obtain the central frequency at 1 kHz. The capacitive sensor shown in Fig. 8c is used for the same. C_x is the capacitance with edible oil sample and C_0 is the free space air capacitor. The experiment has been performed for different samples of the oils. The percentage changes in frequency of the interface for the several identical samples of the oils are measured. The average dielectric constants of the sunflower and the mustard oil are ~3.15, ~2.95 and the percentage frequency changes are ~9.5% and 8.7% respectively. The effectiveness of the circuit for measuring the capacitance of the low value capacitive sensor is visible in the response. However, the circuit is suitable for interfacing the perfect capacitive sensor.

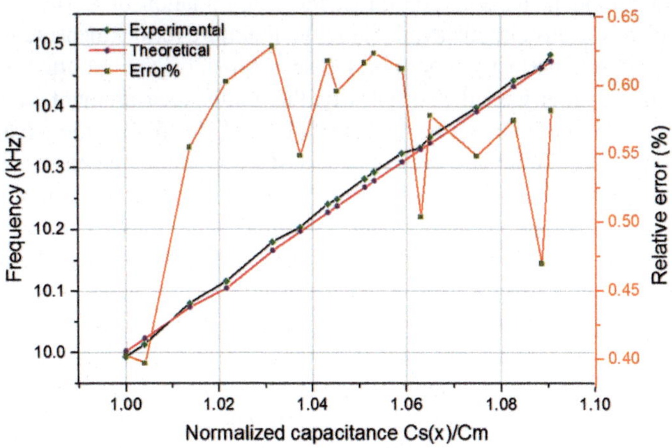

Fig. 19 Frequency output with the capacitance C_x

Table 7 Reading of the circuit with moisture concentration

Moisture (ppm)	Frequency (kHz)
4	9.99
13	10.12
29	10.20
50	10.28
60	10.32
70	10.35
81	10.4
90	10.44
100	10.48

5.5 *An Impedance Measurement Technique for Wide-Range Lossy Capacitive Sensors*

5.5.1 Working of the Interface

A new, simple and accurate impedance measurement technique (IMT) for wide range lossy capacitive sensor ('C_X' in parallel with 'R_X') based on the separation of in-phase and quadrature components of a modified active DeSauty bridge is discussed in this section [44]. The measurement range of the sensor parameters of the basic interface circuit can be improved further using an auto-balancing approach. In the auto balancing bridge, a voltage controlled variable reference capacitor, is connected at one arm of the active bridge, which forcefully nullifies the quadrature component of the output phase shifted signal. The phase difference of the output with respect to the input signal is zero only, when the voltage controlled capacitance is equal to the sensor capacitance, and at the same time, the output voltage of the interface circuit is proportional to the resistance component. Thus, at the quadrature component balance condition, the determination of C_X and R_X is possible. There are various circuits available in the literature for the measurement of the capacitance of the perfect capacitive sensors. Most of the reported works for the lossy sensors suffer from the complexity, the limited measurement range and the lower accuracy. In comparison to other techniques [34, 45–49], the advantages of the present interface circuit are; determination of R_X, C_X simultaneously; wide range of the components values; low base line drift; the stray capacitance immunity; low cost, capability of supplying same frequency signal to the sensor and the simplicity of the electronics. The phase and the amplitude of the output signal change due to the variation of the resistance and the capacitance of the sensor arm. The phase change of the output signal is converted into the time domain pulse wave modulated (PWM) signal using a phase sensitive detector as described in Fig. 14. Because of the lower output impedance of the opamp A1 and the virtual ground of the opamp A2, the output phase difference has negligible effect for the stray capacitances across the sensor arm. The phase difference has been extracted by applying V_i, V_o to the XOR gate through a zero crossing detector. The front-end of the first

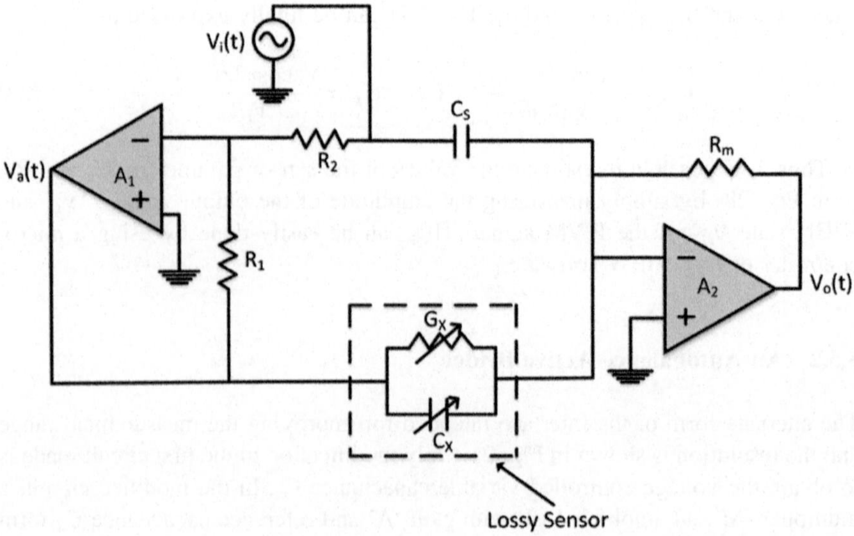

Fig. 20 The first basic impedance measurement circuit

proposed configuration is shown in Fig. 20, which is designed using the following elements: an inverting amplifier (A_1), a reference capacitance (C_s), a lossy capacitive sensor (C_x and G_x) and an amplifier (A_2). If the AC sinusoidal input voltage of the bridge, $V_i(t) = V_i\sin(\omega t)$, is applied to the input, then the output of the opamp A2, V_o can be written as

$$V_O(t) = R_m V_i(t)[(G_x + j\omega C_x) - (j\omega C_s)] \quad (25)$$

The Eq. (25) can be rewritten as

$$V_O[\cos\theta + j\sin\theta] = R_m V_i[(G_x) + j\omega(C_x - C_s)] \quad (26)$$

Separating real and imaginary parts of the Eq. (26), the values of R_x and C_x can be obtained as

$$R_x = \frac{V_i R_m}{V_o \mathrm{Cos}\theta} \; and \; C_x = \frac{V_o \mathrm{Sin}\theta}{V_i\, \omega\, R_m} + C_S \quad (27)$$

The phase difference θ can be converted into the time domain pulse wave signal using a XOR gate. The high output of the XOR gate 't_H' can be related with θ as

$$\frac{t_H}{T} = \frac{\theta}{2\pi}$$

where T is the time period, and the Eq. (27) can be finally expressed as:

$$C_x = \frac{V_0 \mathrm{Sin}(2\pi f \cdot t_H)}{V_i R_m \omega} + C_S, \quad G_x = \frac{V_o \cos(2\pi f \cdot t_H)}{V_i R_m} \tag{28}$$

Thus, it is possible to estimate the values of the sensor parameters (C_x and R_x) from Eq. (28) by simply measuring the amplitude of the output voltage 'V_0' and HIGH state 't_H' of the PWM signal. This can be easily done by using a microcontroller or Lab VIEW software.

5.5.2 An Autobalance Active Bridge

The alternate form of the interface intended for improving the measurement range and the resolution is shown in Fig. 21. Only modification to the first circuit made is to obtain the voltage controlled variable capacitance C_s. In the modified circuit, a multiplier 'M', an amplifier 'A3' with gain 'A' and reference capacitance C_S form one arm of the bridge. The output of the amplifier A3 can expressed as

$$V_3 = \frac{V_i AK}{10} \tag{29}$$

where K is the variable DC voltage obtained from the DAC output. The voltage V_3 drives a current I_S through the reference capacitance C_s, while V_2 (=$-V_i$) causes a current I_x through the sensor arm. The unbalance current at the input of A_2 is converted into a voltage V_o. The output V_o can be written as

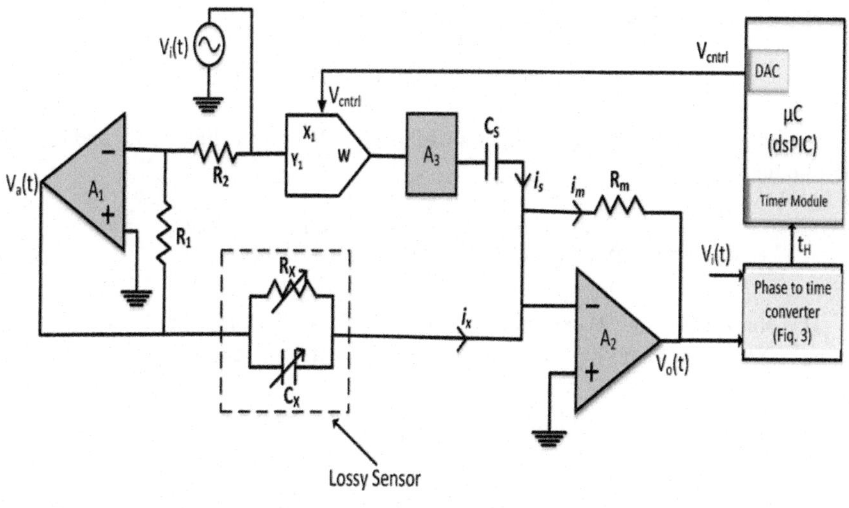

Fig. 21 Auto balancing active bridge for the impedance measurement

$$V_o = V_i R_m = \left[G_x + j\omega \left(C_x - \frac{KAC_s}{10} \right) \right],$$

$$\text{And, } [\theta = \tan^{-1} \left[\omega \left(C_x - \frac{KAC_s}{10} \right) \frac{1}{G_x} \right]$$

The phase difference θ in time domain at the output of the XOR gate can be written as $t_H = \frac{1}{2\pi f} \tan^{-1} \left[\omega \left(C_x - \frac{KAC_s}{10} \right) \frac{1}{G_x} \right]$. At balance condition, when t_H is zero, the C_x and R_x can be determined using the following expression:

$$C_x = \frac{V_{cntrl} A\ C_s}{10} \quad and\ R_x = R_m \frac{V_i}{V_0} \tag{30}$$

Smallest detectable capacitance can be given by $\Delta C_x = \frac{V_{DAC} A\ C_s}{2^n 10}$.

Where, ΔC_x is the smallest detectable capacitance change, n is the number of DAC bits, V_{DAC} is the reference voltage of DAC unit ($=V_{ctrl}$). For example, suppose we have a 10-bit DAC with an operating voltage range of 5 V, if we consider A = 10, and C_s = 10 pF, then the minimum detectable capacitance will be 0.0488 pF. In this sense, the readout operating range and the minimum detectable capacitance are directly related to V_{cntrl} value, which is in-turn depends on the number of DAC bits and the output voltage range of DAC. The electrical limit of the control voltage of the analog multiplier is 10 V, and the saturation level is 9.5 V. Since the output voltage of the multiplier is internally reduced by a factor of 10, so the gain of the non-inverting amplifier A_2 is selected according to the required resolution and the bandwidth.

5.5.3 Experimental Results of the First Electronic Circuit

The circuit shown in Fig. 20 was tested with the values; V_{in} = sin(ωt), f = 1 kHz, C_s = 100 pF, R_m = 480 kΩ. The ultra-low offset operational amplifier OPA-177 was used for A1, A2 and A3, the high slew rate operational amplifier LF351 was used for comparator O1 and O2 and the XOR gate HCF4070 was used for the phase detection purpose. The output of the XOR gate and the interface circuit were acquired in Lab-VIEW software using 'NI-ELVIS II' data acquisition (DAQ) card to simultaneously provide graphical user interface. Initial, discrete capacitive and parasitic resistive components are used for the sensor arm. All passive components have been first measured with an LCR meter (Agilent) and found to be ±2% tolerance for the resistive elements and ±3% tolerance for the capacitive elements.

The experimental results with the circuit for different values of the resistance from 68 to 10 MΩ and the capacitance from 200 pF to 1.1 nF is shown in Table 8. The maximum relative error is found to be 6%.

Table 8 Experimental results with discrete components

True value	R_x = 10 MΩ		R_x = 1 MΩ		R_x = 68 kΩ	
C_x (pF) (LCR meter)	C_x (pF) circuit	% error	C_x (pF) circuit	% error	C_x (pF) circuit	% error
200	195.3	−2.35	192.6	−3.7	189.2	−5.4
300	293.31	−2.23	291.9	−2.7	288.72	−3.76
400	395.12	−1.22	393.4	−1.65	386.8	−3.3
500	493.8	−1.24	491.7	−1.66	489.3	−2.14
600	595.70	−0.72	593.4	−1.1	592.5	−1.25
700	698.10	−0.27	696.43	−0.51	697.2	−0.4
800	803.4	0.43	801.36	0.17	806.64	0.83
900	906.09	0.68	906.48	0.72	919.17	2.13
1000	1009.2	0.92	1017.3	1.73	1035.8	3.58
1100	1114.85	1.35	1123.98	2.18	1147.08	4.28

Table 9 Experimental results with the autobalancing circuit

True value	R_x = 4.7 MΩ		R_x = 680 kΩ		R_x = 100 kΩ	
C_x (pF) (LCR meter)	C_x (pF) circuit	% error	C_x (pF) circuit	% error	C_x (pF) circuit	% error
100	96.8	−3.2	102.1	2.1	104.3	4.3
300	291.9	−2.7	302.79	0.93	312.6	4.2
600	586.32	−2.28	607.08	1.18	622.8	3.8
900	883.44	−1.84	909.09	1.01	934.38	3.82
1200	1178.64	−1.78	1223.04	1.92	1242	3.5
1500	1471.05	−1.93	1522.2	1.48	1556.1	3.74
1800	1781.1	−1.05	1834.56	1.92	1863.54	3.53
2000	2023	1.15	2061.2	3.06	2097.2	4.86

5.5.4 Experimental Results of the Autobalancing Bridge

To implement the autobalancing bridge, an analog multiplier ‘AD633’ [21] and a PIC microcontroller ‘dsPIC33FJ16GS502’ [22] operating at 40 MHz and having an inbuilt 10-bit DAC (3.3 V reference) were employed and the amplifier A3 with gain 30 was used to obtain the autobalance condition. The experiment was conducted with discrete resistance and capacitance having an enhanced range of capacitance (1.6 times of the previous range) and the resistance (3 times of the previous range). The results of the capacitance and resistance measurement are shown in Table 9. The maximum relative errors of the capacitance and the resistance measurement are 5% and 6% respectively. The minimum capacitance value of the autobalance technique is 50 pF. The minimum range can be increased further by using advanced opamps, high signal frequency and the high resolution DAC.

Table 10 Capacitance of the humidity sensor measured by the circuit

RH (%)	11	20	29	39	50	73	85	91
C_X (LCR meter)	70.0	71.9	74.9	81.7	91.1	302.5	576.2	782
Cx (meas)	65.4	66.4	70.7	78.2	87.4	313.6	589.1	798.3
% error	−6.5	−7.6	−5.7	−4.2	−4.2	3.7	2.2	2.1

Finally, the interface has been tested with the lossy capacitive type sensors like humidity sensor. The resistance value of the sensor varies in the range of 35 kΩ–4 MΩ and the capacitance value of the sensor varies in the range of 70–790 pF for the humidity range of 10–90% RH. The result of the humidity measurement is shown in Table 10.

5.6 Current Mode Oscillator Circuit for the Grounded Capacitive Sensors

Voltage mode oscillator combining the advantages of the bridge method of the capacitance measured has been discussed in Sects. 5.3 and 5.4. The voltage mode oscillator has the limited dynamic range because of the frequency dependent gain of the operational amplifiers which are utilized for the implementation of the circuit. This problem can be avoided by a current mode oscillator utilizing the second generation current conveyer (CCII). The current mode oscillator offers large dynamic range, wide bandwidth, high linearity, possibility of designing with low power consumption and simple analog circuit design. An oscillator based interface circuit for the grounded capacitive sensors with wide dynamic range using CCII has been discussed [51]. Its main operation is basically adopted from the voltage mode approach of current integration, and the configuration is designed in such a specific way that minimizes the effect of the parasitic components of CCIIs. The dynamic range has been improved by reducing the effect of the parasitic components at node X. This is done by avoiding the capacitive loads, keeping only grounded resistive load of slightly higher value than the parasitic resistance at the same terminal. In addition, it is also possible to set the circuit sensitivity and the operating frequency range externally, using the passive components. Simulation and the experimental results show a good agreement with the theoretical expectations with good linearity in the wide dynamic frequency range.

5.6.1 Current Conveyer Based Oscillator

Figure 22a shows the symbolic representation of a basic second generation current conveyer. The characteristics of the CCII can be described by the following equations

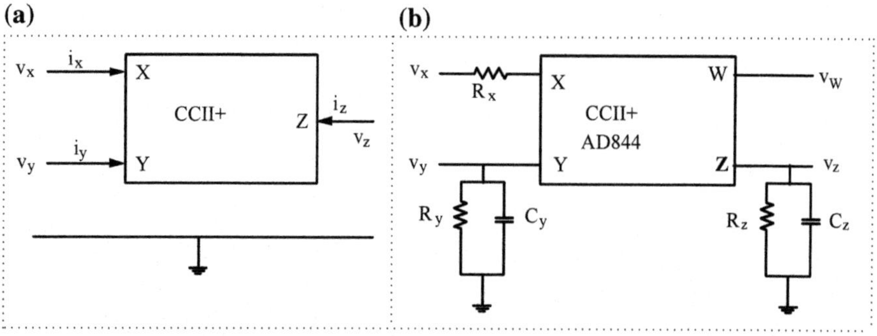

Fig. 22 **a** Symbol of the second generation current conveyer. **b** Analog device IC AD844 as a practical current conveyer

$$i_z = \pm \alpha_i i_x \; and \; v_x = v_y, i_y = 0 \tag{31}$$

The '±' symbol of current transfer ratio a_i indicates, whether the conveyor is framed as non-inverting or inverting circuits, labelled as CCII+ and CCII− respectively. The output current i_z depends on the input current i_x, which may be directly injected by applying the input voltage at X node, or by coping the input voltage from terminal Y. Analog device AD844 can be used as a CCII. The schematic diagram, of the practical AD844 as current conveyer is shown in Fig. 22b. The practical parasitic components values of the device are $C_z = 4.5$ pF, $C_y = 2$ pF, $R_z = 3$ MΩ, $R_y = 10$ MΩ and $R_x = 50$ Ω. The actual interface electronic circuit is shown in Fig. 23a where CCII+'A' works as a voltage to current converter and CCII+'B' works as a hysteresis comparator. The important waveforms at different nodes are shown in Fig. 23b. Considering the ideal conditions of the conveyer, the peak to peak voltage, v_{02} can be written as

$$v_{02}(p-p) = V_s \left(\frac{R_2 - R_1}{R_2} \right) \tag{32}$$

The voltage across the capacitor due to current i_{XB} can be expressed as

$$v_{02}(p-p) = \frac{V_s}{2R_x C_x} T \tag{33}$$

Comparing (32) and (33), the time period of the oscillator output can be written as

$$T = 4R_x C_x \frac{R_2 - R_1}{R_2} \tag{34}$$

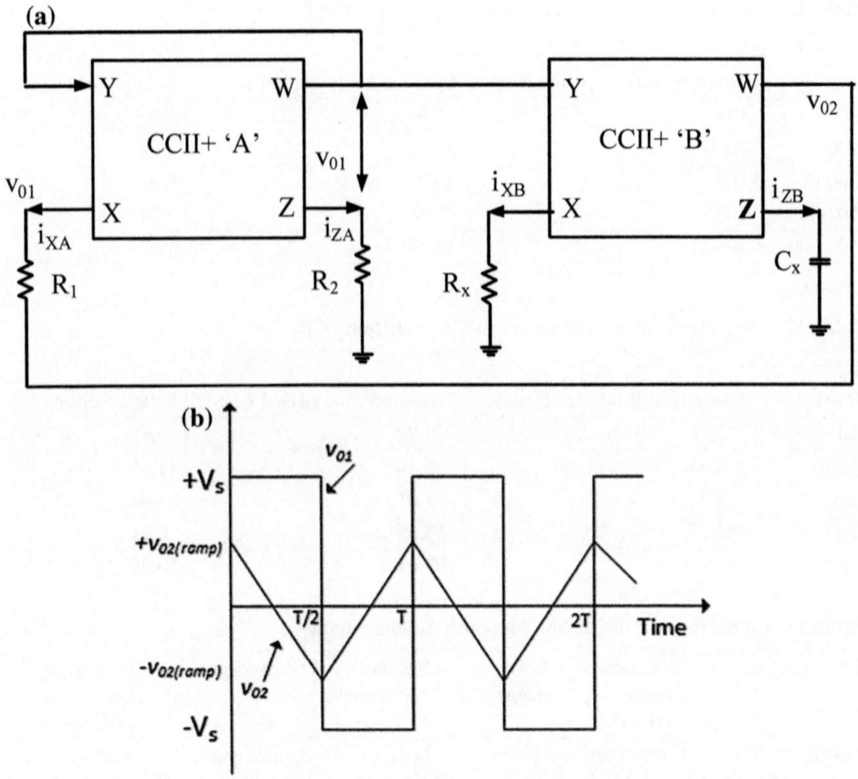

Fig. 23 **a** Current conveyer based interface electronics circuit, **b** waveforms at important nodes

If we consider the parasitic components of the device AD844, the time period can be written as

$$T = 4(R_x + R_{XB})C_x \frac{R_2 - (R_1 + R_{XA})}{R_2} \tag{35}$$

Time period is depending on both the values of R_x and C_x.

5.6.2 Experimental Verification of the Interface Circuit

The circuit shown in Fig. 23a was first simulated by varying C_X and R_X keeping R_1 = 6.8 kΩ and R_2 = 10 kΩ. The simulation results are shown in Tables 11 and 12. The range of the oscillation frequency for the variation of the capacitance from 50 pF to 500 nF is 850 kHz to 99 Hz, very wide.

The circuit was also hardware implemented on the breadboard with the components values R_1 = 6.8 kΩ, R_2 = 10 kΩ and Rx = 40 kΩ. The oscillation frequency in the range of 0–250 kHz was measured with the variation of the

Table 11 Time period of the oscillator with the variation of C_x

$R_x = 20\ k\Omega$			
C_x (pF)	Theoretical time period (μs)	Measured time period (μs)	Relative error (%)
50	1.08	1.14	5.56
500	10.11	10.42	3.74
5000	108	102	5.80
50,000	1080	1020	5.54
500,000	10800	10100	6.48

Table 12 Time period of the oscillator with the variation of R_x

$C_x = 50$ pF			
R_x (kΩ)	Theoretical time period (μs)	Measured time period (μs)	Relative error (%)
10	2.56	2.70	5.18
100	5.31	5.40	1.67
500	27.54	27.01	1.96
1000	57.57	54.05	6.11
2000	119.47	107.41	10.09

Table 13 Comparison of the circuit with other similar circuits

Characteristics	Frequency range (Hz–kHz)	Active components	Passive components	Output signal	Prototyping and experimental
Double CCII(+) based oscillator [51]	100–850	Two	Four	Square and triangular wave	Yes
CCII(+) based design [52]	25–225	Two	Four	Square and triangular wave	Yes
CMOS/CCII(+) differentiation Based [53]	0.128–737	Two	Six	Square	Yes
CMOS based design [54]	25–260	Two	Two	Square	No
CCII(+) based design [55]	100–100	One	Four	Square/Triangular	Yes

capacitance, *Cx* from 100 pF to 200 nF. The maximum relative error of the measured oscillation frequency of the output signal with respect to the theoretical value is nearly 8%. Similarly, the oscillation frequency was also measured with the variation of R_x from 50 kΩ to 1.5 MΩ for the components values of $C_x = 100$ pF, $R_1 = 7.4$ kΩ, $R_2 = 10$ kΩ. The maximum relative error of the actual frequency with respect to the theoretical value is 9%. Finally, a comparison table showing the performance of the current mode oscillator circuit with other existing similar circuits for the capacitive and the resistive sensors is shown in Table 13. The circuit based on charging/discharging effect of resistance-capacitance (RC) cell, is having a simple circuit topology requiring two CCIIs and four passive components only.

6 Conclusions

Capacitive sensors both perfect and lossy are widely used for the measurement of different physical and chemical parameters. The preference and the uses of such type of the capacitive sensors are getting more importance day by day with the needs of new information to be measured efficiently and reliably. The capacitive sensors can be useful sometimes for the hassle free non-contact measurement of the physical parameters. Different types of the capacitive sensors suitable for the measurement of the parameters are discussed. Though the cross-capacitance is used as primary capacitance standards for different national laboratories, its potential is yet to be exploited fully for sensing applications. The capacitive sensor needs extra precautions to avoid the errors due to stray effects, particularly for the low capacitance measurement. Issues related to the development of the interface electronic circuit for the sensors are also discussed. Some advanced interface electronic circuits for both the perfect and the lossy capacitive sensors are discussed. These interface circuits are promising for interfacing different capacitive sensors with the ability to minimize the stray capacitances, where the capacitance values vary widely.

References

1. J. Fraden, *Handbook of Modern Sensors, Physics, Design, and Applications* (Springer, New York, 2003)
2. E. Doeblin, D.N. Manik, *Measurement Systems, Applications and Design*, 5th edn. (Tata Macgraw Hill, New Delhi, 2007)
3. A. De Marcellis, G. Feri, *Analog Circuits and Systems for Voltage Mode and Current Mode Sensor Interfacing Applications* (Springer, 2011)
4. A. Abu-Al-Aish, M. Rehman, Development of a capacitive mass measuring system. Sens. Actuators A **151**, 113–117 (2009)
5. B. Hague, T.R. Ford, *Alternating Current Bridge Methods,* 6th edn. (Pitman Publishing, 1971)
6. Z.M. Rittersma, Recent achievements in miniaturized humidity sensors—a review of transduction techniques. Sens. Actuators A **96**, 196–2002 (2002)
7. T.A. Blanka, L.P. Eksperiandova, K.N. Belikov, Recent trends of ceramic humidity sensors development: a review. Sens. Actuators B **228**, 416–442 (2016)
8. S. Dhanekar, S.S. Islam, T. Islam, A.K. Shukla, Harsh, Organic vapour sensing by porous silicon: influence of molecular kinetics in selectivity studies. Phys. E (Elsevier) **42**(5), 1648–1652 (2010)
9. T. Islam, H. Saha, S.S. Islam, Porous silicon based sensor array for ethanol sensing. Sens. Lett. **7**, 1077–1085 (2009)
10. Golding, Widdies, *Electrical Measurements and Measuring Instruments*, 5th edn. (Reem Publication, 2009)
11. F.N. Toth, D. Bertels, G.C.M. Meijer, A low cost, stable reference capacitor for capacitor sensor systems. IEEE Trans. Instrum. Meas. **45**(2), 526–530 (1996)
12. M. Rehman, S.K. Mukherjee, V.G.K. Murti, Effect of a symmetrically placed dielectric on the capacitance of a cylindrical cross-capacitor. Proc. IEE (England) **129**(5) (1982)

13. W.Ch. Herrens, Applications of capacitance techniques in sensor design. J. Phys. E: Sci. Instrum. **19**, 897–906 (1986)
14. F. Reverter, X. Liand, G.C.M. Meijer, Stability and accuracy of active shielding for grounded capacitive sensors. Meas. Sci. Technol. **17,** 2884–2890 (2006)
15. A.U. Khan, T. Islam, J. Akhtar, An oscillator-based active bridge circuit for interfacing capacitive sensors with microcontroller compatibility. IEEE Trans. Instrum. Meas. (2016)
16. T. Islam, Md.R. Mahboob, S.A. Khan, A simple MOX vapor sensor on polyimide substrate for measuring humidity in ppm level. IEEE Sens. J. **15**(5), 3004–3013 (2015)
17. T. Islam, A.T. Nimal, U. Mittal, M.U. Sharma, A micro interdigitated thin film metal oxide capacitive sensor for measuring moisture in the range of 175–625 ppm. Sens. Actuators B **221**, 357–364 (2015)
18. T. Islam, M.Z.U. Rahman, Investigation of the electrical characteristics on measurement frequency of a thin-film ceramic humidity sensor. IEEE Trans. Instrum. Meas. **65**(3) (2016)
19. A.I. Zia, M.S.A. Rahman, S.C. Mukhopadhyay, et al., Technique for rapid detection of phthalates in water and beverages. J. Food Eng. **116,** 515–523 (2013)
20. T. Yin, W. Wei, L. Yang, X. Gao, Y. Gao, A novel capacitive immunosensor for transferrin detection based on ultrathin alumina sol–gel-derived films and gold nanoparticles. Sens. Actuators B **117**, 286–294 (2006)
21. A. Abu-Al-Aish, R. Mahfoozur, A.H.A. Hassan, M.R. Arshad, Development of an intelligent capacitive mass sensor based on co-axial cylindrical capacitor. J. Sens. Transducers, **105**(6), 1–9 (2009)
22. M. Rehman, V.G.K. Murti, A sensitive and linear pressure transducer. J. Phys. E: Sci. Instrum. (England) **14**, 988–992 (1981)
23. E. Basheer, M.M. Rahman, A novel capacitive rotation speed transducer. Meas. Sci. Technol. **13**, 2027–2031 (2002)
24. Z. Chen, C. Lu, Humidity sensors: a review of materials and mechanisms. Sens. Lett. **3**, 274–295 (2005)
25. S.M. Huangt, A.L. Stottf, R.G. Green, M.S. Beck, Electronic transducers for industrial measurement of low value capacitances. J. Phys. E: Sci. Instrum. **21**, 212–250 (1988)
26. G.C.M. Meijer, *Smart Sensor Systems* (Wiley, Delft, The Netherlands, 2008), pp. 240–241
27. M. Rehman, M.T. Ahmad, V.G.K. Murti, A TRA bridge technique for in-circuit impedance measurement. IEEE Trans. Instrum. Meas. (USA) **IM-33**, 252–256 (1984)
28. A. Ashrafi, H. Golnabi, A high precision method for measuring very small capacitance changes. Rev. Sci. Instrum. **70**(8), 3483–3487 (1999)
29. A. De Marcellis, G. Ferri, E. Palange, A novel analog autocalibrating phase-voltage converter for signal phase-shifting detection. IEEE Sens. J. **11**(2), 259–266 (2011)
30. H. Binxin, J. Wang, G. Song, F. Zhang, A compact wideband precision impedance measurement system based on digital auto-balancing bridge. Meas. Sci. Technol. **27**, 055902–055915 (2016)
31. R. Rybski, J. Kaczmarek, K. Kontorski, Impedance comparison using unbalanced bridge with digital sine wave voltage sources. IEEE Trans. Instrum. Meas. **64**(12), 3380–3386 (2016)
32. A. De Marcellis, G. Ferri, P. Mantenuto, Uncalibrated operational amplifier-based sensor interface for capacitive/resistive sensor applications. IET Circuits Devices Syst. **9**(4), 249–255 (2015)
33. S.N. Nihtianov, G.P. Shterev, B. Iliev, G.C.M. Meijer, An interface circuit for R–C impedance sensors with a relaxation oscillator. IEEE Trans. Instrum. Meas. **50**(6), 1563–1567 (2001)
34. Y. Liu, S. Chen, M. Nakayama, K. Watanabe, Limitations of a relaxation oscillator in capacitance measurements. IEEE Trans. Instrum. Meas. **49**(5), 980–983 (2000)
35. P. Mantenuto, A. Dc Marcellis, G. Ferri, Novel modified De-Sauty autobalancing bridge-based analog interfaces for wide-range capacitive sensor applications. IEEE Scns. J. **14**(5), 1664–1672 (2014)
36. N. Madhu Mohan, V. Jagadeesh Kumar, Novel signal conditioning circuit for push-pull type capacitive transducers. IEEE Trans. Instrum. Meas. **56**(1), 153–157 (2007)

37. D. Wang, Fdc1004: basics of capacitive sensing and applications, Texas Instruments, Dallas, TX, USA, Appl. Rep. SNOA927, 2014
38. T. Islam, K.K. Mistry, K. Sengupta, H. Saha, Measurement of gas moisture in the ppm range by porous silicon (PS) and porous alumina sensors. Int. J. Sens. Mater. **16**(7), 345–356 (2004)
39. T. Islam, Md.F. Khan, S.A. Khan, H. Saha, A sensitive digital moisture detector for Nanostructured thin film sensor. Int. J. Smart Sens. Intell. Syst. **7**(3), 1059–1076 (2014)
40. T. Islam, S.A. Khan, Md.F.A. Khan, S.C. Mukhopadhyay, A relaxation oscillator based transformer ratio arm bridge circuit for capacitive humidity sensor. IEEE Trans. Instrum. Meas. **67**(12), 3414–3422 (2015)
41. T. Islam, A.U. Khan, J. Akhtar, M.Z.U. Rahman, A digital hygrometer trace moisture measurement. IEEE Trans Indus. Electron. **61**(10), 5599–5605 (2014)
42. H. Shibata, M. Ito, M. Asakursa, K. Watanabe, A digital hygrometer using a polyimide film relative humidity sensor. IEEE Trans. Instrum. Meas. **45**(2), 564–588 (1996)
43. T. Islam, L. Kumar, S.A. Khan, A novel sol-gel thin film porous alumina based capacitive sensor for measuring trace moisture in the range of 2.5 to 25 ppm. Sens. Actuators B, **173**, 377–384 (2012)
44. S. Malik, K. Kishore, T. Islam, Z.H. Zargar, S.A. Akbar, A time domain bridge-based impedance measurement technique for wide-range lossy capacitive sensors. Sens. Actuators A **234**, 248–262 (2015)
45. A.D. Ferri, C. Marcellis, V. Di Carlo, A. Stornelli, A. Flammini, D. Depari, E. Marioli, A CCII-based low-voltage low-power read-out circuit for DC-excited resistive gas sensor. IEEE Sens. J. **9**(12), 2035–2041 (2009)
46. S.Y. Yurish, N.V. Kirianaki, Interfacing circuit design for frequency-time domain MEMS sensors, in *International MEMS Conference* (2006)
47. S.C. Bera, J.K. Ray, S. Chattopadhyay, A low-cost non-contact capacitance-type level transducer for a conducting liquid. IEEE Trans. Instrum. Meas. **55**, 778–786 (2006)
48. Analog Devices, Norwood, MZ, USA. Datasheet, AD5933, 12-BitImpedance Converter [Online], http://www.analog.com/media/en/technical-documentation/data-sheets/AD5933.pdf (2013)
49. C. Kolle, P. O'Leary, Low cost, high precision measurement system for capacitive sensors. Meas. Sci. Technol. **9**(3), 510–517 (1998)
50. F. Reverter, O. Casas, A microcontroller-based interface circuit for lossy capacitive sensors. Meas. Sci. Technol. **21**, 65203–65211 (2010)
51. S. Malik, D. Sharma, K. Kishore, M. Maharana, S.A. Akbar, T. Islam, A CCII-based wide frequency range square/triangular wave generator, in *2015 IEEE 2nd International Conference on Recent Trends in Information Systems July 9–11, 2015, Jadavpur University, Kolkata, India*
52. D. Pal, A. Srinivasulu, B.B. Pal, A. Demosthenous, B.N. Das, Current conveyor-based square/triangular waveform generators with improved linearity. IEEE Trans. Instrum. Meas. **58**(7), 2174–2180 (2009)
53. A.D. Marcellis, C.D. Carlo, G. Ferri, V. Stornelli, A CCII-based wide frequency range square waveform generator. Int. J. Circuit Theory Appl. **41**, 1–13 (2013)
54. S.D. Re, A.D. Marcellis, G. Ferri, V. Stornelli, Low voltage integrated astable multivibrator based on a single CCII, *in IEEE Proceedings—Prime* (Bordeaux, France, 2007)
55. M. Abuelma'atti, S. Al-Shahrani, New CFOA-based trinagular/square wave generator. Int. J. Electron. **84**, 583–588 (1998)

A Simple Embedded Sensor: Excitation and Interfacing

Md. Eshrat E. Alahi, A. Nag, N. Afsari Manesh, S.C. Mukhopadhyay and J.K. Roy

Abstract Some of the basic sensors and their operating principle have been explained in this chapter. There is the certain advantage of capacitive/inductive sensors over resistive sensors. The operating principle of capacitive sensors and type of measurement techniques are also explained. Finally, design and development of a low-cost portable sensing system were explained which can measure the impedance of a capacitive sensor and might be useful for different applications. There are some various applications explained by using capacitive sensors.

1 Introduction

The world as we see today is encapsulated with sensors. The phenomenon of analyzing the architect of the environment is done with sensors and circuits. Sensors and transducers have made the understanding of nature much easier. Sensors are incorporated inaccessible and partially accessible locations to get the data without much of human intervention. Human skin can be considered an excellent sensor depending on the dynamicity of applications it senses around it. For example, the sensing it would experience in everyday life would include tactile sensing which would involve the sensing of objects or food items before eating/drinking them. Another type of sensing would involve the weather it senses to determine the attire worn at that moment. The sensors used in the field of science senses the ambient conditions and sent it to the attached conditioning circuit where the raw data is processed and forwarded to the monitoring unit for understanding its significance. The conditioning circuit depends on the type of data collected which as a result depends on the application of the sensor. Figure 1 shows a simple block diagram of a sensor involved in data reception and transmission to the monitoring unit.

Md.E.E. Alahi · A. Nag · N. Afsari Manesh · S.C. Mukhopadhyay (✉)
Macquarie University, Sydney, Australia
e-mail: subhas.mukhopadhyay@mq.edu.au

J.K. Roy
System Advance Technologies Pvt. Ltd, Kolkata, India

B. George et al. (eds.), *Advanced Interfacing Techniques for Sensors*,
Smart Sensors, Measurement and Instrumentation 25,
DOI 10.1007/978-3-319-55369-6_3

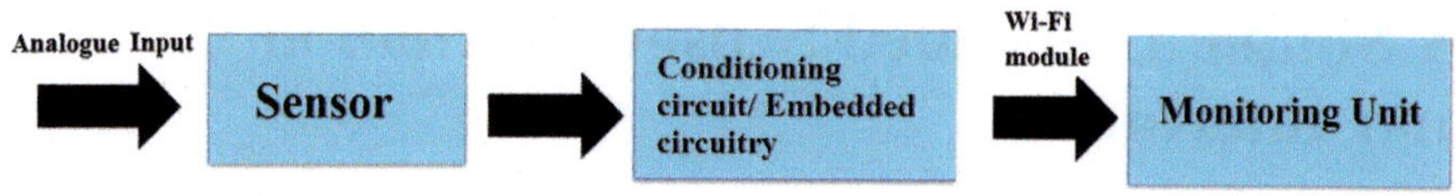

Fig. 1 Schematic diagram for the sensor data transmission

Different types of sensors are available for various applications it covers. Three main categories of sensors, resistive, capacitive and inductive cover most of the sensors commercially available in the market. The differences in the sensors lie in the different parameters of the impedance that are examined. Each type of sensor is briefly explained below with examples.

2 Type of Sensors

Sensors are classified in different types based on the application and circuitry it is associated with. They can also be differentiated depending on the kind of input voltage the circuit is supplied with. For example, resistive sensors need a DC input whereas capacitive and inductive circuits need an AC input. The term DC refers to the application of a fixed input voltage whereas AC refers to the frequency dependency voltage signal as a function of time. Some of the types of sensors are described in the following section.

2.1 Resistive Sensors

The most common type of sensors used for different applications are the resistive sensors. Some of the common examples of resistive sensors are strain gauges, potentiometric and piezo-resistive sensors. The measurement of the resistive sensors is done directly by analysing the change in electrical resistance of a resistive element. The resistance of a material is dependent on the electrical resistivity (ρ), length (l) and area (A) of a material as given below in Eq. (1). The electrical resistivity is the property of a material which determines the amount of electric current flowing through a material. Higher value of resistivity indicates lower current flow through the material. A and l are the cross-sectional area and length of the material whose resistive is evaluated.

$$R = \rho * \frac{l}{A} \tag{1}$$

The resistive sensors analyze the change in resistance which is caused as a result of an analog input to the sensor. As a result, this change in resistance causes a change in the output voltage which is measured to determine the characteristic of the sensed parameter.

2.2 Capacitive Sensors

Capacitive sensors operates on the principle of coupling effect where the displacement current is used as a parameter to transfer a certain amount of energy between oppositely charged poles. Dielectric materials are placed in between the two poles. The dielectric material is an insulator whose charges can be polarized by the influence of an electric field. The intensity with which the electric field can polarize a dielectric material depends on the permittivity of the material. Capacitive sensors are more popular compared to the resistive ones due to their ability to store energy. Different metals and alloys are used to develop these sensors. The type of material used to develop the sensor depends on its application. The major difference between a resistive and capacitive sensor is regarding the frequency dependency of the input voltage to the material. Like resistive sensors, the size of the sensor is one of the parameter that has an effect on the value of capacitance. The capacitance also depends on the permittivity of the dielectric material chosen for the sensor. The basic formula with which the capacitance of a material is calculated is given in Eq. (2).

$$C = \frac{\in_0 * \in_r * A}{d} \tag{2}$$

where,

C is the capacitance of the interdigital sensor,
$\in_o =$ 8.85 is the $\times 10^{-12}$ F m^{-1} is the permittivity of vacuum,
$\in_r$ is the relative permittivity,
A is the effective area, and
d is the effective spacing between electrodes of different polarity.

There are two different types of capacitances—surface capacitance and projectile capacitance. The common one is the surface capacitance where an alternating voltage is applied to one of the electrodes making the other one as ground, thus generating an electrostatic field. A touch on the top layer would create a capacitive effect. The projectile capacitance involves the presence of capacitive layers forming rows and columns from two different layers. The advantages of this mode of capacitance are that it can detect multi—touch activation and also due to the presence of multiple layers, the surface of the material is more resistant compared to surface capacitors.

2.3 Inductive Sensors

When a steady state electric current is fed into a coil with a specific area (A) and length (l), magnetic flux are generated in the coil. This induces an electromotive force (EMF) which is opposite to the direction of the current. The relationship is given in Eq. (3). The induced EMF is exploited in mutual inductance where the current in one conductor induces an EMF in a nearby conductor.

$$V_L = N * \frac{d\varnothing}{dt} = \frac{\mu * N^2 * A}{l} * \frac{di}{dt} \quad (3)$$

where,

V_L is the induced EMF.
N is the number of turns of the coil.
Φ is the magnetic flux.
A is the unit area of the inductor.
l is the unit length of the material.
di/dt is the current per second.

The inductive sensors are useful to determine the position of a metallic object through the generated magnetic field. It is difficult to find specific inductive sensors in the market. The circuit consisting of an inductive element also consists of resistive and capacitive elements. The advantages of using these sensors are the robustness they have towards shock and turbulent environments. These sensors are used in high security environments like aeronautics and space where the precise location of a body is of crucial importance.

3 Types of Resistive Sensors

Resistive sensors were the simplest and elementary devices introduced in the field of sensing technology. The resistive sensors operate just on one parameter, electrical resistance which changes with the change in property of the material. Some of the common types of resistive sensors are explained below.

3.1 Potentiometric Sensors

These types of sensors operate on the principle of potential difference between two analytic materials. These sensors are popular as chemical sensors where the concentration of an analyte is changed to change its potential. The change is measured by comparing it with a reference analyte with a zero or defined potential. The sensor connected to analyte material shows a change in its electrical resistance when the concentration of the gas is changed. These sensors are advantageous because they can used to study the behavior of a large number of analytes together. But there are certain disadvantages associated with these sensors. The circuitry associated with these sensors is complex. Also, the change in potential difference can take place because of many factors, not just the concentration. Thus, there remains an ambiguity with the operation with these sensors.

3.2 *Thermistors*

Thermistors are the devices whose resistance changes with the change in temperature experienced by the sensor. These types of sensors are used as temperature sensors at houses or as self-regulating heating elements. The relation between the changes in resistance with temperature is given in Eq. (4).

$$\Delta R = k * \Delta T \tag{4}$$

where,

ΔR is the change in resistance.
ΔT is the change in temperature.
K is then first order temperature coefficient of resistance.

There are two types of thermistors available in the market—positive temperature coefficient (PTC) where the resistance increases with the rise in temperature and negative temperature coefficient (NTC) where the resistance decreases with the rises in temperature. The use of a particular type of thermistor depends on the application. These types of sensors are generally used in batteries, thermostats, 3D printers etc.

3.3 *Light-Dependent Resistors*

The light dependent resistors or photo-resistors are the devices where the resistor changes depending on the intensity of incident photons. The resistance is inversely proportional to the incident light intensity. The nature of the photo-resistors depends on the intensity of light. For example, these devices behave as an insulator with high resistance in areas with low light intensity. The increase in incident photon intensity beyond a limit causes the transfer of energy of the incident photons to the electrons leading to their excitation. This causes the flow of current making it behave like a semiconductor. These resistors have been used in a lot of applications like compressors or as LEDs. The disadvantages of these devices are the dependency of the resistivity on the ambient temperature as a result of which the time constant (τ) varies to a large extent.

3.4 Piezo-Resistive Sensors

This category of resistive sensor is used as pressure sensors or in accelerometers where the electrical resistance changes when a mechanical stress is applied on the sensor. The highest change takes place when the stress is applied normally on the patch. The electrons in the valence band are energized to travel to the conduction by providing external stress. The relation between the original and changed resistivity is given in Eq. (5). When a mechanical stress is mechanically applied on the sensor, the resistance of the sensor patch decreases. This increases the pressure dependent output voltage.

$$\rho_\sigma = \frac{\frac{\partial \rho}{\rho}}{\in} \tag{5}$$

where,

$\partial\rho$ is the change in resistivity.
ρ is the original resistivity.
ε is the strain on the sensor.

Piezoresistive silicon sensors are a very common in industries due to the advantage of their high elasticity and high resolution and sensitivity. Other types of piezo-resistive sensors are fabricated from germanium and polyesters.

3.5 Resistive Level Sensor

Level sensors are used to detect the level of any solid or liquid in the surface of that form. The criteria for the selection of level sensor are determined by the ambiance parameter like temperature, pressure or dielectric constant of the experimental medium. The level sensor operates in two different forms—point level and continuous level. Point level detection refers to the detection of any physical property like capacitance or admittance of any solid particle like powders or solute particle used in the medium. Different types of point level sensors like vibrating, capacitive, rotating puddle are available in the market. The other one is the continuous level detection sensors where the physical properties like the hydrostatic, magneto-static or the air pressure is calculated of any liquid or gaseous form. The resistive level sensors are mainly used in the large-scale manufacturing industries rather than academic experimental purposes due to their high cost.

3.6 Strain Gages

The operating principle for a strain gauge is similar to that of a piezo-resistive sensor except that face that this device operates on the strain. Strain gauge operates on the principle of Wheatstone bridge. When the gauge is stretched or compressed with a certain stress within its limits, the electrical resistance increases or decreases respectively. The change in strain of the gauge depends on applied stress and the elastic properties of the stressed material. The factor of indicators (G.F.) is an important parameter for the strain which is used to determine the output voltage. The formulas for the strain, G.F., and the output voltage are given are Eqs. (6) and (7).

$$\in = \frac{\partial l}{l} \tag{6}$$

where,

ΔR is the change in resistance of the strain gauge.
RG is the original resistance.

$$G.F. = \frac{\frac{\Delta R}{R_G}}{\in} \tag{7}$$

where,

ΔR is the change in resistance of the strain gauge.
RG is the original resistance.
ε is the strain exerted on the gauge.

$$V_0 = \frac{V_c{*}G.F.{*}\in}{k} \tag{8}$$

where,

V_o is the output voltage.
V_c is the excitation voltage.
G.F. is the Gauge factor.
ε is strain on the sensor and k = 4.

3.7 Limitations of Resistive Sensors and Need of Capacitive/Inductive Sensors

As explained in the previous section, different types of resistive sensors are available and used for different applications. But still there are certain disadvantages of these sensors which make them unsuitable and unreliable to be used for precise measurements. Some of the major disadvantages of resistive sensors are given below:

1. Considerable amount of energy is generated in case of resistive sensors. This energy is released from the system in terms of heat energy which is a loss to the system.
2. Resistive sensors have very basic operating principle. Now, in some applications, the experimental parameter may not vary only in terms of electrical resistance, but also on other aspects like capacitive sensing. For example, in case of LPG gas detection [1, 2] or phthalate detection [3, 4], both the real part (resistive) and imaginary part (capacitive) changes with the change in concentration of the measured parameter.
3. The resistance of some devices like strain gauge changes with ambient temperature.
4. The hysteresis loss in resistive sensors is high.

Each type of capacitive or inductive sensor is suited to particular applications. An inductive sensor is a type of non-contact electric sensor. In order to be sensed by an inductive sensor an object must be conductive. This limits suitable targets to metal objects. In order to be sensed by a capacitive sensor the target doesn't need to be conductive. Capacitive sensors convert changes in position, or properties of dielectric material into an electrical signal. A capacitive sensor will react to an object acting as a dielectric material as well as a conductive object. This makes metal and non-metal objects suitable targets for capacitive sensors.

4 AC Excitation of Capacitive/Inductive Sensors

Inductive sensors are designed based on the operating characteristics and principle of inductors. Inductors store energy in the form of magnetic field that is generated when a voltage is applied the terminals of an inductor. Capacitive sensors are actually variable capacitors that their capacitance varies during the sensing process. In some of them, capacitance change depends on the dielectric constant of the target material, and in some others, it depends on the sensors dielectric constant. Applying a voltage to conductive objects causes positive and negative charges to collect on each conductors, that creates an electric field between them. Capacitive sensors use an alternative voltage which causes the charges to move back and forth between the objects, continuously, creating an alternating current that is detected by the sensor (Fig. 2). Some of the measurement techniques are explained in the following section.

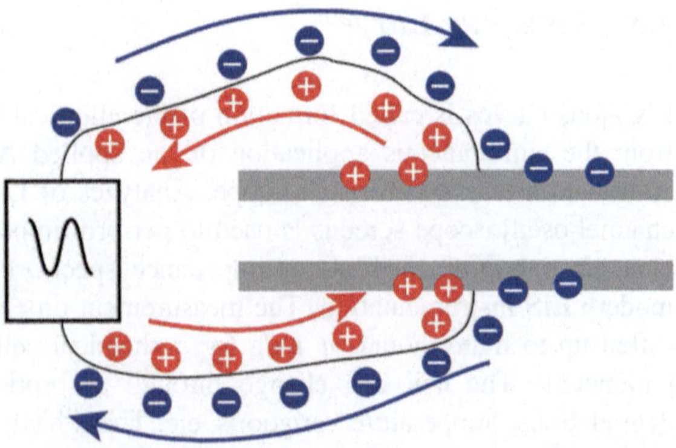

Fig. 2 Applying alternative voltage to the capacitive sensor [5]

4.1 Electrochemical Impedance Spectroscopy

Electrochemical Impedance Spectroscopy (EIS) is a technique to characterize the electrochemical systems. EIS measurement can be done by applying a small amplitude AC excitation signal to the system and measure the response and compute the impedance at each frequency. There are many advantages in using EIS technique: The information content of EIS is much higher than DC techniques or single frequency measurements; EIS provides information on the capacitive behavior of the system and electron transfer, and it may be able to distinguish between two or more electrochemical reactions taking place. This method has been used to study of electrochemistry [6, 7], biomedical applications [8, 9], material science [10] and others. The response of an electrochemical cell to a small amplitude ac perturbation as a function of frequency and it has been reported to estimate the dielectric properties of milk [11], meat inspection [12], quality testing in leather [13], Saxophone reed inspection [14], detection of contaminated seafood with marine biotoxin [15], food endo-toxins [15–19], evaluate electrical properties of drinks, and water [20], and measuring water quality [21].

4.2 AC Bridges

It is the ancient of all the techniques initially used for the measurement of double-layer parameters, principally of the hanging mercury drop electrode. Another use of this method is to measure the electrode impedance in a faradaic reaction to evaluate the dynamic processes at the electrode. Although this method is slow, yet provides an excellent precision of measurements.

4.3 Lissajous Curve

Lissajous curves is called formation of the elliptical figures which can be resulted from the simultaneous application of the applied AC voltage and resulting AC current to a twin beam oscilloscope. Analyzes of Lissajous curves made on twin channel oscilloscope screens is used to perform impedance measurements and had been an approved method for impedance spectroscopy before to the advent of modern EIS instrumentation. The measurement time involved in using this method (often up to many hours) is long for a chemical cell to cause drift in the system parameters. The cell can change through adsorption of impurities, oxidations, degradations, temperature variations, etc. These have been widely used to measure the impedance, but sensitivity to noise and frequency limitations and has limited the use of this technique. Figure 3 shows the formation of Lissajous figures as a consequence of two out of phase signals.

4.4 Fast Fourier Transforms

Fast Fourier Transforms (FFT) is a mathematical tool for evaluating the system impedance. FFT is referred as taking the Fourier transform of the perturbation signal in time domain and generation of corresponding frequency domain data using a computerized algorithm. This method provides an efficient and fast algorithm to compute the Fourier transforms. In practice, only limited length data is transformed, causing the broadening of the computer frequency spectrum, commonly called 'leakage.' Another problem that is called 'aliasing' is linked with the presence of the frequencies larger than one-half of the time domain sampling frequency.

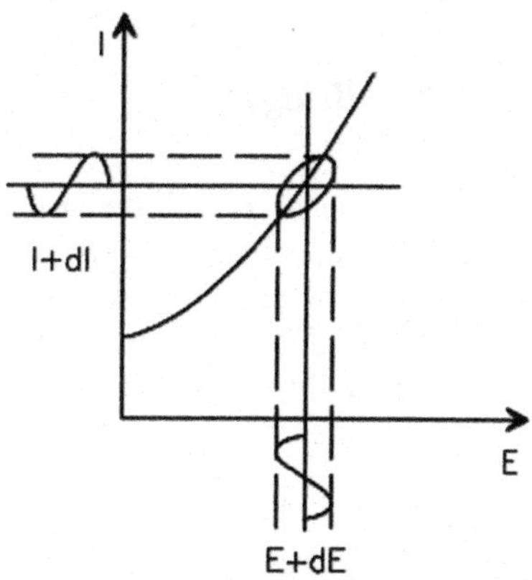

Fig. 3 Formation of Lissajous figure

4.5 Phase Sensitive Detections

Phase sensitive detection (PSD) is used in lock-in amplifiers interfaced with precision potentiostats for system impedance measurements. It contains one time-independent component, depending on the phase difference between two signals and proportional to the amplitude of the measured AC signal. The output signal is given to a low-pass filter that averages the signal component having frequencies above the cut-off frequency. The limitation of the lock-in technique is that it contains the combination of harmonic frequencies present in the input signal.

4.6 Frequency Response Analysis

Frequency Response Analyzer (FRA) is a hi precision technique that determine the frequency response of the system. FRA operates based on correlation of the studied signal with the reference perturbation. The monitored signal is multiplied by the cosine and sine of the reference signal of the same frequency and integrated over one time period. Real and imaginary parts of the signal are recovered, deleting all harmonics. The advantage of the correlation process is that the noise will be reduced, but it is done at the cost of attenuation of the output signal.

FRA is a single sine-wave input measurement method in which a low amplitude AC sine wave of a given frequency is overlaid on a dc bias potential, applied to the working/excitation electrode and measurement of resulting AC current is done. The system behaves pseudo-linear at small amplitude AC potential. This process is repeated for a frequency range, and impedance is computed for five to ten measurements per decade change in frequency. In order to make sure about the system repeatability, linearity, and stability, this technique is rendered viable only for a stable system in equilibrium. For this reason, instantaneous impedance measurements are mandatory for non-stationary systems [22]. A non-linear system contains harmonics (noise) in the measured current response. The drift in the measured system parameters is often observed if the system loses its steady state during the measurement time. The electrochemical cell can change through adsorption, oxidation, coating degradation and temperature variations; to list, these are a few major factors affecting the steady-state condition of the system under test. EIS is used to measure the changes happening in the electrochemical system in and observe the changes in the conductance and capacitance at the sensing surface, interface, and layers, in particular.

5 Simple Embedded Processor Based Excitation System

5.1 Sensing System

The aim of this section is to discuss the design and development of a low-cost sensing system for a different application using the Interdigital sensor. The whole system consisted of a microcontroller, capacitive sensor, power supply, the signal processing circuit, and control circuit. The microcontroller generated the PWM waveform which was filtered to get the sinusoidal excitation signal (input signal) at certain operating frequency (frequency selection was based on the experimental results). This excitation signal was applied to interdigital sensor and output was fed into the signal smoothing circuit to eliminate the noise and amplify the signal. Both the input and output signals were fed into zero cross detector to get the phase difference between them and feedback to the microcontroller.

5.2 Interfacing to Microcontroller

The microcontroller board, Arduino Fio (Fig. 4) which is based on the ATmega328P to perform the signal generation and data acquisition.

Arduino Fio operates at 3.3 v, and it has 14 digital pins and 8 analog pins which are used as input or outputs. It has two external interrupts pins (D2 and D3) and specific pins (D3, D5, D6, D9, D10, D11) for PWM output and controller by corresponding timers. The objective of the interfacing sensor to microcontroller is

Fig. 4 Arduino Fio

to measure the impedance of the sensor. In order to measure both the real and reactive part of the impedance, the input voltage (V1) and the voltage across the series resistance (V2) as shown in Fig. 5 were interfaced to the microcontroller via ADC inputs A0 and A1 respectively. To measure the phase difference between these two signals, the sinusoidal signals were first converted to square wave using zero-crossing detectors (ZCD) and the waveforms from ZCD and then fed into digital inputs D2 and D3 respectively. With the help of microcontroller, the time difference was measured and converted into phase angle in degree.

Figure 6 shows the picture of Printed Circuit Board (PCB) that involves voltage regulator, TLC555IP, the microcontroller, band-pass filter, LM324, LM338, MOSFET for control.

The sinusoidal signal was produced using PWM output and a band-pass filter. Pin 6 of the microcontroller was used to generate the PWM output which was passed through a band-pass filter to reduce the noise and produce a smooth sinusoidal wave. The sinusoidal wave was used as the exciting source of the sensor.

5.3 Power Supply Circuits

Any rechargeable battery (12 V in this project) can be used to provide the power to the embedded circuit. The microcontroller, operational amplifier (LM324) and differential comparator (LM339) were required to provide power at a voltage level at 3.3 V which was necessary to reduce the voltage down to 3.3 V by using a voltage regulator. There are different types of a voltage regulator used for various purposes. The linear voltage regulator is the easiest type to use, and it can be divided into fixed and adjustable value voltage regulator. The advantages of linear

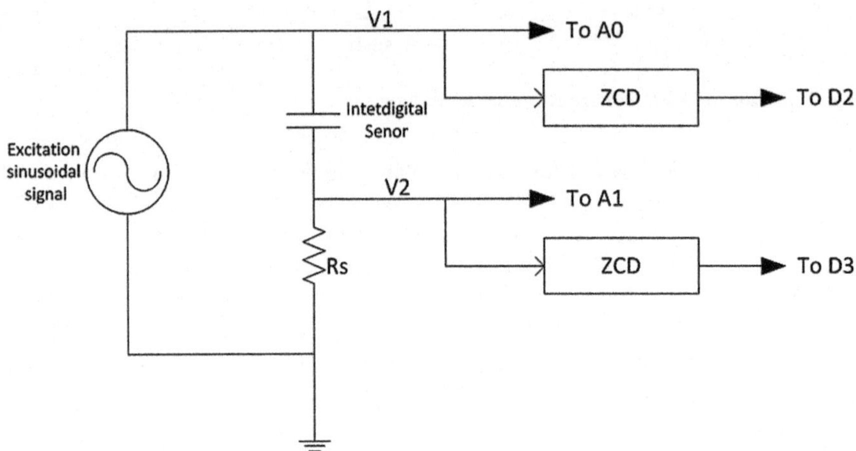

Fig. 5 The interfacing of sensor to microcontroller

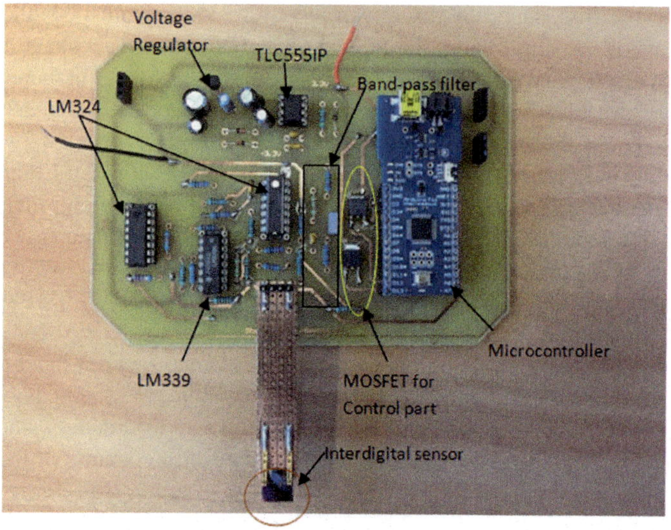

Fig. 6 Picture of system prototype based on the microcontroller

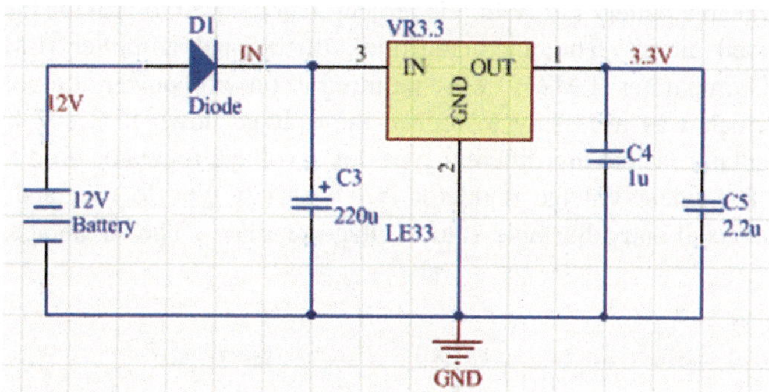

Fig. 7 Circuit for 3.3 V voltage regulator application

regulator include low noise, low output voltage ripple, fast transient response and low cost but the excessive power dissipation is a major limitation of it [23]. L78L33 is a linear fixed voltage regulator with the output voltage of 3.3 V and output current up to 100 mA which is sufficient for those components, so it was employed for the experiment. Figure 7 shows the circuit for the voltage down from 12 to 3.3 V by using L78L33 combined with diode and capacitors. The diode (D1) was used to nullify the inverse voltage in the circuit and allow the current to flow in one direction. Capacitors (C3, C4, and C5) were used as decoupling capacitors to reduce the effect of noise.

On the other hand, the switch-mode power supply (SMPS) is also widely used for the application which requires high efficiency, high power density and low power dissipation [23]. SMPS can generate not only step-down output voltage but also step-up and opposite output voltage. The input signal applied to the sensor requires both the positive and negative magnitude of the alternating source, so the TLC555IP is employed to provide the negative power source. 555 IC is used in a variety of electronic application such as a timer, pulse generation, and oscillator.

Figure 8 shows the circuit that converted the positive input voltage to negative output voltage by using TCL55IP and other essential components. R17 is the pull-up resistor which was connected between VDD and discharging signal to make sure that the discharging signal is in either high or low state. R17, R18 and C6 made up a loop to generate the oscillatory output signal through charging and discharging of C6. When the output signal is high, C8 was charging and D2 conducted the current to the ground but D3 was cut off. When the output signal went to low, the voltage across C8 went to zero as well. In this case, the polarity of D3 was changed so it could conduct the current and C9 started charging. Due to this opposite polarity, it generated the negative voltage.

5.4 Generating an AC Signal

The direct digital synthesis (DDS) method is popularly used to produce wanted output at precise frequencies. DDS is implemented by breaking an analog waveform into discrete digital points which is stored in an array as a lookup table (LUT), and the use of phase accumulator is done to calculate a phase angle for LUT and then generate the digital value at different amplitude [24]. Figure 9 shows the microcontroller code that stored the hexadecimal value of sine waveform in the array as LUT.

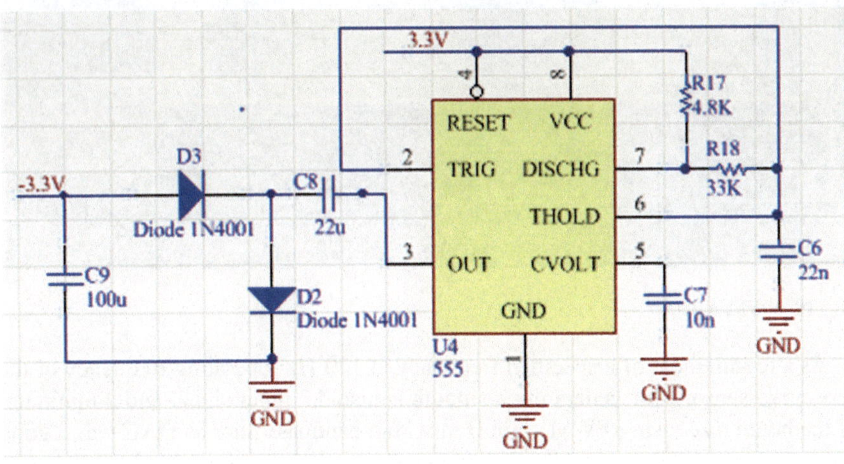

Fig. 8 Use of 555

```
//---------PROGMEM is a variable modifier, store data in flash (program) memory instead of SRAM.
const unsigned char sine[] PROGMEM={
  0x80,0x83,0x86,0x89,0x8c,0x8f,0x92,0x95,0x98,0x9c,0x9f,0xa2,0xa5,0xa8,0xab,0xae,
  0xb0,0xb3,0xb6,0xb9,0xbc,0xbf,0xc1,0xc4,0xc7,0xc9,0xcc,0xce,0xd1,0xd3,0xd5,0xd8,
  0xda,0xdc,0xde,0xe0,0xe2,0xe4,0xe6,0xe8,0xea,0xec,0xed,0xef,0xf0,0xf2,0xf3,0xf5,
  0xf6,0xf7,0xf8,0xf9,0xfa,0xfb,0xfc,0xfc,0xfd,0xfe,0xfe,0xff,0xff,0xff,0xff,0xff,
  0xff,0xff,0xff,0xff,0xff,0xff,0xfe,0xfe,0xfd,0xfc,0xfc,0xfb,0xfa,0xf9,0xf8,0xf7,
  0xf6,0xf5,0xf3,0xf2,0xf0,0xef,0xed,0xec,0xea,0xe8,0xe6,0xe4,0xe2,0xe0,0xde,0xdc,
  0xda,0xd8,0xd5,0xd3,0xd1,0xce,0xcc,0xc9,0xc7,0xc4,0xc1,0xbf,0xbc,0xb9,0xb6,0xb3,
  0xb0,0xae,0xab,0xa8,0xa5,0xa2,0x9f,0x9c,0x98,0x95,0x92,0x8f,0x8c,0x89,0x86,0x83,
  0x80,0x7c,0x79,0x76,0x73,0x70,0x6d,0x6a,0x67,0x63,0x60,0x5d,0x5a,0x57,0x54,0x51,
  0x4f,0x4c,0x49,0x46,0x43,0x40,0x3e,0x3b,0x38,0x36,0x33,0x31,0x2e,0x2c,0x2a,0x27,
  0x25,0x23,0x21,0x1f,0x1d,0x1b,0x19,0x17,0x15,0x13,0x12,0x10,0x0f,0x0d,0x0c,0x0a,
  0x09,0x08,0x07,0x06,0x05,0x04,0x03,0x03,0x02,0x01,0x01,0x00,0x00,0x00,0x00,0x00,
  0x00,0x00,0x00,0x00,0x00,0x00,0x01,0x01,0x02,0x03,0x03,0x04,0x05,0x06,0x07,0x08,
  0x09,0x0a,0x0c,0x0d,0x0f,0x10,0x12,0x13,0x15,0x17,0x19,0x1b,0x1d,0x1f,0x21,0x23,
  0x25,0x27,0x2a,0x2c,0x2e,0x31,0x33,0x36,0x38,0x3b,0x3e,0x40,0x43,0x46,0x49,0x4c,
  0x4f,0x51,0x54,0x57,0x5a,0x5d,0x60,0x63,0x67,0x6a,0x6d,0x70,0x73,0x76,0x79,0x7c
};
```

Fig. 9 Arrays for the hexadecimal of sinusoidal waveform

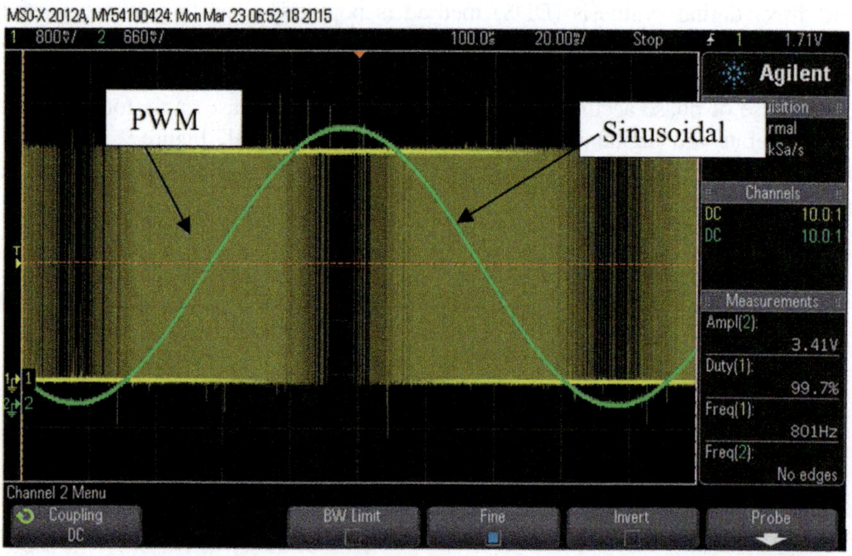

Fig. 10 PWM output

As a requirement of a specified frequency at 120 Hz (operating frequency of the capacitive sensor), for generating a smooth sinusoidal input signal and a minimum of the board size, using PWM combined with a bandpass filter as DAC was a good option. Figure 10 illustrates that the analog amplitude increases with the increase in duty cycle and pulse width.

To generate the smooth sinewave, the band pass filter (Fig. 11) was used only to pass the frequency range between 106 and 338 Hz. The cut-off frequency (f_c) was calculated based on the value of capacitor and resistor, shown in Eq. (9):

$$f_c = \frac{1}{2\pi RC} \tag{9}$$

The band pass filter was combined with a high pass filter (C1 and R1) and a low pass filter (R2 and C2). The capacitor C1 connected to PWM output removed the DC offset from this input signal and combined with R1 only allowed the frequency (f_H) above 106 Hz to pass. The capacitor (C1) value was computed based on rearranged Eq. 5 using the resistor at a value of 10 KΩ:

$$C1 = \frac{1}{2\pi R f_H} = \frac{1}{2 * \pi * 10{,}000 * 106} = 150nF \tag{10}$$

The low pass filter part consisted of R2 and C2, assuming the R2 was also fixed at value of 10 KΩ and cut-off frequency (f_L) is at 338 Hz, the C2 value was calculated using Eq. 6:

$$C2 = \frac{1}{2\pi R f_L} = \frac{1}{2 * \pi * 10{,}000 * 338} = 47nF \tag{11}$$

After filtering, the smooth sinusoidal waveform which had positive and negative analog amplitude was generated as shown in Fig. 12 with a frequency of 120 Hz.

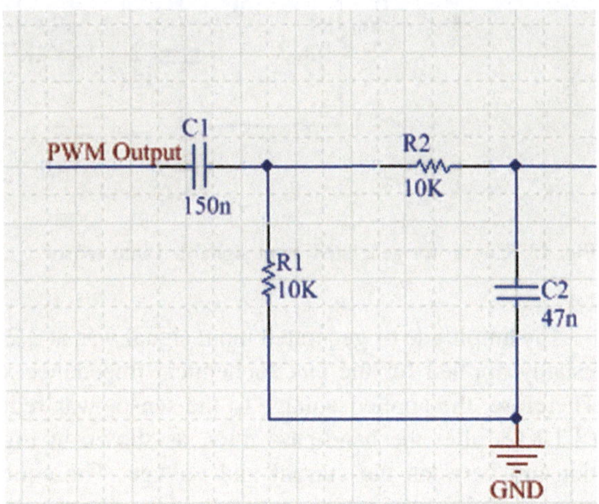

Fig. 11 Bandpass filter for generating smooth sinusoidal waveform

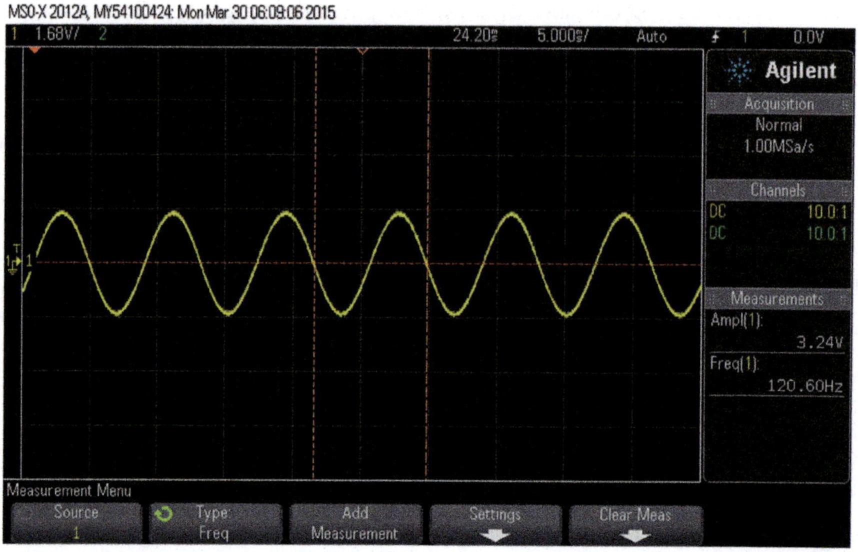

Fig. 12 Smooth sinusoidal waveform produced by combined PWM output and band pass filter

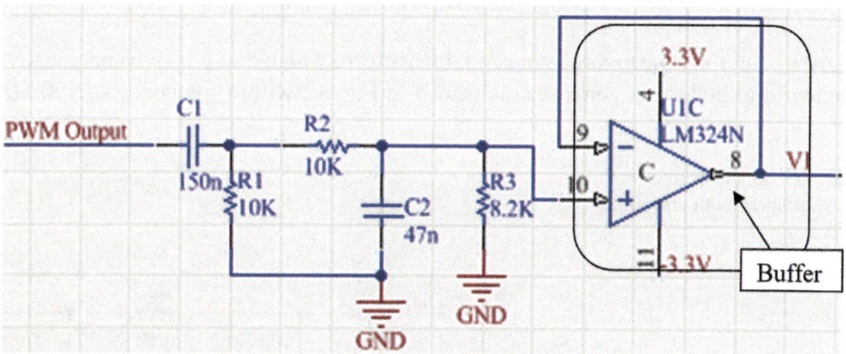

Fig. 13 Circuit for generated input signal fed into sensor

The amplitude of generated input signal was at 3.24 V. A small excitation signal usually applied for the electrochemical impedance spectroscopy (EIS) technique. Therefore, the applied voltage to the sensor was reduced by adding a resistor, R3 (8.2 KΩ) after the band-pass filter, as shown in Fig. 13, to keep the linear relationship between the current and voltage. The equivalent resistance of the filter along with R3 formed a potential divider and reduced the voltage to 860 mV which was suitable for the operation. One amplifier served as the unity gain buffer or voltage follower to reduce any current drawn from the input voltage, and it also played a role in the impedance matching that minimizing the load effect.

5.5 Measurement of the Sensing Voltage

To measure the impedance of the sensor, the voltage applied to the sensor, and the current through the sensor was needed to be measured by the system. The excitation voltage to the sensor is bipolar in nature, so it could not be fed directly to the microcontroller. To avoid the negative excitation voltage, a level shift-up circuit is used as shown in Fig. 14.

A resistor, R_s was connected to the sensor in series to determine current flowing through the sensor as shown in Fig. 14. The value of R_s was significantly small so that the effect of it was negligible. Based on the Ohm's law, the current (I_s) flows through the sensor was calculated by Eq. (12):

$$I_s = \frac{V_{in}}{Z} \tag{12}$$

where:

V_{in}: the voltage applied to the sensor
Z: the total impedance of the circuit

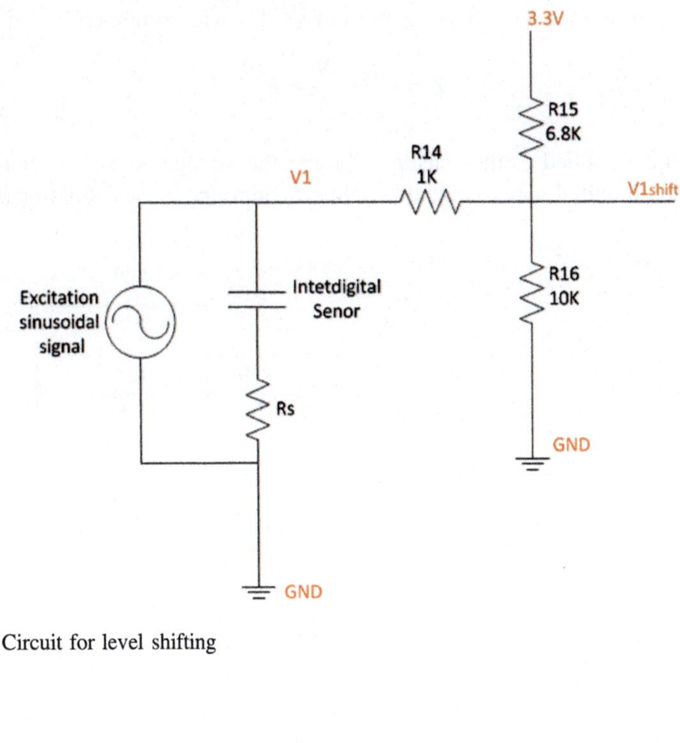

Fig. 14 Circuit for level shifting

Therefore, the voltage across the sensor V_s is rearranged into Eq. 9:

$$V_s = I_s * R_s = \frac{V_{in}}{Z} * R_s \tag{13}$$

where:

- R_s: series resistance, used for measurement of sensing voltage

In Fig. 15, another buffer had been connected to the output of V_s to reduce the signal reflection from the load and transfer max power from the input. Also, the active low-pass filter was employed in order to reduce the noise and amplify the signal with a gain of 11 by using R20 and R21.

The reason to use an active low-pass filter was that the amplitude of the output signal in pass low-pass filter is lower than the amplitude of its input signal [25] and the gain is less than 1. In addition, the voltage across the sensor (V_s) is slight (the peak voltage is around 80 mV) because the total impedance of the sensor was very larger which resulted in a very small current going through the circuit. The resistor value was picked at any value and gain was changed but the value of 10 and 1 KΩ is a standard resistor value which is available. However, it resulted in the distortion of the sinusoidal waveform which affect the accurate measurement of V_s. Therefore, using the active low-pass filter with gain control reduced the noise across the signal and amplify the V_s.

Rearranging the Eq. (12) into the term of Z, this was computed by Eq. (13):

$$Z = \frac{V_{in}}{I_s} = \frac{V_{in}}{V_s} * R_s \tag{14}$$

The voltage applied to the sensor (V_{in}) and the voltage across the sensor (V_s) were measured and Z was calculated by microcontroller. Series Resistor (R_s) was

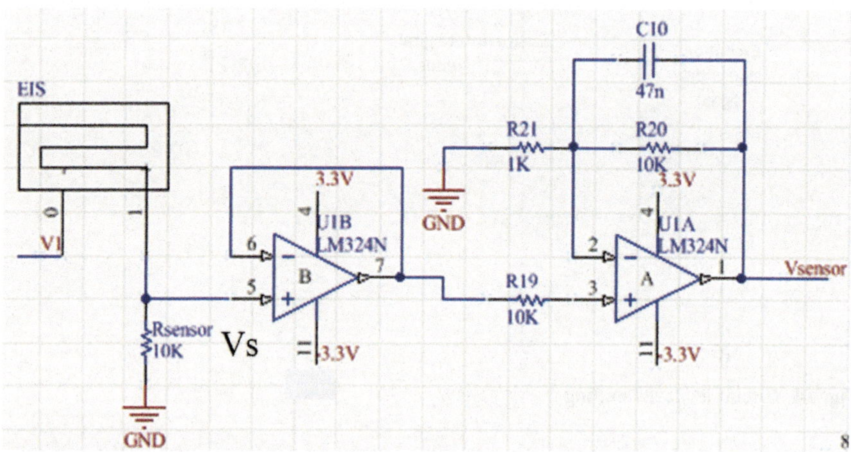

Fig. 15 Circuit for processing the sensing voltage

given at 10 KΩ. Figure 16 shows the precision rectifier circuit which was composed of two operational amplifiers for reducing the voltage loss because the V_{sensor} is very small. The output of voltage (V_{sensor}) from the sensor was full sine wave which was fed into the first operational amplifier. When $V_{sensor} > 0$, only positive half wave (V_{pos}) was generated from the first op-amp and $V_{pos} = V_{sensor}/2$. The second op-amp worked as a subtractor, $V_{pos} = V_{full}/2$ (where $V_{sensor} = V_{full}$) when $V_{sensor} > 0$, $V_{inmicro} = V_{full} - V_{pos} = V_{full}/2$; when $V_{sensor} < 0$, $V_{inmicro} = 0 - (-V_{full}) = V_{pos}$; Therefore the output of $V_{inmicro}$ was a full wave rectified wave. The $V_{inmicro}$ was then fed into the analogue input pin of the microcontroller.

The voltage applied to the sensor (V_{in}) and the voltage across the sensor (V_s) were all fed into microcontroller ADC input pins for calculation of total impedance. The compare output B register of Timer 1 was used as the trigger source for ADC conversion. 160 sample points of the whole sinusoidal waveform were taken with the sample rate at 19.2 kHz and each point was compared to get the maximum and minimum of the value. Figure 17 illustrates the microcontroller code for collecting and storing 160 sample values into an array. Figure 18 indicates how to obtain the amplitude of voltage.

5.6 *Measurement of the Sensor Impedance*

The total impedance has two parts: real part (R) and imaginary part (X) of impedance. Following equations can calculate them:

$$R = Z^* cos\emptyset \tag{15}$$

$$X = Z^* sin\emptyset \tag{16}$$

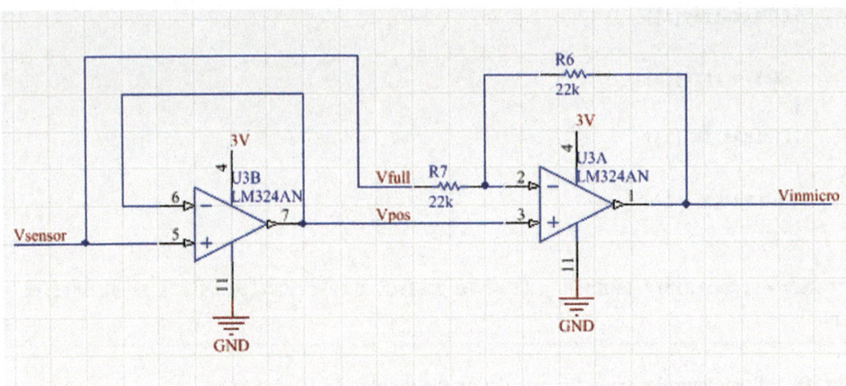

Fig. 16 Precision rectifier circuit

```
ISR(ADC_vect)
{
  TIFR1 = (1<<OCF1B);       //clear Timer compare match flag
  if (adcon ==1)
  {
    array[i] = ADCW;       //ADCW takes care of ADCL and ADCH
    i++;
  }
}
void check()
{
  i = 0;
  while(i<160)
  {
    SMCR |=0x02;           //SMCR - Sleep Mode Control Register
                           //bit 3:1, 001, ADC Noise Reduction Mode
                           //ADC noise reduction mode
    adcon = 1;
  }
  adcon = 0;
  SMCR &= 0xFD;            //idle mode
}
```

Fig. 17 Microcontroller code for ADC data collection

```
void collect()
{
  i=0;
  Max = 0;                 //Peak value after level shifter
  Min = array[0];
  for(i=0;i<160;i++)
  {
    if(Max<array[i])
    {
      Max = array[i];
    }
    if(Min>array[i])
    {
      Min=array[i];
    }
  }
  Vmax = ((Max-Min)/2)+Min;        //To collect the Peak value before level shifter
}
```

Fig. 18 Microcontroller code for calculating the amplitude of voltage where: R and X are the resistance and reactance of the circuit, ϕ: phase angle between V_{in} and I_s. Therefore, the resistive part of the sensor (R_{sensor}) was calculated by Eq. 13:

$$R_{sensor} = R - R_s \tag{17}$$

To measure the phase angle (ϕ) between V_{in} and I_s, the Schmitt Trigger was used as zero crossing detector (ZCD). The Schmitt Trigger is based on the hysteresis comparator with zero reference voltage is shown in Fig. 19. The purpose of employing Schmitt Trigger was to convert the sinusoidal waveform to square waveform which can be easily red by the microcontroller and can reduce undesirable transitions caused by the noisy input signal. The time delay has been calculated inside the microcontroller. The time delay depends on the nature of the impedance of the sensor. From the time difference (T_d) and the period of one square wave (Period), the phase angle can be calculated by Eq. (18):

$$phase\,angle = \frac{T_d}{Period} * 360° \tag{18}$$

Finally, real part and imaginary part of the impedance can be calculated from Eqs. (15) and (16) and can be stored in the microcontroller for further use.

(a)

(b)

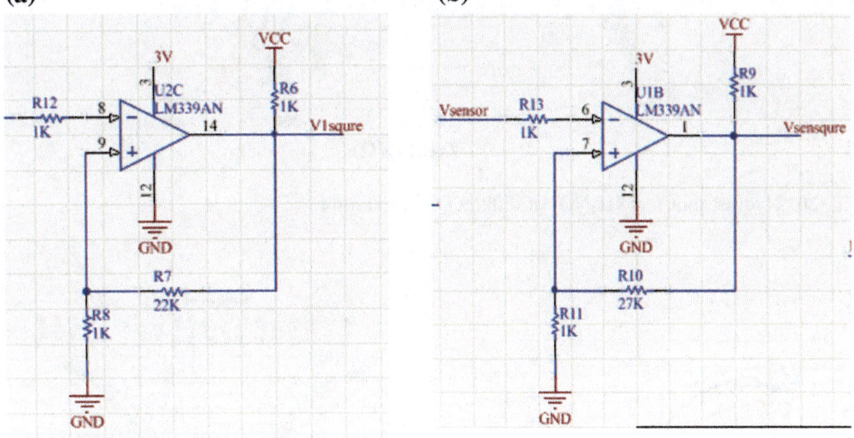

Measurement of the input signal.

Measurement of the output signal.

Fig. 19 Zero cross detector circuit

6 Result and Discussion

Some of the application of capacitive sensor will be explained in this part to understand the range of implementation of a capacitive sensor.

In the first application, one interdigital sensor was used to measure the nitrate concentration in water. The concentration level was 0.1–0.5 mg/L and measurement was done by HIOKI LCR meter. The LCR meter provided the changing impedance with changing concentration. The impedance measurement result is represented in the Nyquist plot as shown in Fig. 20 with a frequency range from 1 Hz to 100 kHz. It illustrates that the total impedance is reduced when the concentration of the water solution was increased.

Due to the presence of ionic salts in the solution, there was a significant change in the real part of the impedance compared to the imaginary part. From Fig. 21, it is seen that the real part of the impedance changes with the change in frequency.

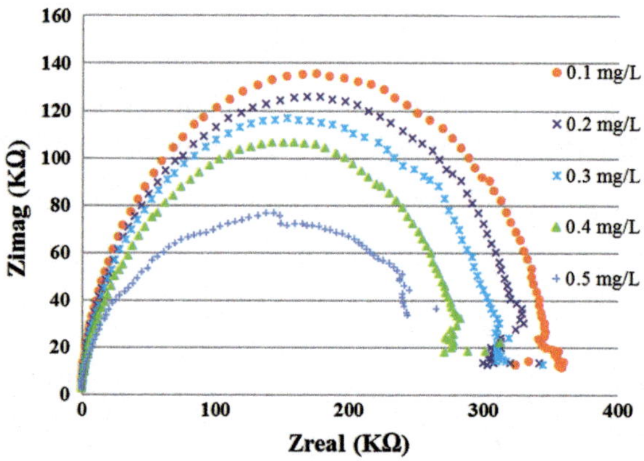

Fig. 20 Nyquist plot for NH_4NO_3 at different concentration

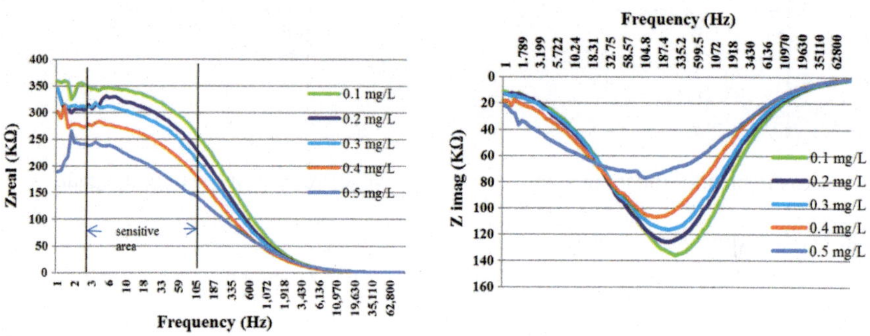

Fig. 21 Real part and imaginary part of impedance versus frequency at different concentration (NH_4NO_3)

Different concentrations of solution were considered to determine the sensitive region (from 5 to 150 Hz) of the sensor under consideration. It also shows the corresponding change in the imaginary part of impedance with the change in frequency. It is seen from the figure that there is not much change in the reactance values for the complete frequency range as compared to the change in the resistive (Real part of Impedance) values.

In the next application, A real-time, non-invasive, and label-free sensing technique for the early detection of bone turnover by employing an EIS technique has been reported. The proposed system incorporated antibody-antigen-based functionalization by employing streptavidin agarose as a cross-linker for binding CTx-I peptides-a bone loss biomarker. A planar capacitive interdigital sensor has been used in conjunction with a FRA algorithm to measure the electrochemical impedance of the samples. Four known concentration samples (0.147, 0.437, 0.798, and 1.693 ng/mL) were measured in the developed sensing system. The standard solution with zero concentration of analyte was considered as the control. Experiments were performed at room temperature (21 °C) at a humidity level of 31%. The developed sensing system performed tests on the samples immediately after preparing the sample solutions. Figure 22a represents the real part of impedance in the frequency domain for all the analyte concentrations. As shown in this figure, the real part of impedance (X) shows clear changes, especially at lower frequencies between 150 Hz and 750 Hz, with a change in analyte concentration that is attributed to the dielectric properties of the sample. The resistive of impedance versus frequency for different concentrations of CTx-I is plotted in Fig. 22b. The change in the real part of the impedance can be seen only at very low frequencies (up to 150 Hz), that is mainly due to the ionic properties of the sample. The sensitivity obtained from the imaginary part is also higher than the resistive part. Therefore, the reactance was used to evaluate the concentration of the analyte (CTx-I) in the sample solutions.

Figure 23 illustrates the Nyquist plot for the impedance spectrum obtained for all four concentrations of CTx-I in the frequency range of 42 Hz–100 kHz. It was seen that the diameter of the semicircle increases by increasing the analyte concentration

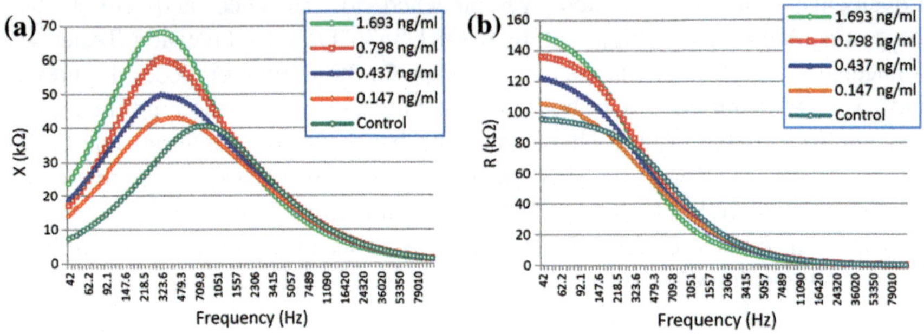

Fig. 22 **a** Imaginary part of the impedance versus frequency; and **b** real part of the impedance versus frequency

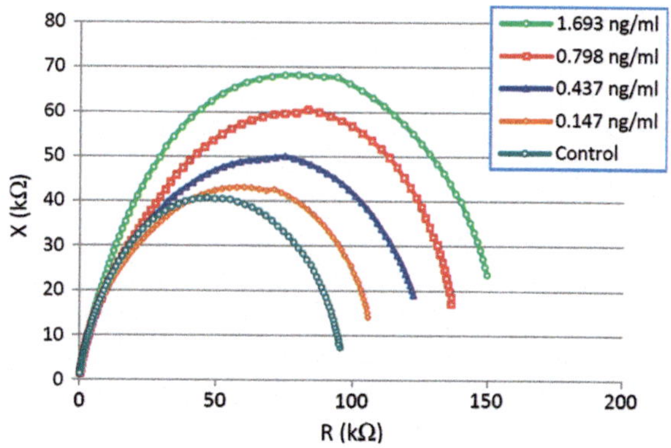

Fig. 23 Nyquist plot for different CTx-I concentrations

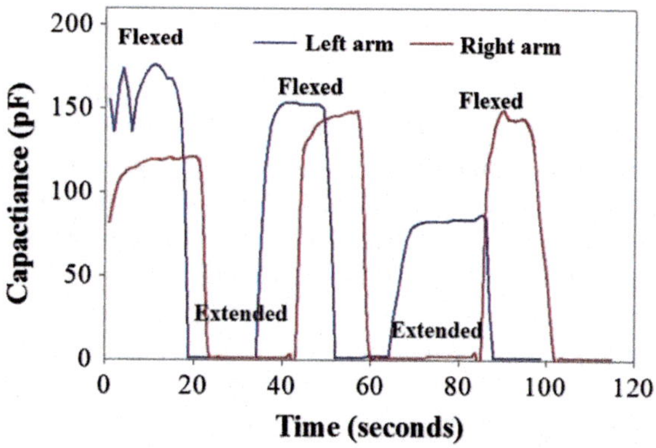

Fig. 24 Movement of *left* and *right* arm

depicting the increase in charge transfer resistance due to the presence of the more concentration of CTx-I attached to the sensing surface.

The flexible, strain—sensitive sensors were used for determining physiological movements. The sensor patch was attached to the knee and elbow with bio-compatible tapes. A high testing HIOKI 3536 LCR Hi Precision Tester was connected from the other end via alligator clips for monitoring purposes. Figures 24 and 25 show the monitoring of both the limbs in an oscillatory motion. The reading was analyzed based on the change in capacitance for two different conditions. Flexed referred to a condition when the limb was bent whereas extended relates to a condition when the limb was stretched. The bending of the limb led an increase in the capacitance value compared to the stretched one. The two different conditions were easily identified for both the limbs using the flexible sensors.

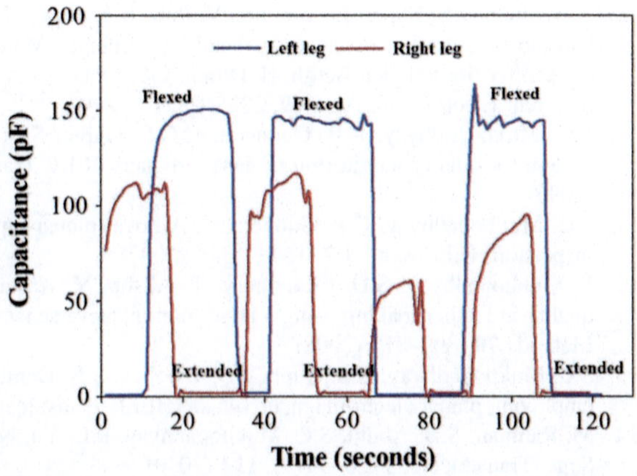

Fig. 25 Movement of *left* and *right* leg

7 Improvement

The embedded sensing device is a bit large compared to different ICs as impedance analyzer which is available in the market. Though the explained system is easy to develop and low cost which is one of the major advantage in different applications.

References

1. A. Nag, A.I. Zia, X. Li, S.C. Mukhopadhyay, J. Kosel, Novel sensing approach for LPG leakage detection: part I—operating mechanism and preliminary results. IEEE Sens. J. **16**, 996–1003 (2016)
2. A. Nag, A.I. Zia, X. Li, S.C. Mukhopadhyay, J. Kosel, Novel sensing approach for LPG leakage detection—part II: effects of particle size, composition, and coating layer thickness. IEEE Sens. J. **16**, 1088–1094 (2016)
3. A.I. Zia, S. Mukhopadhyay, I. Al-Bahadly, P. Yu, C.P. Gooneratne, J. Kosel, Introducing molecular selectivity in rapid impedimetric sensing of phthalates, in *2014 IEEE International Instrumentation and Measurement Technology Conference (I2MTC) Proceedings* (2014), pp. 838–843
4. A.I. Zia, A.M. Syaifudin, S. Mukhopadhyay, P. Yu, I. Al-Bahadly, C.P. Gooneratne et al., Electrochemical impedance spectroscopy based MEMS sensors for phthalates detection in water and juices, in *Journal of Physics: Conference Series* (2013), p. 012026
5. M. Khafaji, S. Shahrokhian, M. Ghalkhani, Electrochemistry of levo-thyroxin on edge-plane pyrolytic graphite electrode: application to sensitive analytical determinations. Electroanalysis **23**, 1875–1880 (2011)
6. J. Fischer, H. Dejmkova, J. Barek, Electrochemistry of pesticides and its analytical applications. Curr. Org. Chem. **15**, 2923–2935 (2011)
7. C. Valero Vidal, A. Igual Muñoz, Effect of physico-chemical properties of simulated body fluids on the electrochemical behaviour of CoCrMo alloy. Electrochim. Acta **56**, 8239–8248 (2011)

8. C. Wolner, G.E. Nauer, J. Trummer, V. Putz, S. Tschegg, Possible reasons for the unexpected bad biocompatibility of metal-on-metal hip implants. Mater. Sci. Eng., C **26**, 34–40 (2006)
9. C. Xhoffer, K. Van den Bergh, H. Dillen, Electrochemistry: a powerful analytical tool in steel research. Electrochim. Acta **49**, 2825–2831 (2004)
10. S.C. Mukhopadhyay, C.P. Gooneratne, G.S. Gupta, S.N. Demidenko, A low-cost sensing system for quality monitoring of dairy products. IEEE Trans. Instrum. Meas. **55**, 1331–1338 (2006)
11. S.C. Mukhopadhyay, C.P. Gooneratne, A novel planar-type biosensor for noninvasive meat inspection. IEEE Sens. J **7**, 1340–1346 (2007)
12. S. Mukhopadhyay, S.D. Choudhury, T. Allsop, V. Kasturi, G. Norris, Assessment of pelt quality in leather making using a novel non-invasive sensing approach. J. Biochem. Biophys. Methods **70**, 809–815 (2008)
13. S.C. Mukhopadhyay, G.S. Gupta, J.D. Woolley, S.N. Demidenko, Saxophone reed inspection employing planar electromagnetic sensors. IEEE Trans. Instrum. Meas. **56**, 2492–2503 (2007)
14. M. Rahman, S.B. Abdul, S.C. Mukhopadhyay, P.L. Yu, Novel sensors for food inspections. Sens. Transducers (1726–5479), **114** (2010)
15. A. Mohd Syaifudin, S. Mukhopadhyay, P. Yu, C.H. Chuang, H.P. Wu, C.P. Gooneratne et al., Characterizations and performance evaluations of thin film interdigital sensors for Gram-negative bacteria detection (2011), pp. 181–186
16. A. Mohd Syaifudin, S. Mukhopadhyay, P. Yu, M.J. Haji-Sheikh, C.H. Chuang, H.P. Wu, Detection of natural bio-toxins using an improved design interdigital sensors (2011), pp. 1028–1031
17. A. Mohd Syaifudin, S. Mukhopadhyay, P. Yu, Modelling and fabrication of optimum structure of novel interdigital sensors for food inspection. Int. J. Numer. Model. Electron. Networks Devices Fields **25**, 64–81 (2012)
18. M.S. Abdul Rahman, S.C. Mukhopadhyay, P.-L. Yu, J. Goicoechea, I.R. Matias, C. P. Gooneratne et al., Detection of bacterial endotoxin in food: new planar interdigital sensors based approach. J. Food Eng. **114**, 346–360 (2013)
19. A.I. Zia, A.R. Mohd Syaifudin, S.C. Mukhopadhyay, I.H. Al-Bahadly, P.L. Yu, C. P. Gooneratne et al., MEMS based impedimetric sensing of phthalates, in *2013 IEEE International Instrumentation and Measurement Technology Conference (I2MTC)* (2013), pp. 855–860
20. M.E. Alahi, L. Xie, A. I. Zia, S. Mukhopadhyay, L. Burkitt, Practical nitrate sensor based on electrochemical impedance measurement, in *2016 IEEE International Instrumentation and Measurement Technology Conference Proceedings (I2MTC)* (2016), pp. 1–6
21. B.-Y. Chang, S.-M. Park, Electrochemical impedance spectroscopy. Ann. Rev. Anal. Chem. **3**, 207–229 (2010)
22. H.J. Zhang, *Basic Concepts of Linear Regulator and Switching Mode Power Supplies* (2013), http://cds.linear.com/docs/en/application-note/AN140fa.pdf
23. C.S. Eva Murphy, *All About Direct Digital Synthesis* (2004), http://www.analog.com/library/analogdialogue/archives/38-08/dds.pdf. Accessed 03 Jan 2017
24. E. Tutorials, *Active Low Pass Filter*, http://www.electronics-tutorials.ws/filter/filter_5.html. Accessed 03 Jan 2017
25. L.P. Mark Kretschmar, *Capacitive Sensor* (2016), http://www.sensorsmag.com/sensors/position-presence-proximity/capacitive-sensor-operation-part-1-the-basics-5844. Accessed 3 Jan 2017

Advanced Techniques for Directly Interfacing Resistive Sensors to Digital Systems

Ferran Reverter, Fernando Vidal-Verdú and José A. Hidalgo-Lopez

Abstract This chapter reviews advanced techniques for the direct connection of resistive sensors to digital systems without using any analogue circuit, such as an amplifier or an analogue-to-digital converter, in the signal path. The sensor electronic interfaces proposed herein rely on the following operating principle: the digital system measures through an embedded digital timer the charging/discharging time of an RC circuit formed by the resistive sensor and a known capacitor. The chapter first explains how resistive sensors with a single, differential or bridge topology can be directly measured using a low-cost microcontroller. The uncertainty sources involved in the measurement (such as the mismatch of the internal resistances, quantisation and trigger noise) and the performance in some applications are reported. Next, the chapter deals with the direct connection of resistive sensor arrays to field-programmable gate arrays, where different resistances of the array are measured in parallel through a set of timers running simultaneously. The new uncertainty sources (mainly, crosstalk) and the applications are also reported. Although the proposed sensor interfaces are quite simple in terms of operating principle, their linearity and resolution are quite remarkable provided that the design rules indicated along this chapter are followed.

F. Reverter (✉)
Universitat Politècnica de Catalunya - BarcelonaTech, Barcelona, Spain
e-mail: ferran.reverter@upc.edu

F. Vidal-Verdú · J.A. Hidalgo-Lopez
Universidad de Málaga - AndalucíaTech, Málaga, Spain
e-mail: fvidal@uma.es

J.A. Hidalgo-Lopez
e-mail: jahidalgo@uma.es

B. George et al. (eds.), *Advanced Interfacing Techniques for Sensors*,
Smart Sensors, Measurement and Instrumentation 25,
DOI 10.1007/978-3-319-55369-6_4

1 Introduction

In the society of the 21st century, many people have smart home appliances inside smart buildings located in smart cities whose streets are full of smart cars. Almost everything is getting smart thanks to the proliferation of *information and communication technology* and the deployment of technologies such as *wireless sensor networks* and the *internet of things*. To become smart, it is essential in the first place to monitor through sensors what is happening in and/or around the smart *thing*. The data collected is processed and then a smart decision is taken with the aim of improving the safety, efficiency, sustainability, mobility, etc. of the smart *thing* and, hence, the people's quality of life.

Real-time monitoring systems employ sensors to acquire information, the same as human beings use their senses. The information acquired can be very diverse, for instance: the carbon monoxide (CO) concentration in air in a smart city, the tire pressure in a smart car, the vibration level in a smart building, or the laundry weight in a smart washing machine. The magnitude of the measurand (e.g. CO concentration, pressure, vibration and weight) determines the magnitude of the electrical signal (e.g. resistance, capacitance, voltage or current) provided at the sensor output. Anyhow, such an electrical signal is generally of low amplitude and carries some noise and, therefore, an electronic interface is required between the sensor and the processing digital system so as to correctly extract the information of interest.

A classical block diagram of a sensor electronic interface is shown in Fig. 1 [1]. The sensor output signal is first processed in the analogue domain by a signal conditioning circuit that generally relies on operational amplifiers (OpAmp). The main functions of this block are level shifting and amplification so as to match the sensor output span to the input span of the ensuing analogue-to-digital converter (ADC) and, hence, to make good use of the ADC dynamic range. Other common tasks of the signal conditioning circuit are: sensor output-to-voltage conversion, filtering, linearization and/or demodulation. The resulting analogue signal is then digitized via the ADC. Finally, a digital system acquires, stores, processes, controls, communicates (to other devices or systems) and/or displays the digital value with information about the measurand. Nowadays, the most popular digital systems are microcontrollers (μC) and Field-Programmable Gate Arrays (FPGA).

The sensor electronic interface shown in Fig. 1 can be implemented in various ways, for example: (i) each block has its own integrated circuit (IC) and then those are interconnected in a printed circuit board (PCB); or (ii) an application-specific IC

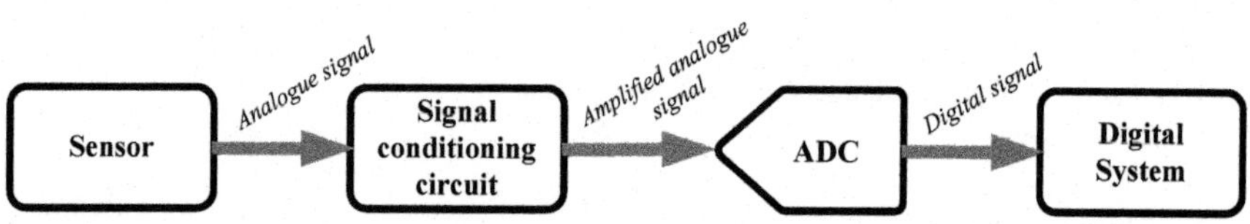

Fig. 1 Classical block diagram of a sensor electronic interface

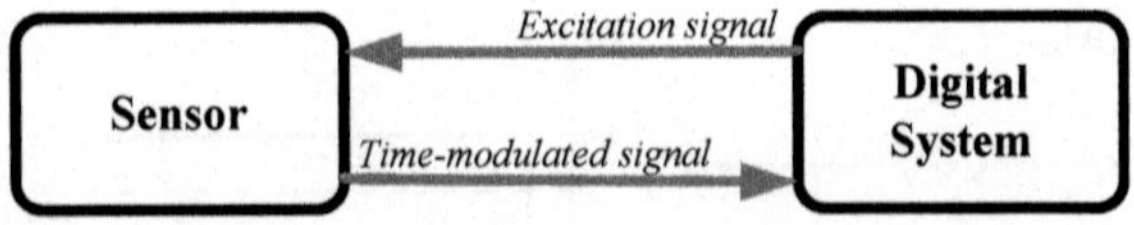

Fig. 2 Direct interface circuit

(ASIC) including the electronics of the four blocks shown in Fig. 1 is designed. Intermediate solutions are also offered by the main semiconductor companies through commercial ICs that include: (i) some signal conditioning circuit, the ADC and the digital system (e.g. MSC1210 from Texas Instruments, TI); (ii) the sensor, its signal conditioning circuit and the ADC (e.g. ADXL312 from Analog Devices, AD); and (iii) the signal conditioning circuit and the ADC to measure a specific type of sensor (e.g. ADS1232 from TI for bridge-type resistive sensors and AD7745 from AD for capacitive sensors). These chips including the sensor together with analogue and digital electronics are commonly known as integrated *smart sensors* [2].

An alternative approach to reading some sensors (e.g. resistive [3–5], capacitive [6–8], inductive [9, 10] and voltage-output [11] sensors) is shown in Fig. 2. This circuit topology is known as *direct interface circuit* since the sensor is directly connected to the digital system without using either the signal conditioning circuit or the ADC [12, 13]. The digital system excites the sensor to get a time-modulated signal that is directly measured in the digital domain through a digital timer embedded into the digital system. In comparison with the sensor electronic interface shown in Fig. 1, a direct interface circuit is simpler and needs fewer components. Actually, it can be implemented with a common general-purpose 8-bit μC which is a low-cost (say, 1 $) and low-power (say, about 1 mA in active mode and less than 1 μA in power-down mode [14]) device. Therefore, a direct interface circuit offers advantages in terms of cost, physical space and power consumption, which is of major interest, for instance, in autonomous sensors powered by either batteries or energy harvesters. Furthermore, as will be shown along this chapter, the performance of such circuits in terms of accuracy and resolution is quite remarkable taking into account their simplicity.

This chapter reviews most of the research work carried out about direct interface circuits for resistive sensor and is organized as follows. Section 2 describes the operating principle of such circuits. Section 3 explains how a μC can be applied to measure different topologies of resistive sensors. Section 4 does the same but using FPGA applied to resistive sensor arrays. Finally, Sect. 5 takes some conclusions and forecasts the future research work about this topic.

2 Operating Principle

Direct interface circuits for resistive sensors rely on measuring the charging or discharging time of an RC circuit, i.e. a circuit with a resistance (R) and a capacitance (C), as shown in Fig. 3a. The digital system excites the RC circuit and then

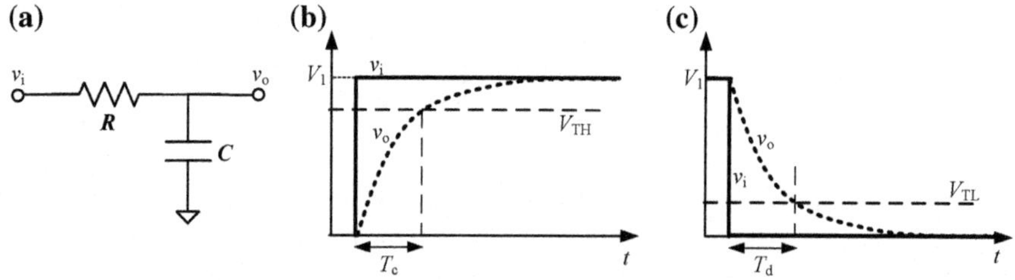

Fig. 3 **a** RC circuit; **b** measurement of the charging time; and **c** measurement of the discharging time

measures the time interval needed to charge or discharge the capacitance C to a given threshold voltage through the sensor resistance.

The basics of the operating principle for the measurement of the charging time and the discharging time are explained by means of Fig. 3b, c, respectively. In Fig. 3b, assuming C initially discharged, if a step of amplitude V_1 is applied to the input of the RC circuit, then the transient response of the output voltage is

$$v_o(t) = V_1\left(1 - e^{-\frac{t}{RC}}\right), \tag{1}$$

and the time required to charge C from 0 to a given high threshold voltage (V_{TH}) is

$$T_c = RC\ \ln\left(\frac{V_1}{V_1 - V_{TH}}\right), \tag{2}$$

which is proportional to R. On the other hand, in Fig. 3c, assuming C already charged to V_1, if a step towards ground is applied to the input, then the transient response of the output voltage is

$$v_o(t) = V_1 e^{-\frac{t}{RC}}, \tag{3}$$

and the time needed to discharge C from V_1 to a given low threshold voltage (V_{TL}) is

$$T_d = RC\ \ln\left(\frac{V_1}{V_{TL}}\right) \tag{4}$$

which again is proportional to R. Therefore, in an RC circuit, changes of resistance are proportionally converted to changes of time interval.

The RC circuit in Fig. 3a can be directly connected to a digital system using the circuit topology shown in Fig. 4a, where R has been replaced by R_x (i.e. a resistive sensor). Two input/output digital ports (pins 1 and P) are employed to excite the RC circuit and to monitor through a Schmitt trigger (ST) buffer embedded into Pin 1 the exponential charging or discharging voltage represented before in Fig. 3b, c,

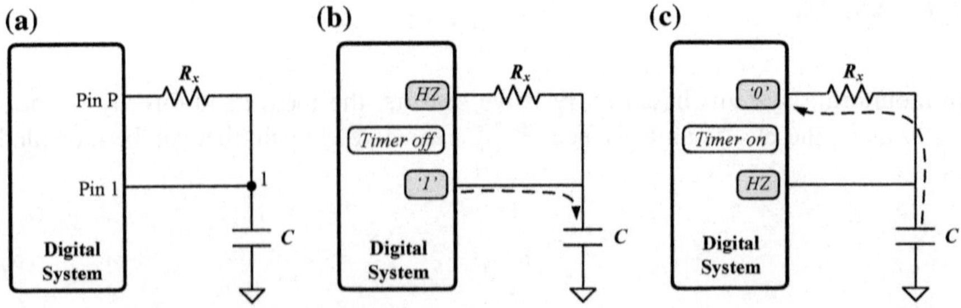

Fig. 4 **a** Basic topology of a direct interface circuit for a resistive sensor; **b** pin configuration during the charging stage; and **c** pin configuration during the discharging and measurement stage

respectively. This circuit can measure either the charging time or the discharging time, but the measurement of the latter is preferable since it has lower variability. This is because the discharging-time measurement uses the V_{TL} of the ST buffer, which is less noisy than the V_{TH} used for the charging-time measurement [15]. For this reason, the rest of the chapter always assumes that the direct interface circuit measures the sensor resistance through the discharging time.

The circuit in Fig. 4a involves two operation stages: charging stage, and discharging and measurement stage. During the charging stage, Pin 1 is set as an output providing a digital '1', whereas Pin P is set as an input offering high impedance (HZ), as shown in Fig. 4b. Therefore, the capacitor C is quickly charged to the analogue output voltage (V_1) corresponding to a digital '1', which is generally equal to the supply voltage (V_{DD}) of the digital system. During the discharging and measurement stage, Pin 1 is set as a HZ input and Pin P is set as an output providing a digital '0', as shown in Fig. 4c. Consequently, C is discharged towards ground through R_x while a digital timer (embedded into the digital system) measures the time interval required to do so. When the exponential discharging voltage crosses the V_{TL} of the ST buffer embedded into Pin 1, the timer is read and a digital number proportional to R_x (see Eq. (4)) is achieved.

3 Interfacing Resistive Sensors to Microcontrollers

The operating principle explained in Sect. 2 can be implemented by a µC to measure resistive sensors with a single, differential or bridge topology. Next, we discuss the main features of both the sensor and the µC, and then we explain how to join them to build a direct interface circuit. The uncertainty sources involved in the measurement and the application of the proposed circuits are also reported.

3.1 Sensor

In monitoring systems based on resistive sensors, the measurand directly or indirectly alters the electrical resistance (R) of a resistive element that can be modelled as

$$R = \rho \frac{l}{A}, \quad (5)$$

where ρ is the resistivity of the material, and l and A are the length and cross-sectional area of the conductor, respectively. Any of the three parameters involved in Eq. (5) can be altered by the measurand, thus causing a change of resistance.

Resistive sensors can be classified according to the number of sensing elements that make up the sensor and how these are interconnected, with the following three types:

(a) **Single resistive sensors**, with one sensing element whose resistance (R_x) changes with the measurand, as shown in Fig. 5a. Such a resistance can be modelled as

$$R_x = R_0 \pm \Delta R = R_0(1 \pm x_R) \quad (6)$$

where R_0 is the nominal resistance at a reference value of the measurand, ΔR is the change of resistance due to (and, for some sensors, proportional to) the measurand, and x_R is the relative change of resistance (i.e. $x_R = \Delta R/R_0$). These sensors are commonly employed to measure temperature (e.g. platinum sensors and thermistors), light (e.g. light-dependent resistors, LDR), gas (e.g. tin dioxide gas sensors) and humidity.

(b) **Differential resistive sensors**, with two sensing elements (R_{x1} and R_{x2}) that share a terminal, as shown in Fig. 5b, and undergo opposite changes: a change of the measurand causes an increase of R_{x1} and a decrease of R_{x2}, or vice versa. Such resistances can be modelled as

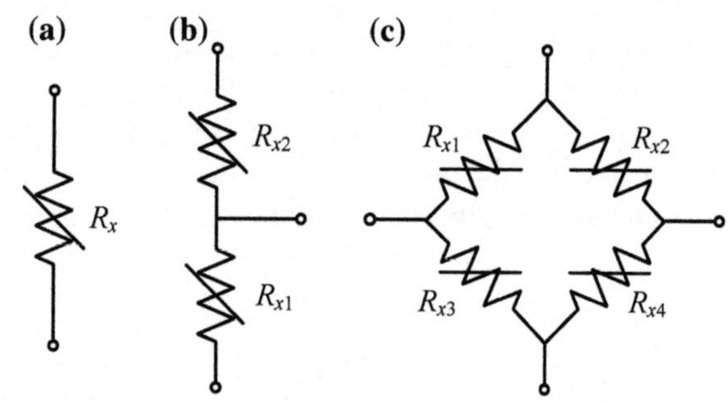

Fig. 5 Resistive sensor with **a** single, **b** differential, and **c** bridge topology

$$\begin{aligned} R_{x1} &= R_0(1 \pm x_R) \\ R_{x2} &= R_0(1 \mp x_R) \end{aligned}, \tag{7}$$

where x_R is assumed to be equal in magnitude but opposite in direction for R_{x1} and R_{x2}. Such a differential topology is quite often implemented through potentiometric sensors that are applied to measure linear or angular position/displacement, pressure (e.g. sensors based on Bourdon tubes) and liquid level (e.g. float-based sensors).

(c) **Bridge-type resistive sensors**, with one, two or four sensing elements in a Wheatstone bridge, thus resulting in a quarter-bridge, half-bridge or full-bridge sensor, respectively. For the full-bridge topology shown in Fig. 5c, which is the most popular since it provides the highest sensitivity, the four sensing elements undergo the same x_R but with opposite signs as follows

$$\begin{aligned} R_{x1} &= R_{x4} = R_0(1 \pm x_R) \\ R_{x2} &= R_{x3} = R_0(1 \mp x_R) \end{aligned}, \tag{8}$$

These sensors are commonly used to measure weight (e.g. load cells based on metal strain gages), pressure (e.g. sensors based on semiconductor strain gages) and magnetic field [e.g. Anisotropic (AMR) and Giant (GMR) Magnetoresistive Sensors].

3.2 Microcontroller

A μC is a programmable processor-based digital IC widely used in control and measurement electronic systems. It has three main blocks embedded: (i) a central processing unit (CPU), which executes instructions sequentially; (ii) a memory, which saves the instructions to be executed and data to be processed; and (iii) peripherals, which enable the μC to interact with the off-chip world. The peripherals can be digital (e.g. a timer/counter), analogue (e.g. an analogue comparator), or mixed (e.g. an ADC). However, the direct interface circuits of interest exclusively need digital peripherals, to be precise: input/output digital ports (if possible, with a ST buffer embedded) and a digital timer (if possible, of 16 bits). With regard to the number of bits of the CPU, 8 bits is enough for direct interface circuits, with the corresponding benefits in terms of power consumption.

The tasks of the digital system shown in Fig. 4 can be implemented by a μC following the operating principle represented in Fig. 6. First of all, the start of the discharging-time measurement is synchronized with the timer. Once the measurement has been started, the timer increases by one at every rising edge of its reference oscillator whose period equals T_S. Then, when the exponential discharging voltage crosses the V_{TL} of the ST buffer embedded into Pin 1, the timer stops.

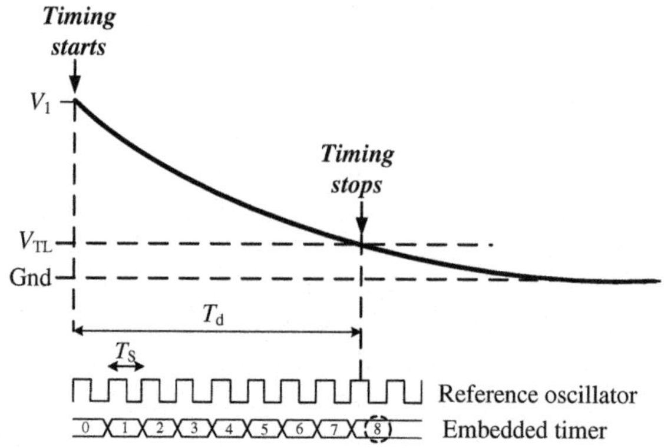

Fig. 6 Discharging-time measurement carried out by the μC

In Fig. 6, the measurement result is the digital number 8, which has information about the value of the sensor resistance included in the RC circuit.

In order to have an accurate measurement of the discharging time shown in Fig. 6, the μC should have the following:

- A crystal oscillator as a reference for the embedded timer, whose temperature coefficient and time drifts are very low.
- A reference oscillator of high frequency (nowadays, it can be up to tens of MHz) to reduce the quantisation error in the discharging-time measurement. The higher the frequency, the better the resolution, but also the higher the power consumption.
- A capture module associated to Pin 1 (see Fig. 4) to automatically capture the value of the timer when the voltage-threshold crossing occurs, regardless of the instruction being executed by the CPU.
- A CPU with a power-down (or sleep) mode to suspend its activity and, hence, to reduce the noise during the discharging-time measurement [16], provided that the timer and the interrupt system keep working in this operating mode. This feature is also of interest to decrease the power consumption.
- An appropriate decoupling capacitor between the power supply pins and a suitable layout of the ground and supply tracks of the PCB to have a clean supply voltage and, consequently, a clean V_{TL} [15].

The measurement of time-modulated signals with a slow slew rate (i.e. a slow transition from '1' to '0', or vice versa) is very susceptible to noise. In the case shown in Fig. 6, the comparison between the two voltages (i.e. the discharging voltage and V_{TL}) can be erroneously triggered due to noise superimposed on either of the two voltages, thus resulting in a wrong value of the digital number. Therefore, any initiative promoting the reduction of trigger noise in the circuit (e.g. power supply noise or CPU-activity noise) will improve the resolution of the measurement.

Nowadays, there are many commercial μCs from different semiconductors companies but with quite similar features that can be employed to build a direct interface circuit. Some examples are: PIC16 family from Microchip Technology, MSP430 family from Texas Instruments, and AVR family from Atmel. Low-power versions of these µCs (e.g. PIC16 with extreme low power technology, or MSP430 with ultra-low power technology) are also available. Direct interface circuits have been implemented using different commercial µCs, but the performance seems to be fairly independent of the µC employed.

3.3 *Interface Circuits*

The resistive sensor topologies shown in Fig. 5 can be directly measured by a µC through the interface circuits proposed in Fig. 7. In comparison with the circuit shown in Fig. 4, the circuits in Fig. 7 have two additional resistors: R_i between Pin 1 and Node 1, which improves the rejection of power supply noise/interference [17] at the expense of a longer charging stage; and R_s between Node 1 and the sensor, which ensures that the discharging current is lower than the maximum output current sunk by a port pin even when the sensor resistance is very low.

The direct interface circuit proposed for **single resistive sensors** is shown in Fig. 7a [3], which applies the three-signal auto-calibration technique to have a measurement result insensitive to both multiplicative and additive errors of the circuit [18]. In order to apply such a technique, three measurements are performed sequentially: (1) sensor measurement, which is intended to measure the discharging time through R_x; (2) reference measurement, which is intended to measure the discharging time through a reference resistor (R_{ref}) whose value is known; and (3) offset measurement, which is intended to measure the discharging time through the internal resistance (R_p) of the port pins of the µC. The waveform of the voltage across C in a whole measurement is shown in Fig. 8. The state of pins 2, 3 and 4 in Fig. 7a during the discharging stages and the resulting discharging time for each of the three measurements is summarised in Table 1, where $k_R = C \cdot \ln(V_1/V_{TL})$.

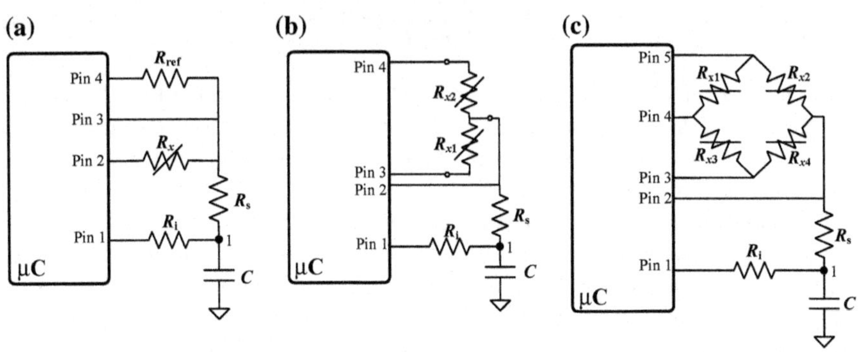

Fig. 7 Interface circuit for a **a** single, **b** differential, and **c** bridge-type resistive sensor

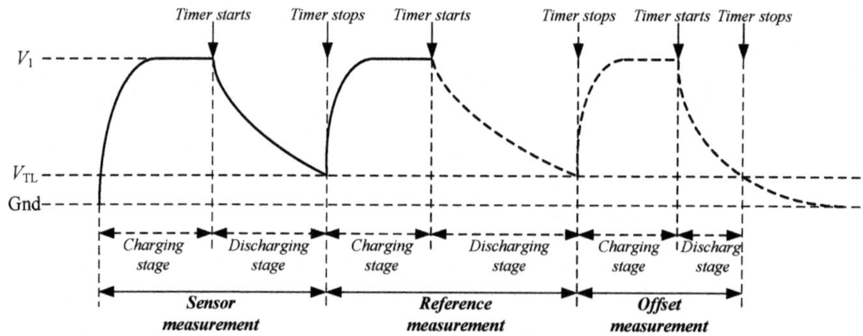

Fig. 8 Waveform of the voltage across C during the charge-discharge process for each of the three measurements involved in the circuit shown in Fig. 7a

Table 1 Pins configuration and discharging times for the circuit in Fig. 7a

Measurement	Pin 2	Pin 3	Pin 4	Discharging time
Sensor	'0'	HZ	HZ	$T_x = k_R(R_s + R_x + R_p)$
Reference	HZ	HZ	'0'	$T_{ref} = k_R(R_s + R_{ref} + R_p)$
Offset	HZ	'0'	HZ	$T_{off} = k_R(R_s + R_p)$

Table 2 Pins configuration and discharging times for the circuit in Fig. 7b

Measurement	Pin 2	Pin 3	Pin 4	Discharging time
Sensor #1	HZ	'0'	HZ	$T_1 = k_R(R_s + R_{x1} + R_p)$
Sensor #2	HZ	HZ	'0'	$T_2 = k_R(R_s + R_{x2} + R_p)$
Offset	'0'	HZ	HZ	$T_{off} = k_R(R_s + R_p)$

Using the three discharging times (T_x, T_{ref} and T_{off}), the sensor resistance can be estimated by

$$R_x^* = \frac{T_x - T_{off}}{T_{ref} - T_{off}} R_{ref}, \tag{9}$$

which is insensitive to the tolerance and low-frequency variability of C, V_1 and V_{TL}. A circuit similar to that shown in Fig. 7a but including diodes and switches has also been proposed to measure remote resistive sensors cancelling the effects of the connecting lead resistances [19].

For **differential resistive sensors**, we propose the direct interface circuit shown in Fig. 7b [4], which also carries out three measurements: (1) sensor measurement #1, (2) sensor measurement #2, and (3) offset measurement, which are intended to measure the discharging time through R_{x1}, R_{x2} and R_p, respectively, applying the pins configuration indicated in Table 2. Using the three discharging times (T_1, T_2 and T_{off}), the parameter x_R of the differential sensor can be estimated by

Table 3 Pins configuration and discharging times for the circuit in Fig. 7c

Measurement	Pin 2	Pin 3	Pin 4	Pin 5	Discharging time
Sensor #1	HZ	'0'	HZ	HZ	$T_1 = k_R[R_s + (R_{x4} \| (R_{x1} + R_{x2} + R_{x3})) + R_p]$
Sensor #2	HZ	HZ	'0'	HZ	$T_2 = k_R[R_s + ((R_{x3} + R_{x4}) \| (R_{x1} + R_{x2})) + R_p]$
Sensor #3	HZ	HZ	HZ	'0'	$T_3 = k_R[R_s + (R_{x2} \| (R_{x1} + R_{x3} + R_{x4})) + R_p]$
Offset	'0'	HZ	HZ	HZ	$T_{off} = k_R(R_s + R_p)$

$$x_R^* = \frac{T_1 - T_2}{T_1 + T_2 - 2T_{off}}. \tag{10}$$

Note that here it is better to estimate the measurand by means of x_R rather than R_{x1} (or R_{x2}), since R_{x1} (or R_{x2}) can also be altered by undesired inputs such as temperature, thus causing multiplicative errors. Moreover, unlike the measurement of single resistive sensors, here x_R can be estimated without using any reference resistor.

Resistive sensors in a **bridge topology** can be directly connected to a μC using the interface circuit shown in Fig. 7c [5, 20]. This circuit measures four discharging times (T_1, T_2, T_3 and T_{off}) applying the pins configuration indicated in Table 3, where the symbol "||" means in parallel. For a full-bridge topology, the parameter x_R of the sensor can be estimated by

$$x_R^* = \frac{T_1 - T_3}{T_2 - T_{off}}. \tag{11}$$

For other bridge topologies, x_R can be estimated using other time-based equations [5]. Furthermore, for sensors whose output is temperature dependent (e.g. piezoresistive pressure sensors), the result obtained from Eq. (11) can be easily corrected by estimating the temperature through the sensor itself [21].

3.4 Uncertainty Sources

The direct interface circuits proposed in Fig. 7 measure three (or four) discharging times and then use them to estimate the sensor resistance or the relative change of resistance through Eqs. (9)–(11). This operating principle involves the following uncertainty sources:

(a) **Mismatch of the internal resistances**: For a CMOS μC, R_p corresponds to the channel resistance of the NMOS transistor embedded into the output buffer that provides a digital '0'. In Sect. 3.3, we have assumed that R_p was the same for all the port pins of the μC. However, there is a mismatch between those internal resistances that brings about systematic errors in the measurement. If this mismatch is considered, the estimated value (R_x^* and x_R^*) can be expressed in function of the actual value (R_x and x_R) for the single [3], differential [4] and bridge [20] topology as follows, respectively,

$$R_x^* \approx (R_x + \Delta R_{23})\left(1 - \frac{\Delta R_{43}}{R_{\text{ref}}}\right) \tag{12}$$

$$x_R^* \approx \left(x_R - \frac{\Delta R_{43}}{2R_0}\right)\left(1 + \frac{\Delta R_{23} + \Delta R_{24}}{2R_0}\right) \tag{13}$$

$$x_R^* \approx \left(x_R + \frac{\Delta R_{35}}{R_0}\right)\left(1 + \frac{\Delta R_{24}}{R_0}\right), \tag{14}$$

where $\Delta R_{23} = R_{p2} - R_{p3}$, $\Delta R_{43} = R_{p4} - R_{p3}$, $\Delta R_{24} = R_{p2} - R_{p4}$ and $\Delta R_{35} = R_{p3} - R_{p5}$. According to Eqs. (12)–(14), we have $R_x^* = R_x$ and $x_R^* = x_R$ only when the internal resistances are matched, otherwise there are offset and gain errors. If the mismatch between internal resistances (ΔR_p) is a few tenths of ohm [3] and the sensor resistance is higher than 1 kΩ, the resulting $\Delta R_p/R$ is very low. Therefore, offset and gain errors due to internal resistances are expected to be in the range of 0.01%.

(b) **Quantisation**: The starting point of the discharging-time measurement is synchronized with the program executed by the μC and, hence, there is no quantisation error at this point. However, the stopping point does suffer from quantisation effects, as shown in Fig. 6. Because of these, the relation between the discharging time to be measured (T_d) and the measurement result (T_q) has a quantisation error (i.e. $T_q - T_d$) that ranges from $-T_S$ to 0 [4]. Equations (9)–(11) can compensate for offset and gain errors obtained during the discharging-time measurements, but not for the non-linearity error caused by quantisation. This non-linearity error in the discharging-time measurement causes offset, gain and/or non-linearity errors in the estimation of R_x and x_R [4]. However, such errors are very low when a high-value capacitance (C in Fig. 7) is employed, but at the expense of a longer measuring time.

(c) **Trigger noise**: The use of a high-value capacitance to reduce the effects of quantisation can cause some non-desired secondary effects. The higher the capacitance, the lower the slew rate at the stopping point of the discharging-time measurement and, hence, the higher the effects of noise coming from the supply voltage or the activity of the CPU. Due to this noise, the discharging-time measurement shows some variability, i.e. different digital numbers are obtained in the digital timer for the same value of the measurand, thus limiting the resolution. Accordingly, the measurement resolution is not only limited by quantisation but also by the trigger noise [22]. Nevertheless and unlike what happens with quantisation effects, trigger noise effects can be reduced by averaging provided that the noise is random, but again at the expense of a longer measuring time.

(d) **Others**: There are other uncertainty sources affecting the measurement but their effects are expected to be less significant. A first example is the *input leakage current* of the pins set as a HZ input during the discharging stage. However, in modern low-power microcontrollers (e.g. PIC16F with extreme low power technology), such a leakage current is 5 nA that is a million times lower than the operating current of the circuit when measuring resistances of units of kiloohm and, therefore, their effects are negligible. Another minor uncertainty source is the *dielectric absorption* (DA) of the capacitor of the RC circuit. To cope with that, it is not advisable the use of electrolytic capacitors whose DA is higher than 10%, but the use of "poly" type capacitors, such as polycarbonate or polypropylene, whose DA is lower than 0.1%.

3.5 Applications

The direct interface circuits for resistive sensors shown in Fig. 7 have been applied to measure many physical and chemical quantities, for example: temperature [3], magnetic field [5], atmospheric pressure [23, 24], gas [25, 26], light [27] and

Table 4 Applications of the direct interface circuits for resistive sensors shown in Fig. 7

Reference	[3]	[4]	[5]
µC	PIC16F873[a]	AVR ATtiny2313[b]	MSP430F123[c]
Supply voltage	5 V	5 V	3 V
Ref. oscillator	20 MHz	20 MHz	4 MHz
Sensor	Temperature sensor (Pt1000)	Potentiometric sensor (1 kΩ)	Magnetoresistive sensor (HMC1052[d])
Topology	Single	Differential	Full-bridge
Interface circuit	Figure 7a	Figure 7b	Figure 7c
Capacitor (C)	2.2 µF	470 nF	2.2 µF
Other components	R_{ref} = 1470 Ω R_s = 330 Ω	R_s = 470 Ω R_i = 100 Ω	R_i = 120 Ω
Meas. range	[−45, 120] °C	[−100, 100]%[e]	[75, 600] µT
Max. NLE[f]	0.01% FSS[g]	0.01% FSS	1.8% FSS
ENOB[h] (measuring time)	11 b (5 ms) or 12.5 b (50 ms)	11.5 b (1 ms) or 13 b (100 ms)	7 b (50 ms)

[a]From Microchip Technology. [b]From Atmel. [c]From Texas Instruments. [d]From Honeywell. [e]Such a range means that the movable common terminal of the potentiometric sensor moves from one end to the other. [f]NLE stands for non-linearity error. [g]FSS stands for Full-Scale Span. [h]ENOB stands for Effective Number Of resolution Bits

respiratory rate [28]. The performance of these circuits in some of the previous applications using different commercial μCs is summarised in Table 4.

Taking into account the simplicity of the proposed interface circuits, the values of non-linearity and resolution shown in Table 4 for the first two cases [3, 4] are quite remarkable. In these cases, the non-linearity error is mainly due to the effects of quantisation in the discharging-time measurement, whereas the resolution is determined by the effects of both quantisation and noise affecting the voltage-threshold crossing. The experimental results for the third case in Table 4 [5], however, are not as excellent as the previous ones. On the one hand, this is due to the non-linearity of the commercial sensor tested; in other words: if the direct interface circuit in Fig. 7c measures a bridge circuit emulated by resistors instead of such a sensor, the maximum non-linearity error of the circuit is about 0.1% FSS. On the other hand, the lower value of resolution is due to the low sensitivity of the commercial sensor. As a rule of thumb, direct interface circuits are able to detect changes of resistance of about 0.1 Ω, which is a very low value when measuring a temperature sensor [3] but not when measuring such a magnetoresistive sensor whose dynamic range is around ±6 Ω.

4 Interfacing Resistive Sensor Arrays to FPGAs

As the complexity of the system and the number of sensors to be measured increase (e.g. array sensors composed by a high number of sensing units), μCs may not have enough resources to implement the techniques described in Sect. 3. In such a case, FPGAs can be a good alternative since they have a high number of I/O pins and also reconfigurable hardware resources to build timer-capture modules operating in parallel. Next, we explain the main features of array sensors and FPGAs, and how to join them to build a direct sensor-to-FPGA interface circuit.

4.1 Array Sensor

Array sensors are composed of many sensing units. These arrays are built to obtain spatial patterns (for instance, a pressure map in tactile sensors) or exploit redundancy to improve sensitivity or selectivity (for example, in electronic noses). Many array sensors are small and implemented with microelectromechanical technologies or conventional technologies for ICs. This is the case of smart vision chips or arrays of thermopiles for infrared imaging. These array sensors commonly incorporate signal conditioning circuitry on the same substrate and, hence, the concept of direct interfacing is not the best choice for them. However, there are discrete arrays of sensors, such as arrays of MOX gas sensors [29], whose interface with the

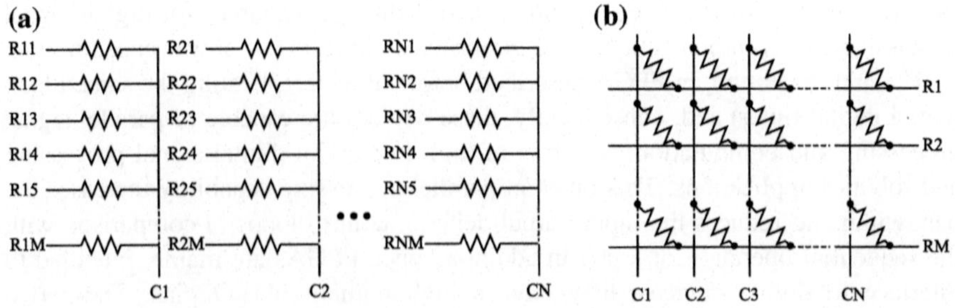

Fig. 9 Architecture of an array sensor **a** with one selection track per sensing unit, and **b** addressed in rows and columns

processing electronics could be noticeably simplified through direct connection. Moreover, direct interfacing is especially suitable for large-size sensors of different shapes, for instance those made with printable electronics. Conductive polymer gas sensor arrays [30], thermal imaging sensors [31] and tactile sensors [32] have been implemented with these technologies.

Regarding its architecture, an array sensor with *M* rows and *N* columns can be built with either one selection track per sensing unit in the array or organized in a row and column fashion where many sensing units share the selection tracks, as shown in Fig. 9a, b, respectively. The latter is obviously advantageous in terms of cost and complexity of hardware. However, shared connections create parasitic current paths that may originate crosstalk between sensing units.

4.2 FPGA

FPGAs are close to ASICs in terms of performance for real-time computing, although ASICs exhibit better dynamic response-power consumption trade-off. The main advantage of FPGAs is that they are programmable, thus allowing rapid system prototyping at low cost. FPGAs are basically composed of cells that have local memory such as flip-flops and are able to perform logic functions. These logic cells are connected through switches to vertical and horizontal routing channels, so the hardware is configurable. The same routing matrix connects the logic cells to a high set of I/O pins. Besides the distributed local memory, FPGAs usually have memory blocks and may incorporate more complex blocks such as multipliers. Advanced versions of FPGA also implement processors as embedded cores, thus resulting in powerful devices called Programmable Systems-on-Chip. The embedded processors can be programmed in high-level languages, while configurable logic is programmed with graphical tools such as circuit schematics or with hardware description languages (HDL). The main vendors of FPGAs are Xilinx and Altera providing a large portfolio of devices with different technologies, number of

I/O pins, number of logic cells, memory and dedicated resources for digital signal processing and communications, and performance in terms of power consumption.

The architecture of an FPGA described before allows replicating the same block (e.g. a digital timer) and, consequently, these devices are capable of parallel signal processing and computation, which is of high interest in digital signal processing and robotics applications. This inherent parallel processing capability increases the bandwidth and reduces the input-output delay in control loops in comparison with the sequential operation of a μC. In addition, since FPGAs are mainly intended to interface to digital devices, they have a high number of I/O pins. These two characteristics make them especially suitable to implement direct interfaces for array sensors. Note, however, that FPGAs are more power demanding than μCs. Moreover, FPGAs are much more limited than μCs regarding the interface to analogue electronics because they do not commonly include ADCs or analogue comparators.

Regarding the issues involved in the concept of direct interfacing depicted in Fig. 4, FPGAs can easily be configured to have embedded digital timers measuring the discharging time of the RC circuit. They also have enough I/O digital pins to control the charge/discharge process of the RC circuit. These I/O pins, however, do not commonly have a ST buffer. In any case, the flexibility of the FPGA allows building dedicated hardware capture modules with similar or better performance than that of ST buffers in terms of noise rejection. I/O drivers are flexible and can be set in different modes such as 'high impedance' and 'strong drive'. The current sunk or sourced by an I/O pin is obviously limited, as in a μC. I/O pins also have non-zero output impedance, which may cause crosstalk errors in the measurement circuit. The clock block in FPGAs allows a flexible managing of the clock, for instance to increase the frequency of the reference clock signal and reduce the quantisation error. Current FPGAs can run at several hundreds of MHz.

Finally, the same requirements as those mentioned in Sect. 3.2 to have an accurate measurement of the discharging time apply here. Specifically, a stable reference oscillator, careful layout design, and decoupling capacitors are necessary.

4.3 Interface Circuits

If the measurement system has a large set of resistive sensors in an array topology, the straightforward approach to measuring them is to use a replica of the circuit in Fig. 7a for each sensing unit. Since the circuits for each sensor would be independent of each other, this approach would not suffer from crosstalk between sensors. However, a number of connections at least as high as two times the number of sensors would be required to address the array sensor. Moreover, a capacitor and a timer-capture module would also be needed per sensor.

A step in the direction of reducing the cost and complexity of the hardware is to share the capacitor and the timer-capture module. This can be done with the architecture of Fig. 9a and the interface circuit shown in Fig. 10. The array sensor is

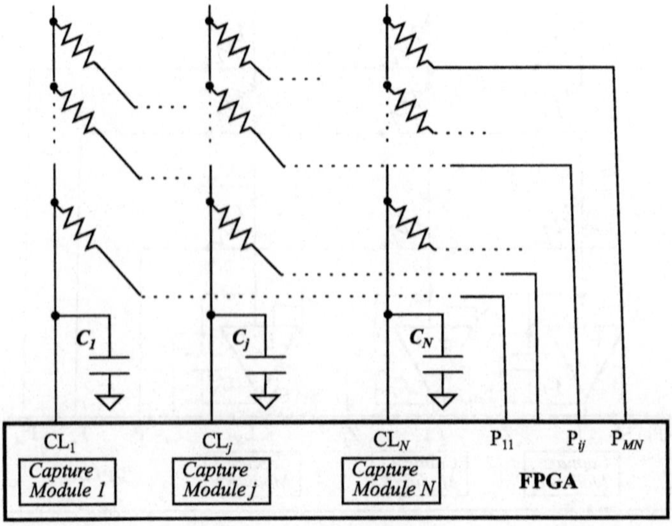

Fig. 10 Interface circuit for an array sensor with one selection track per sensing unit

read as follows. First, the capacitors C_j with $1 \leq j \leq N$ are charged by setting pins CL_j to '1' and the remaining I/O pins to HZ. Then, a whole row is selected by setting its corresponding I/O pins to '0'. For instance, pins of the ith row P_{ij} with $1 \leq j \leq N$ are set to '0' while the remaining pins P_{kj} with $k \neq i$ are set to HZ. The capacitors are then discharged through the sensing resistances of that row and the exponential discharging voltages across them are monitored independently by pins CL_j, which are set to HZ. A set of timers are started at the beginning of the discharging phase and they are stopped when V_{TL} is reached at the related column pins. Therefore, a whole row is read in parallel. Techniques similar to those described in Sect. 3.3 can also be applied here to improve the accuracy by adding reference resistors. In this case, a row of reference resistors can be added to carry out the three-signal auto-calibration technique.

Large array sensors addressed in rows and columns, as shown in Fig. 9b, can be interfaced to the FPGA using the circuit shown in Fig. 11, where passive integrators are replaced by active ones implemented by OpAmps. The basics of this circuit are explained through Fig. 12 involving two stages. In the charging stage shown in Fig. 12a, the OpAmp is shut-down by pin Sh and C is charged to V_1. In the discharging stage shown in Fig. 12b, the timer starts, the OpAmp is turned on and the current through R_x is integrated into C. Therefore, the voltage at Pin 1 linearly decreases until V_{TL} is reached and then the timer stops, as shown in Fig. 12c. The discharging time can be expressed as

$$T_d = R_x C (V_1 - V_{TL}) / V_1 \tag{15}$$

As in Fig. 10, the array sensor in Fig. 11 is scanned so that all the resistances in a row are measured at the same time. In a first phase, the selection pins P_i with

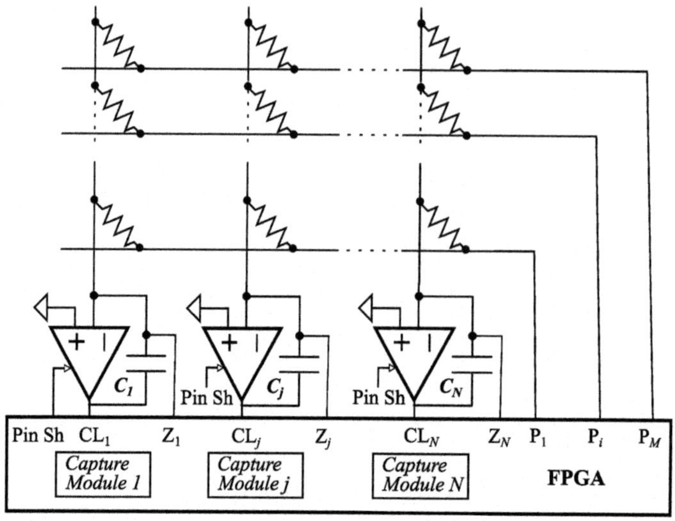

Fig. 11 Interface circuit for an array sensor addressed in rows and columns

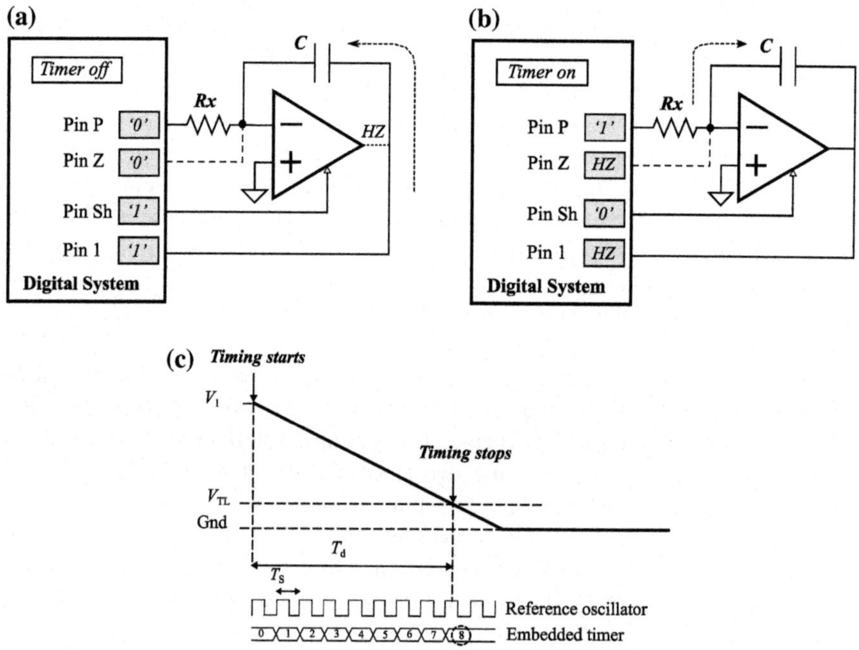

Fig. 12 Basic topology of the interface circuit with an active integrator employed in Fig. 11. Pin configuration during **a** the charging stage, and **b** the discharging stage, and **c** the resulting discharging time

$1 \leq i \leq M$ are set to '0', CL_j with $1 \leq j \leq N$ are set to '1', pins Z_j are set to '0', Pin Sh is set to '1' and the capacitors C_j are charged to V_1. In the second phase, a row is selected and the set of timers start counting. For instance, pin P_i is set to '1', thus resulting in a voltage drop V_1 across the resistances R_{ij}. The OpAmps are turned on by setting pin Sh to '0', and pins CL_j and Z_j are now at HZ. Therefore, currents $i_{Dj} = V_1/R_{ij}$ flow into the integrators and, consequently, the voltages at pins CL_j decrease until V_{TL} is reached at every pin CL_j, with the corresponding stop of the timer. At this time, Z_j is set to '0' so as to avoid that the voltage at the inverting input of the OpAmp grows and interferes the measurement of other resistances. Note that the columns in Fig. 11 are virtually grounded thanks to the negative feedback loop of the OpAmp, thus following a common strategy [33] to short circuit the non-selected resistances and, hence, to avoid any contributing parasitic current to the output.

The total number of I/O pins dedicated to address the array sensor is $(M+1) \times N$ in Fig. 10, whereas is $2 \times N + M + 1$ in Fig. 11. For instance, an array of 8 × 8 resistances requires 72 I/O pins in Fig. 10, but 25 in Fig. 11.

4.4 Uncertainty Sources

The interface circuits for array sensors shown in Figs. 10 and 11 suffer from uncertainty sources similar to those described in Sect. 3.4. However, additional errors arise due to the I/O features of the FPGA and to the array nature of the sensors. These novel uncertainty sources are described next.

(a) **Trigger noise**: As said in Sect. 3.4, trigger noise alters the threshold voltage and the discharging voltage signal, so the discharging time is affected in consequence. The use of an I/O pin with a ST buffer in μCs reduces this uncertainty because crosses of the discharging signal with the threshold after the first one are ignored thanks to the hysteresis of the buffer. Unfortunately, FPGAs generally do not have ST input buffers so the contribution of the trigger noise is significant in a straightforward realization where the output of the input buffer is used to stop the timer. Embedded resources of the FPGA can be configured to build smart capture modules [34] that detect the first change of logical value at the input buffer when the input signal reaches the threshold (label F in Fig. 13). This can be done by adding positive feedback in digital circuits to achieve the memory of the hysteresis cycle or with a level triggered latch. In addition, the flexibility of the storage elements in the FPGA to be synchronized with both edges of the clock signal, and also the detection of not only the first (label F in Fig. 13) but the last (label L in Fig. 13) transition at the output of the input buffer can be exploited to carry out averaging. This actually filters part of the trigger noise and achieves more precision without losing bandwidth.

(b) **Crosstalk**: An additional source of uncertainty in the sensor topologies shown in Fig. 9 with respect to those described in Sect. 3 is crosstalk. The sharing of circuit

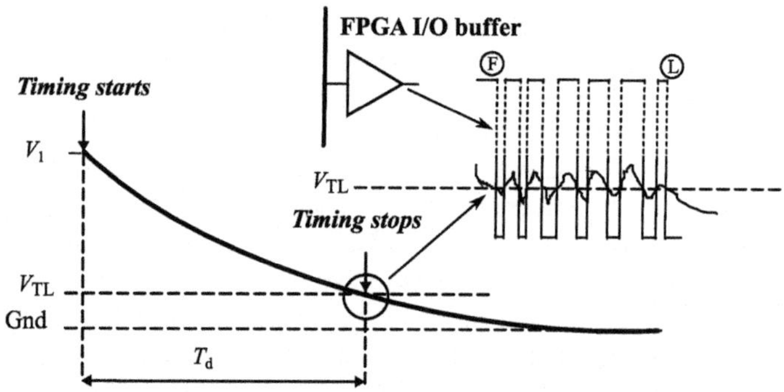

Fig. 13 Effects of trigger noise when a ST buffer is not employed

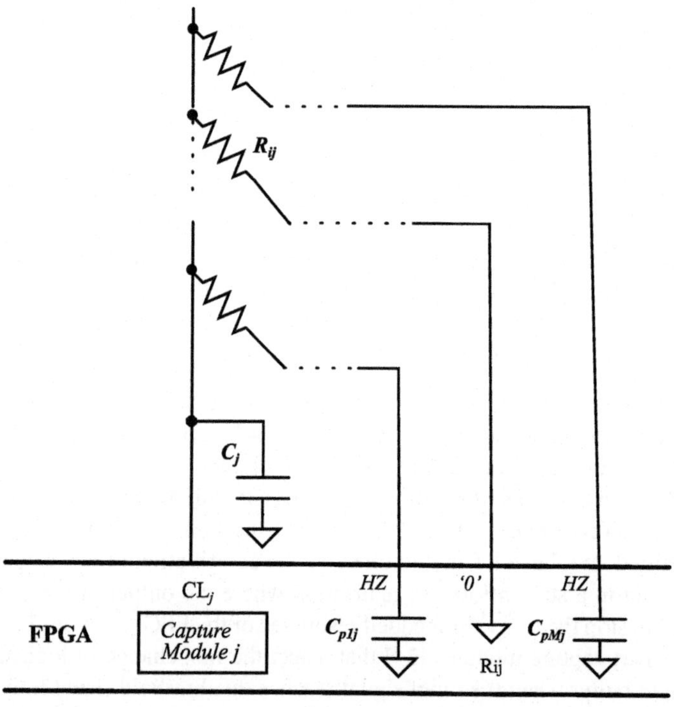

Fig. 14 Crosstalk due to the parasitic capacitances in the interface circuit shown in Fig. 10

components to lower the cost has the drawback of introducing such an error. As a consequence, the timing does not depend only on the value of the resistance that is being measured but also on the value of other resistances in the array.

Regarding the interface circuit in Fig. 10, crosstalk is mainly due to parasitic capacitances associated to connection tracks and I/O buffers set at HZ. If they are taken into account, the discharging circuit is not that in Fig. 4c but the one depicted in Fig. 14 for the column j, where C_{pkj} with $1 \leq k \leq M$ and $k \neq i$ are such parasitic capacitances. For a given R_{ij}, the higher the resistance of the non-selected rows, the shorter the discharging time T_{dij}, while its maximum value will be registered when all these resistances are minimum. Therefore, a higher range of resistances increases the difference between the minimum and maximum values of T_{dij} due to crosstalk and the uncertainty in T_{dij} in consequence. A worst case estimation can be done for a maximum range. In this case, for $R_{kj} \to \infty$ with $1 \leq k \leq M$ and $k \neq i$, T_{dij} is minimum and takes the value

$$T_{dij(min)} = R_{ij} C_j \ln\left(\frac{V_1}{V_{\mathrm{TL}j}}\right). \tag{16}$$

On the other hand, for $R_{kj} \to 0$, T_{dij} is maximum and results in

$$T_{dij(max)} = R_{ij} C_{eqij} \ln\left(\frac{V_1}{V_{\mathrm{TL}j}}\right), \tag{17}$$

where $C_{eqij} = C_j + \sum_{\substack{k=1 \\ k \neq i}}^{M} C_{pkj}$. Therefore, $T_{dij(min)} \leq T_{dij} \leq T_{dij(max)}$ and its actual value depends on the specific values of the remaining resistances in that column. For a uniform distribution of these resistances, the relative standard uncertainty of T_{dij} generated by crosstalk is given by

$$\frac{u(T_{dij})}{T_{dij}} = \frac{\sum\limits_{\substack{k=1 \\ k \neq i}}^{M} C_{pkj}}{\sqrt{12} C_j}, \tag{18}$$

From (18), the higher the aggregated parasitic capacitance with respect to C_j, the higher the relative uncertainty. Such an aggregated parasitic capacitance increases with M and depends on the circuit layout and technology used to implement it. The relative uncertainty in (18) can be reduced with a higher value of C_j, although there is a tradeoff with the measuring time and with the trigger noise since both increase with C_j. On the other hand, a high value of C_j reduces the uncertainty due to quantisation. The aggregation of the errors caused by all these sources determines the resolution of the measurement.

The circuit in Fig. 11 is subjected to more crosstalk error sources than that in Fig. 10. The three main error sources are: (i) the internal resistance of the output buffer that sources current to the selected row (i.e. R_{pi} in Fig. 15), (ii) the internal resistance of the output buffer that sinks current from the column once the discharging process has ended (i.e. R_{nj} in Fig. 15), (iii) the input offset voltage of the OpAmps (i.e. V_{off} in Fig. 15). Because of these, the discharging time does not

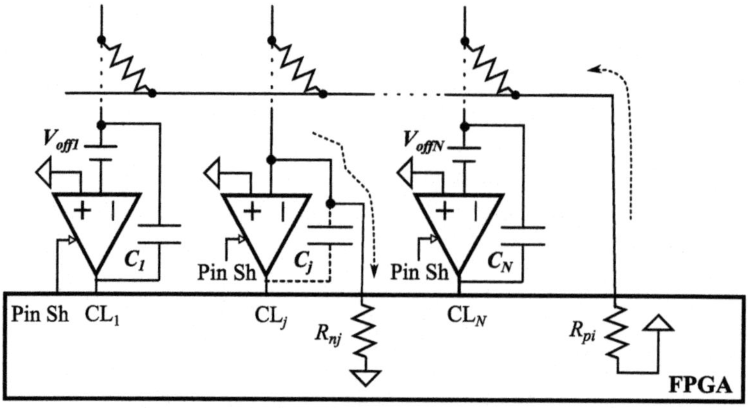

Fig. 15 Crosstalk due to the I/O driver impedance and the input offset voltage of the OpAmp in the interface circuit shown in Fig. 11

depend only on the resistance being measured but also on other resistances in the array.

As for the effects of R_{pi} in Fig. 15, the discharging time in (15) is modified as

$$T_{dij} = R_{ij} C_j \frac{V_1 - V_{TLj}}{V_1} \frac{R_{pi} + R_{eqi}}{R_{eqi}}, \tag{19}$$

where R_{eqi} is the equivalent parallel resistance of the resistances in the row selected. Note from (19) that the lower R_{eqi} with respect to R_{pi}, the higher the crosstalk. This imposes a lower limit of the range of resistances and also limits the number of columns in the array for a given accuracy. The addition of known reference resistors in a new reference column is proposed in [35] to obtain the following expression

$$T_{dij} = \frac{R_{ij}}{R_{ic}} \frac{C_j}{C_c} \frac{V_1 - V_{TLj}}{V_1 - V_{TLc}} T_{dic}, \tag{20}$$

where R_{ic}, C_c, V_{TLc} and T_{dic} are the row resistance, the capacitor, the input buffer threshold voltage and the discharging time associated to the reference column, respectively. Since the other resistances in the array are not in (20), the crosstalk caused by R_{pi} is cancelled. Moreover, the use of known reference resistors in another new reference row provides an expression of the discharging time that only depends on the known value of the reference resistors and the measured discharging times associated to these resistors and R_{ij} [35].

The crosstalk caused by R_{nj} in Fig. 15 is related to the role of pin Z to clamp the voltage at the column to zero. As said in Sect. 4.3, since non-selected rows are also driven by a zero voltage signal, parasitic resistive paths are in principle short

Table 5 Applications of the interface circuits shown in Figs. 10 and 11

Reference	[37]	[35]
FPGA	Spartan3AN[a]	Spartan3AN[a]
OpAmp	–	TLV2475N[b]
Supply voltage	3.3 V	3.3 V
Ref. oscillator	50 MHz	50 MHz
Technology	PCB, insertion sockets, axial-lead discrete resistors	
Sensor topology	Figure 9a	Figure 9b
Sensor size	M = 7, N = 8	M = 8, N = 6
Interface circuit	Figure 10	Figure 11
Capacitor (C)	47 nF	47 nF
Meas. range	[200, 7350] Ω	[556, 3159] Ω[c]/[3296, 9975] Ω[d]
ENOB (scan time)[e]	12.2 b (5 ms)	7.9/8.6 b (5 ms)
Max. NLE	0.028% FSS	0.038% FSS/0.037% FSS

[a]From Xilinx. [b]From Texas Instruments. [c]The OpAmp output does not saturate, i.e. Eq. (21) is satisfied. [d]The voltage at column *j*th is clamped to zero voltage with pin Z_j when the threshold is reached. [e]Scan time is the time to read the whole array

circuited and then crosstalk is reduced. However, if the sensor resistance to be measured is low, the current sunk by Pin Z is high, thus generating a voltage rise at the column that causes parasitic currents and then crosstalk errors. This error can be neglected for high enough values of the sensor resistance (see Table 5 in Sect. 4.5). A minimum resistance is also required to accomplish with the maximum current that is able to source the pin that selects the row and to sink the pin Z. A direct strategy to overcome this limitation and reduce the crosstalk errors consists in adding external resistors in series with R_{pi} to limit the current. In this case, note that Eq. (20) is still valid since the measurements obtained from the reference column provide indirect estimations of these resistances.

Another alternative to reduce the crosstalk caused by R_{nj} consists in ensuring that OpAmps always work in the linear region so the negative feedback imposes always a voltage close to ground. This is achieved if the range of resistances values is set to guarantee that the longest discharging time corresponding to the highest resistance is short enough to avoid that the output of the OpAmp of the column that reads the smallest resistance (i.e. shortest discharging time) saturates. This is achieved if

$$R_L \le R_{ij} \le \frac{V_1}{V_1 - V_{Tj}} R_L, \tag{21}$$

where R_L is the lowest resistance value in the array. A strategy to increase the range imposed by (21) is the reading of two rows at the same time, the one being scanned and a reference row with known resistances. In this way, the reference resistors are in parallel with the ones being read and the equivalent maximum resistance is lowered, so Eq. (21) is met by the equivalent parallel resistances but the actual range of the resistances in the array is much higher [35].

The input offset voltage of the OpAmps in Fig. 15 is also a source of crosstalk error because it changes the voltage at the corresponding column with respect to ground, thus causing parasitic currents in the array. Input bias currents of the OpAmps also introduce error since they are added to the current being integrated and change the discharging time. A procedure is proposed in [35] to reduce these second-order effects where two reference rows and one reference column with known resistances are added to the array. The target resistance R_{ij} can be expressed as a function of the known resistors and the associated discharging times only, and the effects of the offset voltages and bias currents are cancelled.

4.5 Applications

The interface circuits presented before have been applied to the measurement of tactile array sensors in [36] using an FPGA as a digital system. In addition, [37] and [35] report results for the interface circuits shown in Figs. 10 and 11 applied to the measurement of a generic array sensor made of discrete-lead axial resistors in insertion sockets on a PCB. Table 5 shows a summary of the results for different resistive ranges. Note that the time to read the whole array is 5 ms in all cases so a very fast scanning is achieved thanks to the parallel acquisition. Moreover, a resolution as high as 12.2 ENOB for the circuit in Fig. 10 was experimentally measured for two times the standard deviation as error estimation and 500 samples. The circuit in Fig. 11 requires much less I/O pins to address the array at the expense of more complex circuitry with more uncertainty sources. A resolution around 8 bits was reached, although it can increase for higher resistance values and wider measurement ranges.

5 Conclusions

After explaining such circuits and techniques for the direct connection of resistive sensors to digital systems, the following conclusions can be drawn:

- Direct interface circuits clearly simplify the measurement chain because neither an amplifier nor an ADC are needed between the sensor and the digital system. The key element is a digital timer that measures the charging/discharging time of an RC circuit formed by the resistive sensor and a known capacitor.
- A common low-cost general-purpose 8-bit μC can be the core of these sensor interfaces without requiring any on-chip ADC, OpAmp or analogue comparator. For resistive sensor arrays, the use of an FPGA is more advisable because different resistances of the array can be measured in parallel through a set of timers running simultaneously.

- In spite of their simplicity and low cost, these sensor interfaces have a satisfactory performance in terms of linearity and resolution and, therefore, they are very attractive for medium-accuracy, medium-resolution applications. A measuring time of around units or tens of millisecond can be their main limitation if the quantity to be measured changes quite fast.
- Since very similar results have been obtained when using different commercial digital systems from different manufacturers, the design of these sensor interfaces can be considered independent of any specific device or IC from any manufacturer.

From the authors' point of view, future research work on direct interface circuits could be focussed on the following directions:

- The analysis of the limitations when measuring resistive or capacitive sensors subjected to dynamic changes, and not quasi-static changes as considered so far.
- The direct measurement of other types of sensor. For instance, the use of digital timers to directly measure sensors providing an amplitude-modulated sinusoidal voltage signal.
- The use of new digital peripherals, such as configurable logic cells, embedded into the new generation of μCs to improve the performance of the proposed sensor interfaces and/or to develop novel operating principles.

Acknowledgements This work has been partially funded by the Spanish Government and by the European ERDF program funds under contract TEC2015-67642-R.

References

1. R. Pallàs-Areny, J.G. Webster *Sensors and Signal Conditioning*, 2nd edn. (Wiley, New York, 2001)
2. J.H. Huising, Smart sensor systems: Why? Where? How?, in *Smart Sensor Systems*, ed. by G.C.M. Meijer (Wiley, Chichester (UK), 2008), pp. 1–21
3. F. Reverter, J. Jordana, M. Gasulla, R. Pallàs-Areny, Accuracy and resolution of direct resistive sensor-to-microcontroller interfaces. Sens. Actuators, A **121**, 78–87 (2005)
4. F. Reverter, O. Casas, Interfacing differential resistive sensors to microcontrollers: a direct approach. IEEE Trans. Instrum. Meas. **58**(10), 3405–3410 (2009)
5. E. Sifuentes, O. Casas, F. Reverter, R. Pallàs-Areny, Direct interface circuit to linearise resistive sensor bridges. Sens. Actuators, A **147**, 210–215 (2008)
6. F. Reverter, O. Casas, Direct interface circuit for capacitive humidity sensors. Sens. Actuators, A **143**, 315–322 (2008)
7. F. Reverter, O. Casas, Interfacing differential capacitive sensors to microcontrollers: a direct approach. IEEE Trans. Instrum. Meas. **59**(10), 2763–2769 (2010)
8. F. Reverter, O. Casas, A microcontroller-based interface circuit for lossy capacitive sensors. Meas. Sci. Technol. **21** (065203), 8 pp (2010)
9. Z. Kokolanski, J. Jordana, M. Gasulla, V. Dimcev, F. Reverter, Direct inductive sensor-to-microcontroller interface circuit. Sens. Actuators, A **224**, 185–191 (2015)
10. N. Ramadoss, B. George, A simple microcontroller based digitizer for differential inductive sensors, in *Proceedings IEEE I2MTC*, 11–14 May 2015, Pisa, Italy (2015), pp. 148–153

11. L. Bengtsson, Direct analog-to-microcontroller interfacing. Sens. Actuators A **179**, 105–113 (2012)
12. F. Reverter, R. Pallàs-Areny, *Direct Sensor-to-Microcontroller Interface Circuits. Design and Characterization* (Marcombo, Barcelona, Spain, 2005)
13. F. Reverter, The art of directly interfacing sensors to microcontrollers. J. Low Power Electron. Appl. **2**, 265–281 (2012)
14. F. Reverter, Power consumption in direct interface circuits. IEEE Trans. Instrum. Meas. **62**(2), 503–509 (2013)
15. F. Reverter, J. Jordana, R. Pallàs-Areny, Internal trigger errors in microcontroller-based measurements, in *Proceedings XVII IMEKO World Congress*, 22–27 June 2003, Dubrovnik, Croatia (2003), pp. 655–658
16. F. Reverter, R. Pallàs-Areny, Uncertainty reduction techniques in microcontroller-based time measurements. Sens. Actuators, A **127**, 74–79 (2006)
17. F. Reverter, M. Gasulla, R. Pallàs-Areny, Analysis of power-supply interference effects on direct sensor-to-microcontroller interfaces. IEEE Trans. Instrum. Meas. **56**(1), 171–177 (2007)
18. G.C.M. Meijer, Interface electronics and measurement techniques for smart sensor systems, in *Smart Sensor Systems*, ed. by G.C.M. Meijer (Wiley, Chichester (UK), 2008), pp. 23–54
19. R.N. Ponnalagu, B. George, J. Kumar, A microcontroller sensor interface suitable for resistive sensors with large lead resistance, in *Proceedings of the International Conference on Sensing Technology*, 2–4 Sept 2014, Liverpool, UK (2014), pp. 327–331
20. E. Sifuentes, O. Casas, F. Reverter, R. Pallàs-Areny, Improved direct interface circuit for resistive full- and half-bridge sensors, in *Proceedings ICEEE 2007*, 5–7 Sept 2007, Mexico City, Mexico (2007), pp. 197–200
21. F. Reverter, G. Horak, V. Bilas, M. Gasulla, Novel and low-cost temperature compensation technique for piezoresistive pressure sensors, in *Proceedings XIX IMEKO World Congress*, 6–11 Sept 2009, Lisbon, Portugal (2009), pp. 2084–2087
22. F. Reverter, R. Pallàs-Areny, Effective number of resolution bits in direct sensor-to-microcontroller interfaces. Meas. Sci. Technol. **15**, 2157–2162 (2004)
23. J. Jordana, R. Pallàs-Areny, A simple, efficient interface circuit for piezoresistive pressure sensors. Sens. Actuators, A **127**, 69–73 (2006)
24. S.Y. Yurish, A simple and universal resistive-bridge sensors interface. Sens. Transducers J. **10**, 46–59 (2011)
25. J. Courbat, D. Briand, L. Yue, S. Raible, N.F. Rooij, Drop-coated metal-oxide gas sensor on polyimide foil with reduced power consumption for wireless applications. Sens. Actuators, B **161**, 862–868 (2012)
26. S. Knobelspies, B. Bierer, A. Ortiz Perez, J. Wöllenstein, J. Kneer, S. Palzer, Low-cost gas sensing system for the reliable and precise measurement of methane, carbon dioxide and hydrogen sulfide in natural gas and biomethane. Sens. Actuators B **236**, 885–892 (2016)
27. E. Sifuentes, Sensor autónomo para detectar vehículos estáticos. Ph.D. thesis, Universitat Politècnica de Catalunya, Spain (2009) (in Spanish)
28. E. Sifuentes, J. Cota-Ruiz, R. González-Landaeta, Respiratory rate detection by a time-based measurement system. Revista Mexicana de Ingeniería Biomédica **37**(2), 91–99 (2016)
29. L.C. Wang, K.T. Tang, C.T. Kuo, C.L. Ho, S.R. Lin, Y. Sung, C.P. Chang, Portable electronic nose system with chemiresistor sensors to detect and distinguish chemical warfare agents. J. Micro/Nanolith. MEMS MOEMS **9**(3), 302–306 (2010)
30. M. Camara, F. Molina-Lopez, E. Danesh, G. Mattana, A. Bontempi, D. Teyssieux, L. Thiery, P. Breuil, C. Pijolat, K. Persaud, D. Briand, N.F. de Rooij, Printed micro-hotplates on flexible substrates for gas sensing, in *Proceedings TRANSDUCERS'13*, 16–20 July 2013, Barcelona, Spain (2013), pp. 1059–1062
31. http://www.pstsensors.com
32. http://www.tekscan.com
33. T. D'Alessio, Measurement errors in the scanning of piezoresistive sensors arrays. Sens. Actuators, A **72**, 71–76 (1999)

34. Ó. Oballe-Peinado, F. Vidal-Verdú, J.A. Sánchez-Durán, J. Castellanos-Ramos, J.A. Hidalgo-López, Smart capture modules for direct sensor-to-FPGA interfaces. Sensors **15** (12), 31762–31780 (2015)
35. Ó. Oballe-Peinado, F. Vidal-Verdú, J.A. Sánchez-Durán, J. Castellanos-Ramos, J.A. Hidalgo-López, Improved circuits with capacitive feedback for readout resistive sensor arrays. Sensors **16**(2), 149 (2016)
36. F. Vidal-Verdú, Ó. Oballe-Peinado, J.A. Sánchez-Durán, J. Castellanos-Ramos, R. Navas-González, Three realizations and comparison of hardware for piezoresistive tactile sensors. Sensors **11**(3), 3249–3266 (2011)
37. Ó. Oballe-Peinado, F. Vidal-Verdú, J.A. Sánchez-Durán, J. Castellanos-Ramos, J.A. Hidalgo-López, Accuracy and resolution analysis of a direct resistive sensor array to FPGA interface. Sensors **16**(2), 181 (2016)

Interfaces for Autarkic Wireless Sensors and Actuators in the Internet of Things

Hubert Zangl, Stephan Muehlbacher-Karrer and Raiyan Hamid

Abstract Wireless devices need to operate without connection to an energy source autarkically see below over a long time. As energy harvesting is limited this also implies that power and energy management are a major concern. Thus, the devices are often extremely reduced in terms of capabilties and availability. Nevertheless, accessing such devices from the internet should still be easy and hassle-free. We show how the ISO/IEC/IEEE 21450-2010 smart transducer standard can be used for this purpose. In addition, we provide an overview of common concepts that might be used in a similar way. Furthermore, we discuss requirements with respect to various energy modes.

1 Introduction

In the Internet of Things (IoT) many devices covering a huge range of applications from home automation over automotive to industrial applications are interconnected. The scale ranges from a single constrained device up to massive cross-platform deployments of embedded technologies and cloud systems exchanging information in real-time.

Numerous legacy communications protocols exist and numerous emerging communication protocols are developed in order to tie everything together. With respect to standardization, many alliances and coalitions have been created in order to unify the somewhat unstructured organic growth of the IoT landscape. That limit the potential applications of the IoT. The fragmentation between the protocols utilized for communication within and across resource-constrained devices and resource-rich devices is not foreseen to change in the near future [1].

Sensors and actuators will play an important role in the IoT and embedded devices are expected to be dominant in the IoT [2]. Many of these devices may be

H. Zangl (✉) · S. Muehlbacher-Karrer · R. Hamid
Institute of Smart Systems Technologies, Alpen-Adria-Universitaet Klagenfurt,
Klagenfurt, Austria
e-mail: hubert.zangl@aau.at

B. George et al. (eds.), *Advanced Interfacing Techniques for Sensors*,
Smart Sensors, Measurement and Instrumentation 25,
DOI 10.1007/978-3-319-55369-6_5

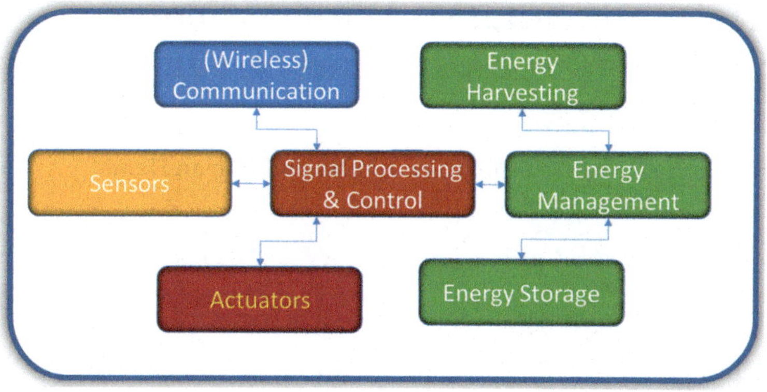

Fig. 1 A typical transducer node architecture consists of several main blocks: a power supply unit, a sensing or actuating subsystem, a processing and storage unit and a communication subsystem

wireless. True wireless sensors and actuators are either battery powered or come with an energy harvesting system. However, if the operation should be over a long time, power and energy considerations are a major concern. From the user perspective the access of the sensors should be as intuitive as possible to provide a high usability. This access will mainly be realized with software interfaces using certain protocols. A typical configuration of an autarkic, i.e. self-powered wireless sensor/actuator device is depicted in Fig. 1.

Even though the general structure is similar, the actual implementation of such a wireless device may be quite different depending on the field of application. For outdoor weather monitoring systems, large photo voltaic cells may be used and the available energy may be quite sufficient to power long range wireless communication using cellular networks. In other situations, e.g. indoor sensors for Heating, Ventilation and Air Conditioning (HVAC) systems, the size of the devices may be limited. Considering photo voltaic energy harvesting, indoor light intensities can also very low, such that we obtain a very energy constraint system. In this contribution we focus on such constrained situations, where energy management becomes a major concern. Such systems will be in sleep mode as much as possible and limit transmission of data as much as possible.

Some typical IoT configurations are shown in Fig. 2. As we are interested in wireless sensors (and actuators) that should operate over a long time or survive solely by energy harvesting, we may prefer designs, where the sensors are not directly accessible using IPv4 or IPv6 protocols. Instead, we use a gateway (no power constrains) that can be accessed directly from the internet. The communication between the gateway and the sensor does not have to be Internet Protocol (IP) based (yet it could). Gateway and server may be the same device (as in example (b)) or may be separated devices that would then also communicate using the IPv4/IPv6 network.

This chapter is structured as follows: In Sect. 2 we discuss energy management and energy saving mechanisms, in Sect. 3 we analyse requirements with respect to

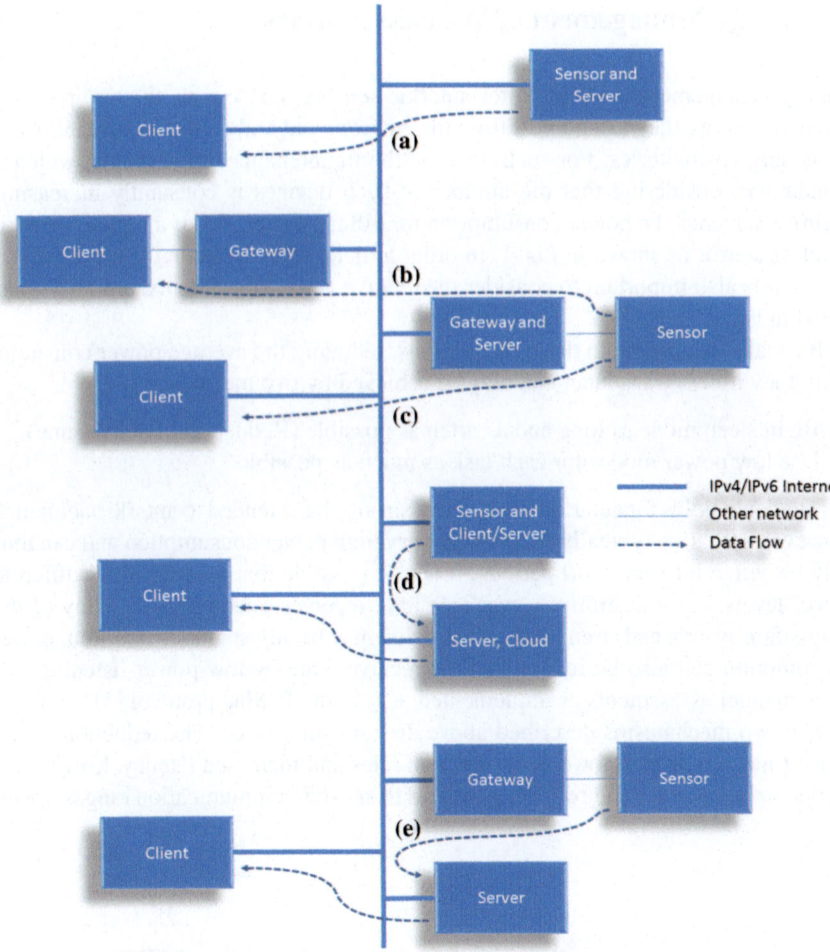

Fig. 2 Examples (non-exhaustive) for the communication between a sensor ("data source") and a client ("data sink") in the IoT. **a** Both client and sensor are linked to the IPv4/IPv6 internet, the sensor directly acts as a server. **b** Both client an sensor do not directly use IP but are linked to the internet by means of gateways. **c** As in (**b**) but here the client connects directly using IPv4/IPv6. **d** Both sensor and client communicate directly using IPv4/IPv6. However, an additional server, e.g. a cloud service or a message broker, relays the information. The server may actually store and preprocess the data of many sensors. As the processing is moved to a powerful server, the complexity of the sensor node can be reduced. **e** Same as (**d**) but with an additional gateway

autarkic wireless sensors and actuators, in Sect. 4 we provide an overview of current approaches used in the Internet of Things. Finally, in Sects. 5–7 we specifically discuss the use of ISO/IEC/IEEE 21450-2010 for autarkic wireless sensors.

2 Energy Management of Wireless Sensors

Energy management is essential for autarkic sensors, i.e. sensors that are powered solely by energy they can harvest from the environment and do not require exchange or recharge of batteries. For such devices, the maintenance effort is low, which is mandatory considering that the number of such devices is constantly increasing. Figure 3 sketches the power consumption for different subtasks of a typical autarkic wireless sensor as shown in Fig. 1. In order to determine the actual energy requirements, it is also important to consider the relative active times, as exemplarily illustrated in Fig. 4 [3].

Basically, according to these illustrations, reducing the average power consumption of a wireless transducer can only be achieved by two mechanisms:

1. Be in sleep mode as long and as often as possible (Reduction of active time).
2. Use low power modes for each task as much as possible.

This not only holds for autarkic sensors but can also be extended to autarkic actuators. However, actuation typically comes with very high power consumption and can thus only be active for very short periods. It is also possible that modules offer different power levels, e.g. transmitter power (see [4]), receiver sensitivity, accuracy of the acquisition system and strength of an actuation may be adjustable. In addition, power consumption can also be reduced on the receiver side by low power listening and clear channel assessments as implemented, e.g. in the B-Mac protocol [5].

The two mechanisms described above do come at a price. The reduction of the active time may lead to lower measurement rates and increased latency. Low power modes for transmitter and receiver may lead to shorter communication ranges, lower

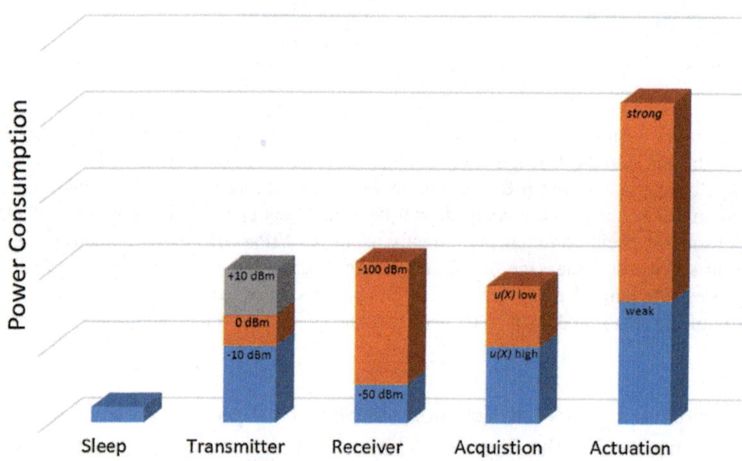

Fig. 3 Example of the power consumption of modules within a wireless transducer. The bar's annotations correspond to examples for transmission power (*transmitter*), receiver sensitivity (*receiver*), measurement uncertainty (*acquisition*) and actuation intensity (*actuation*)

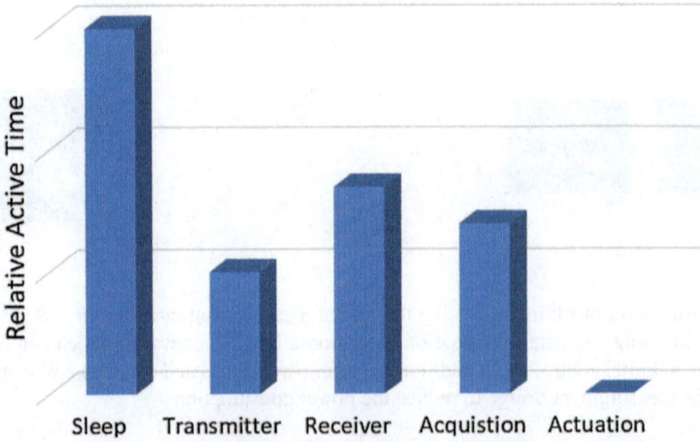

Fig. 4 Example of active time of different modules of a wireless transducer. As transmission is triggered by the transducer itself, it is only in active mode as long as needed. In contrast, if bi-directional, asynchronous communication the receiver needs to be in active mode for much longer times than for actual data transmission. The active time of sensors and actuators varies of large scales depending on the application and implementation but may be significant

data rates, higher probability of packet loss and higher latency. Low power modes for data acquisition modules may lead to higher uncertainty. Lowering power for an actuator may lead to slow responses or even to the failure to perform a desired action. For practical applications it is therefore important that the concepts are comparatively simple [6]. This has to be taken into account for the definition of the transducer interface.

If we use a networked sensor, the user (or client) should somehow be notified of these energy related properties of the device. While classically, only the state of charge of the battery is reported, there are many more aspects that could be considered. A universal sensor interface should provide mechanisms that this information can be conveyed.

As mentioned above, one of the most common means to reduce the average power consumption of wireless sensors is duty-cycling, i.e. to reduce the active time as much as possible, e.g. by a reduction of the measurement rate. In the following we present two other examples how reduction of average power consumption of wireless sensors can be reduced. These can be applied, e.g. when a very accurate wireless sensor is used in an application where high accuracy is not needed at all or at least only under certain circumstances. Then we may save energy and thus extend operation time by reducing the measurement accuracy.

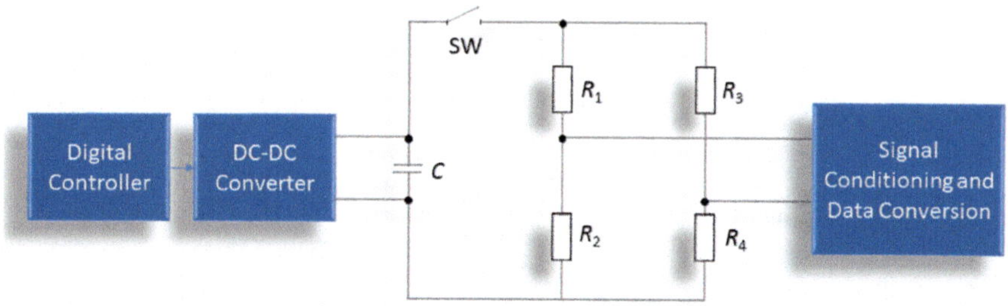

Fig. 5 Measurement block incorporating the bridge measurement circuitry (R_1, R_2, R_3 and R_4), a signal conditioning and data conversation block and a DC-DC converter which can be powered on and off by a digital controller. In addition, a buffer capacitor C and a switch SW is used to turn on and off the measurement bridge to reduce the power consumption

2.1 *Energy Efficient Acquisition: Trading Energy Versus Accuracy*

In this Section we illustrate by means of a simple example that different modes of the data acquisition submodule can be beneficial. Basically, we are trading energy and power consumption versus accuracy [3].

The concept is exemplarily depicted in Fig. 5 for a bridge measurement circuitry. Bridge resistors, e.g. for strain gauges or pressure sensors, typically have comparatively low impedances in the order of $k\Omega$.

In order to achieve reasonable Signal to Noise Ratios (SNRs) it is thus necessary to use substantial currents. While the SNR increases with the current in the resistors, so does the power consumption

$$P = I_t^2 * R_t \tag{1}$$

where I_t represents the total current and R_t the total resistance. Reducing the current by a factor of 10 will decrease the SNR by 20 dB but also the power consumption is reduced to, actually by a factor of 100.

Calculating the power consumption from the resistors and the auxiliary bridge voltage we obtain

$$P = \left(\frac{1}{R_1 + R_2} + \frac{1}{R_3 + R_4} \right) * V_{in}^2 \tag{2}$$

This means that the power consumption is nonlinear in the voltage. If we assume an adjustable voltage, we obtain a nonlinear relation between the parameter voltage and power consumption. However, in order to predict the power consumption for a certain choice of the voltage, this nonlinear relation should be provided in an electronic data sheet of the device. Additionally, it should be described how the choice of the voltage affects the uncertainty of the measurement result, which can easily be determined for changing parameters using software tools [7]. An approach for such a

Table 1 Example for power consumption versus combined standard uncertainty for a bridge circuit as shown in Fig. 5 [3]. With less power used for the bridge supply voltage, the standard uncertainty of the measurement result for the unknown bridge resistor R_1 increases. However, when the higher uncertainty can be accepted in the application, significant power savings can be achieved

Supply voltage V_{in}	Standard uncertainty $u(R_1)$	Power consumption P
V	mΩ	mW
0.1	400.3	0.01
0.8	52.9	0.6
1.6	30.4	2.6
3.3	21.1	11

description based on ISO/IEC/IEEE 21450-2010 is provided in Sect. 5 [3]. Example data is provided in Table 1.

2.2 *Energy Efficient Data Transmission: Trading Energy Versus Accuracy*

Communication has often a major share of the total energy consumption of a wireless sensor. Consequently, reducing the number of transmissions can be a useful approach. Actually, the costs to transmit a single byte or a small payload may not be that much different, as we typically have a minimum costs due to start up period, synchronization, collision avoidance etc. So up to a certain packet size, the costs of a transmission do not significantly increase. Depending on the permitted latency, we may make use of a buffered transmission, i.e. collect several measurements and transmit it at once.

This idea can be extended if we can compress the data. Basically, any lossless data compression method could be used. However, in a resource constrained device, the algorithms must remain simple. A simple yet lossy approach is to only transmit data when a significant change has occurred or the elapsed time since the last transmission would otherwise exceed the maximum permitted latency time (this ensures a heartbeat signal). Again, it is necessary that the client and the sensor can exchange information, i.e. on the one hand what changes are significant for the client and on the other hand how much energy can be saved on the sensor side.

The concept is illustrated in Fig. 6. Note that data reduction not only helps to save power but also helps to significantly reduce the load in the communication channel. Similar approaches have been used in many applications ranging from data compression (e.g. [8]) to process control (e.g. [9]). Another (in terms of power consumption) similar approach suggested in [10] could also be treated in a similar way.

It should be noted that triggered transducer configuration may also have an influence on the acquisition system. For instance, the trigger may be implemented in the

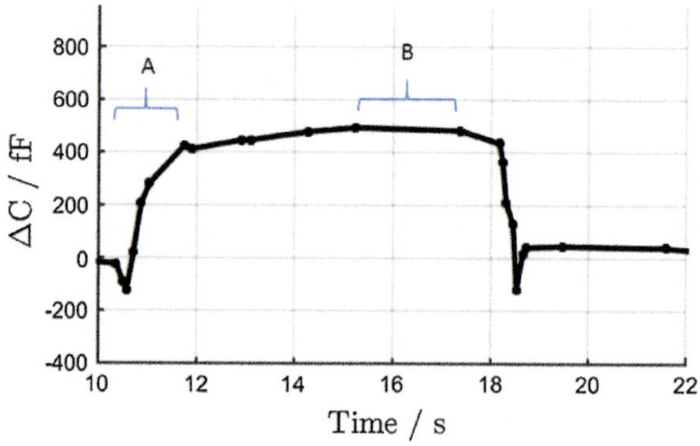

Fig. 6 Example for a simple data reduction approach [11]. Here, a capacitive hand detection sensor for a steering wheel only transmits data when the signal changes significantly (the level for significance can be adjusted) or the maximum latency (2 s) has expired. Consequently, during the transition of the signal (*region "A"*) many samples are transmitted, whereas in the steady phase ("B") only a heartbeat signal is transmitted. In comparison to the lossless compression in [12] (where every signal change is compressed and transmitted) this approach discards data which does not contain new information to the receiver. However, there is no information loss due to the fact that the receiver knows that the signal stays at a certain level as long as it does not get a new package from the transmitter. This approach trades energy savings against an increase of the uncertainty of the data

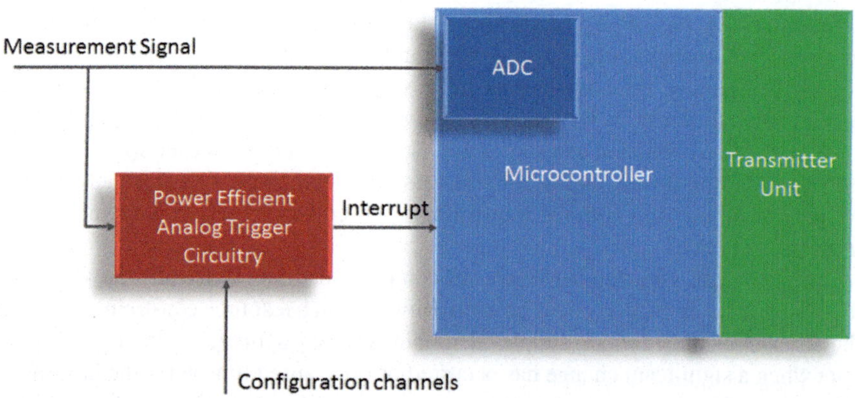

Fig. 7 Sketch of a trigger circuitry. The microcontroller including an ADC and transmitter unit are in sleep mode as long as the measurement signal is within a predefined range of 2 Δ (see Fig. 6). If the measurement signal is not in the range, the analog trigger circuitry fires a wake up interrupt to the microcontroller and the measurement data is transmitted via the transmitter unit. The configuration channels of the analog trigger circuitry are used to set the value of Δ

analog domain with highly power efficient analog comparators and Digital to Analog Converters (DACs). Such a configuration is shown in Fig. 7. Note that we can apply the same strategy as described in Sect. 2.1. Two embedded actuator channels may be used to assign the limits for the change identification. The power consumption for the acquisition module will now be different for the triggered and the non-triggered mode.

3 Requirements on Protocols and Standards with Respect to Autarkic Wireless Sensors

Our aim is to provide quick and easy access to autarkic wireless devices. This implies:

- No specific software should be required to access sensors and actuators, read and understand sensor data, write actuator data and set configurations.
- Sensor responses should be human readable and also machine readable.
- All necessary information (i.e. the electronic datasheet) should be stored within the devices, not just references to some server. This allows that such local devices can be used and configured without the need to have direct access to the internet. Additionally, no long term support (hosting of information) from the manufacturer or a vendor is needed.
- Energy and power related aspects should be covered, i.e. different situation dependent configurations (with varying performance and energy requirements) should be possible. Standardized descriptions of the capabilities should be provided.

Analyzing above requirements, we find that the ISO/IEC/IEEE 21450-2010 [13] standard covers several of these. It provides a Hypertext Transfer Protocol (HTTP) Application Programming Interface (API) that allows to access sensors solely by a web browser. It defines Transducer Electronic Datasheets (TEDS), suitable to be stored in the device itself. With different data formats provided (html, xml and text) the responses are both human and machine readable. Energy and power related aspects of wireless sensors are not been directly addressed in the standard but can be implemented, e.g. in the manufacturer specific part of the TEDS as proposed in [3]. More details on ISO/IEC/IEEE 21450-2010 are discussed in Sect. 5.

4 Overview of Current IoT Approaches

In the following we provide a brief overview of related protocols and standards, with a focus on sensors.

4.1 Infrastructure Internet

In Fig. 2 the IPv4/IPv6 internet connects all the devices. We consider this network as the backbone of the IoT. We refer to the "Internet" as the global network, where all connected devices could (if not prevented by firewalls) communicate with each other using IPv4/IPv6 by knowing the address or an associated name.

- *IPv4*: Internet Protocol version 4 (IPv4) is the fourth version of the IP and is described in Internet Engineering Task Force (IETF) publication RFC 791 (September 1981) [14]. As of 2016 it still routes most internet traffic today despite the ongoing deployment of a successor protocol, IPv6 [15]. IPv4 is a connectionless protocol for packet-switched networking on the network layer according to the Open Systems Interconnect (OSI) model [16]. The protocol does not ensure packet delivery, correct packet sequencing and packets may be duplicated. If needed, these aspects have to be handled on next upper transportation layer and their protocols, e.g. TCP, UDP etc.
  IPv4 uses a 32-bit address space and is thus theoretical limited to 4.29×10^9 addresses, which has been considered insufficient for the future internet. This has led to the development of IPv6.
- *IPv6*: [17] is intended to replace IPv4 on the OSI network layer. Besides other technical benefits, a main advantage is the larger address space. IPv6 uses 128-bits for device addressing, usually represented as eight groups of four hexadecimal digits, separated by colons, e.g. 2001:0db8:0000:0042:0000:8a2e:0370:7334. With about 3.4×10^{38} addresses it is believed that this will be sufficient for the future internet, even though not all of the addresses can be used as some of them are reserved for special use. Google statistics report that as of June 2016 about 12% of accesses to their servers are over IPv6.
- *6LoWPAN*: is an acronym of IPv6 over Low power Wireless Personal Area Networks. 6LoWPAN is an open standard defined in RFC 6282 [18] by the IETF. 6LoWPAN introduces an adaptation layer to be able to transmit IPv6 datagrams over IEEE 802.15.4-Based wireless networks [19]. It allows IPv6 packets to be transported within small link layer frames as those defined by IEEE 802.15.4. It has also been adapted and used over a variety of other networking media such as sub-1 GHz low power Wi-Fi, Bluetooth Smart, and Power Line Control (PLC) [20].
- *TCP*: The Transmission Control Protocol (TCP) provides reliable, ordered, and error-checked delivery of a stream between applications running on hosts communicating over an IP network [21] on the transportation layer. In order to achieve this, it uses acknowledgment and retransmission of lost packets.
- *UDP*: In contrast to TCP, the User Datagram Protocol (UDP) [22] is a connectionless transmission model on the transportation layer. With it, a device can send messages to other hosts using an IP network without the need of a prior connection setup. Thus it is simple, yet it does not guarantee delivery, ordering or duplicate removal. It avoids overhead of acknowledgment and retransmission and is thus favorable, e.g. for real-time or streaming applications, where retransmission of lost packets is not useful.

Table 2 IoT application protocols (adapted from [23])

	REST	Transport	Publish/ Subscribe	Request/ Response	Security	QoS
HTTP	x	TCP		x	SSL	
CoAP	x	UDP	x	x	DTLS	x
MQTT		TCP	x		SSL	x
MQTT-SN			x		SSL	x
AMQP		TCP	x		SSL	x
XMPP		TCP	x	x	SSL	x
DDS		TCP/UDP	x		SSL/DTLS	x
STOMP		TCP	x		SSL	

4.2 Data Protocols

While IPv4/IPv6 in combination with TCP or UDP ensure the data transmission between two end points (transportation-oriented), it is also necessary to define how this data communication is used, i.e. how requests and responses are described and how the data is transported (application-oriented). Table 2 provides a comparison of a selection of common protocols. Besides Message Queue Telemetry Transport for Sensor Networks (MQTT-SN) these have in common that they are based on TCP or UDP and thus using the IPv4/IPv6 internet. Most protocols provide a publish/subscribe approach, HTTP only provides a request/response approach. Security is provided using the standard Secure Socket Layer (SSL) or Datagram Transport Layer Security (DTLS) protocols.

- *REST HTTP*: REST stands for Representational State Transfer [24]. It constraints the client server interaction to be stateless, i.e. each request from client to server must contain all of the information necessary to understand the request by the server. The server does not store and context, states are entirely kept by the client. REST is usually used with HTTP and provides a simple approach to use HTTP request to read, write or delete data.
- *CoAP*: The Constrained Application Protocol (CoAP) [25] is an application layer protocol that is intended for use in resource-constrained internet devices, such as Wireless Sensor Networks (WSN) nodes. It realizes a subset of REST common with HTTP but optimized for Machine-to-Machine (M2M) applications. CoAP can be used for refashioning simple HTTP interfaces into a more compact protocol. It also offers features for M2M such as built-in discovery, multicast support, and asynchronous message exchanges.
- *MQTT*: It is a lightweight publish-subscribe-based messaging protocol for M2M communication on top of the TCP/IP protocol [26]. It requires a message broker

such as *Mosquito* [27], which is responsible for distributing messages to interested clients (compare Fig. 2d).

- *MQTT-SN*: It is a lightweight publish/subscribe middleware specifically designed for embedded devices on non-TCP/IP networks, such as Zigbee. MQTT-SN is close to MQTT, but redesigned to deal with the requirements of WSN such as high link failures, low bandwidth, etc. and resource limited devices in terms of energy, processing power, memory, etc. [28] (compare Fig. 2e).
- *XMPP*: Extensible Messaging and Presence Protocol represents an open technology for real-time communication, which aims to power a wide range of applications including instant messaging, presence, multi-party chat, voice and video calls, collaboration, lightweight middleware, content syndication, and generalized routing of XML data. The core specifications for XMPP are developed at the IETF [29]. An extension of XMPP specifically addresses sensor data in the IoT [30]. As of 2016, the status of the extension is "experimental".
- *DDS*: Data-Distribution Service for Real-Time Systems represents a middleware standard developed by the Object Management Group (OMG). It directly addresses data centric publish-subscribe communications for real-time and embedded systems [31]. In particular it provides control of Quality of Service (QoS) parameters, including reliability, bandwidth, delivery deadlines, and resource limits. It uses a background discovery protocol to automatically find data. DDS systems are typically more contained, i.e. not spread across Wide-Area Network (WAN).
- *AMQP*: Advanced Messaging Queuing Protocol is an open internet protocol for business messaging [32], often used for server to server communication, yet it also finds applications in the IoT. Originating from banking industry, a main focus is on not losing messages, regardless of failures or reboots.
- *STOMP*: Simple (or Streaming) Text Oriented Message Protocol [33] is a simple text based interoperable protocol designed for asynchronous message passing between clients via mediating servers (brokers), similar to MQTT. STOMP is a frame based and assumes a reliable 2-way streaming network protocol (such as TCP) underneath. Being text-based, it is possible to directly communicate with a server, e.g. using a telnet connection.

Beside this selection of common protocols numerous other protocols have been developed and all of them come with certain advantages and disadvantages. The best choice will depend on the number of devices, required response times, etc. Servers often support multiple protocols so that clients using different protocols can access the same data.

4.3 Semantics

Semantics in this context means that sensors are described in a standardized, typically machine-interpretable representation. As much of the IoT potentials is due to the capabilities of low-cost and energy-efficient sensors (and actuators) with mature wireless communication capacities and the interests in integrating the physical into the cyber worlds such semantics can help to handle the heterogeneity of things and to infer new knowledge together with other intelligent processing techniques [34]. Additionally, if the relation between performance and power consumption can be included in such a device description, an intelligent adaptive energy management becomes possible (compare Sect. 2).

Besides TEDS defined in [13] (compare Sect. 7), many other approaches to describe sensors have been suggested. In the following we briefly describe some prominent examples.

- *SensorML*: It provides standard models and an XML encoding for describing sensors and measurement processes [35]. It originates from geospatial sensing and thus includes location information but it is not limited to such applications. Definitions of metrological terms are partially different compared to the field of metrology as defined in [36, 37].
- *EDDL*: The Electronic Device Description Language is used in several industrial standards. An Electronic Device description (EDD) based on EDDL is usually provided by the manufacturer of a device. It is not stored in the device itself and it needs an interpreter to be executed. It is rather a programming language, e.g. for user interfaces in the Industrial Internet of Things (IIoT) and Industry 4.0 [38].
- *Semantic Sensor Net Ontology—W3C*: This ontology describes sensors and observations they make of the physical world using definitions of classes and properties (e.g. measurement range, latency, attached system) [39]. With respect to the 'quality of results' is uses the term measurement accuracy, which [36] defines as non-quantitative.
- *Wolfram Language*: Connected Devices provide symbolic representations of devices [40]. The devices can be accessed with a standard set of Wolfram Language functions like DeviceRead, DeviceExecute, DeviceReadBuffer and DeviceReadTimeSeries. It has a large number of devices in the database, yet not all device descriptions are very detailed. Currently, the support with, e.g. respect to measurement uncertainty appears limited.

4.4 Power Efficient Lower Layer Wireless Sensor Protocols

Numerous wireless technology standards such as WIFI, NFC, ANT, IEEE 802.15.4, ZigBee, WirelessHART, Bluetooth SMART, LoRaWAN etc. as well as proprietary solutions are used for the lower layer of wireless communication between sensors and the internet. From the perspective of the sensor interface as seen from the internet, it should not make a difference how the data is actually transferred. Consequently, all these technologies allow to cope with the requirements given in Sect. 3. The choice for a protocol will mainly depend on the requirements with respect to range, data rate, routing and multi-hop capabilities and of course power consumption.

4.5 Security

Security is a crucial topic for any communication network. In our approach we have three aspects:

- Security of the internet communication. This can be achieved by the usual means that are used for internet traffic.
- Security of the wireless link. This has to be addressed by the wireless communication protocol and depends on the respective standards.
- Security of the server.

The server in an architecture, where it, e.g. acts as a message broker may not just relay the information provided by the sensor but may also perform some processing, e.g. calibration or archiving for historical data access. Consequently, in this situation it must be able to read the data. Therefore, an approach using, e.g. message broker can not provide an end-to-end encryption.

As security is one of the major topics for IoT it actually is a topic on its own. A survey of existing protocols and open research issues can be found, e.g. [41].

5 Using ISO/IEC/IEEE 21450-2010 with Autarkic Wireless Sensors

The ISO/IEC/IEEE 21450-2010 [13] (prepared as IEEE 1451.0-2007) is part of a set of smart transducer interface standards developed by the Institute of Electrical and Electronics Engineers (IEEE) Instrumentation and Measurement Society's Sensor Technology Technical Committee describing a set of open, common, network-independent communication interfaces for connecting transducers (sensors or actuators) to microprocessors, instrumentation systems, and control/field networks. To go beyond the previous definitions, an ISO/IEC/IEEE 21450-2010 smart transducer is defined as a smart transducer that provides functions beyond those necessary for generating a correct representation of a sensed or controlled quantity. This functionality allows for simplifying the integration of the transducers into applications in a networked environment. One of the key elements of these standards is the definition of TEDS for each transducer. The TEDS is a memory *inside* the transducer, which stores transducer identification, sensor and actuator channel descriptions, optional calibration data, etc. The goal of this family of standards is to allow the access of transducer data through a common set of interfaces whether the transducers are connected to systems or networks via a wired or wireless means. This means ISO/IEC/IEEE 21450-2010 smart transducers have capabilities for self-identification, self-description, self-diagnosis, self-calibration, location-awareness, time-awareness, data processing, reasoning, data fusion, alert notification (report signal), standard-based data formats, and communication protocols [42].

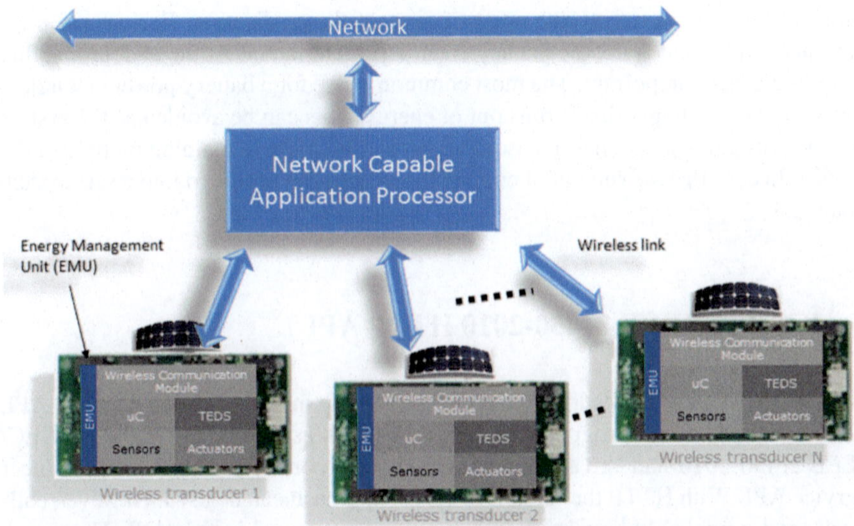

Fig. 8 Architecture in style of IEEE 1451-5. The network capable application processor, which is accessible from the network (e.g. IPv4) acts as an interface to the autarkic wireless devices. This corresponds to architecture (**c**) in Fig. 2. Energy and power management is important for both, the wireless link and the other components inside the transducer

Using ISO/IEC/IEEE 21450-2010 with wireless sensors, a so called Network Capable Application Processor (NCAP) acts as a wireless base station and connects the transducer to a network. Energy management of NCAPs is not considered here, as we assume that these devices are not limited in terms of power or energy. Therefore, the usual architecture is that the transducers connect in a star-like topology to a NCAP. Such topologies are fairly common, e.g. in wireless communication in vehicles or aircraft and also in industry [43]. Distances are usually short and the main backbone of the system is wired (at least for the power supply), e.g. based on Ethernet. An example configuration is illustrated in Fig. 8.

In this approach the NCAP acts both as a server and a gateway. Even when using 6LoWPAN for communication between the NCAP and the server, the sensors would not directly be accessed by their IPv6 address but only through the NCAP. From the client point of view it does not matter which wireless communication is used. It may follow one of the IEEE-1451-5 wireless standards such as ZigBee of Bluetooth or may make us of proprietary protocols of manufacturers or researchers. The different behaviour of the communication interface might find an abstraction in the PHY-TEDS, which is a mandatory part of the TEDS.

As energy management is crucial for wireless devices, a standard on smart transducers should also consider this aspect. A TEDS may provide information on average power consumption in different operation modes. The idea is to simply provide the information on how much power each mode requires, no matter how it is implemented. Therefore, we are not looking into the implementation details of each protocol but look at it from the point of view of the application. Besides the measurement and update rate, the latency is an important factor. At the end, probability of failure is most important. The most common cause for a battery powered wireless device to fail is simply that it runs out of energy. This can be avoided as the system that utilizes a transducer has a reasonable prediction of the remaining lifetime of the device. In case that we run out of energy, the system can safely go into a safe system state.

6 ISO/IEC/IEEE 21450-2010 HTTP API

Besides the standard transducer service API and the module communications API, which may be used in programming languages such as Java or C++, the ISO/IEC/IEEE 21450-2010 standard also defines HTTP APIs corresponding to the transducer service API. With HTTP the request/response approach can be used. However, callbacks and a publish/subscribe approach are not supported in this API. The server may directly be implemented on the NCAP (compare Fig. 8) or may run on a separate device. In the following we assume that the NCAP also hosts the HTTP server.

The methods provided in the API can be classified into four groups:

- Discovery: Methods for applications to discover available Tranducer Interface Modules (TIM) (i.e. wireless sensor nodes) and TransducerChannels organized in this interface.
- TransducerAccess: Access to sensor and actuator TransducerChannels will be by use of methods on this interface.
- TransducerManager: Applications that need more control over TIM access can use methods on this interface. An example are locks on TIMs for exclusive use.
- TEDSManager: Applications can use methods on this interface to read (and write) TEDS.

The HTTP APIs focuses mainly on accessing transducer data and TEDS using the HTTP 1.1 protocol. Users can send a HTTP request (see Table 3) to a server on the NCAP and get a response in the following way:

(a) A user or client sends a HTTP request to the HTTP server on the NCAP.
(b) The HTTP server on the NCAP receives the HTTP request, processes it, communicates with the transducers if necessary and gets the corresponding results.
(c) The HTTP server returns the corresponding HTTP response to the user.

In the standard it is suggested that internally in the NCAP the standard APIs of ISO/IEC/IEEE 21450-2010 should be used between the HTTP server and the NCAP. However, as this can not be seen from the client side it is also be possible that the HTTP server directly processes the request and communicates with the wireless devices. In our implementation we actually follow this approach.

Table 3 Commands in the HTTP API

Discovery API	TIMDiscovery	1451/Discovery/TIMDiscovery
	TransducerDiscovery	1451/Discovery/TransducerDiscovery
Transducer Access API	ReadData	1451/TransducerAccess/ReadData
	StartReadData	1451/TransducerAccess/StartReadData
	MeasurementUpdate	1451/TransducerAccess/MeasurementUpdate
	WriteData	1451/TransducerAccess/WriteData
	StartWriteData	1451/TransducerAccess/StartWriteData
TEDS Manager API	ReadTeds	1451/TEDSManager/ReadTeds
	ReadRawTeds	1451/TEDSManager/ReadRawTeds
	WriteTeds	1451/TEDSManager/WriteTeds
	WriteRawTeds	1451/TEDSManager/WriteRawTeds
	UpdateTedsCache	1451/TEDSManager/UpdateTedsCache
Transducer Manager API	SendCommand	1451/TransducerManager/SendCommand
	StartCommand	1451/TransducerManager/StartCommand
	CommandComplete	1451/TransducerManager/CommandComplete
	Trigger	1451/TransducerManager/Trigger
	StartTrigger	1451/TransducerManager/StartTrigger

A HTTP request can be made using any web browser by typing a command with the following syntax in the address field:

`http://<host>:<port>/<path>?<parameters>`

`<host>` represents the domain name or internet address of the NCAP, e.g. for IPv4 this could be "191.168.0.101". `<port>` is optional and only needed if the HTTP server on the NCAP does not use the standard port (80) for HTTP servers. `<path>` indicates the ISO/IEC/IEEE 21450-2010 path including the command (e.g. "1451/TransducerAccess/ReadData"), parameters associated with the command are passed using `<parameters>` e.g.

`timId=1\&channelId=2\&timeout=14\&samplingMode=`
`continuous\&format=xml`).

The available commands are provided in Table 3.

Using the format parameter, it can be selected if the response should come as plain text, HTML or XML. An example schema for a XML response on a ReadData request is shown in Table 4.

Table 4 Example schema for a response of the HTTP server in XML format (command: ReadData)

```
<?xml version="1.0" encoding="UTF-8"?>
<xs:schema xmlns:xs="http://www.w3.org/2001/XMLSchema"
xmlns:stml=http://grouper.ieee.org/groups/1451/0/1451HTTPAPI
<xs:complexType name="ReadDataHTTPResponse">
<xs:sequence>
<xs:element name="errorCode" type="stml:UInt16"/>
<xs:element name="timId" type="stml:UInt16"/>
<xs:element name="channelId" type="stml:UInt16"/>
<xs:element name="transducerData" type="stml:ArgumentArrayType"/>
</xs:sequence>
</xs:complexType>
</xs:schema>
```

Table 5 General format of ISO/IEC/IEEE 21450-2010 TEDS

Field	Description	Type
-	TEDS length	UInt32
1 to N	Data block	Variable
-	Checksum	UInt16

7 TEDS Structure and Energy Related Extensions

In contrast to most description languages described in Sect. 4.3, which are text based, the ISO/IEC/IEEE 21450-2010 TEDS represents a comparatively compact binary description. However, with its clear structure, it can easily be translated into other formats, e.g. into XML. There is not just a single TEDS for one transducer. Actually, four TEDS are mandatory. The Meta-TEDS for all information needed to gain access to any TransducerChannel, plus information common to all TransducerChannels, the TransducerChannel TEDSs for all information concerning the TransducerChannels, the Transducer Name TEDS to provides a place to store the name by which the system or the end user will know this transducer and the PHY TEDS for all information needed to with respect to communication (this depends on the type of communication and is thus not described in ISO/IEC/IEEE 21450-2010).

The general structure of a TEDS is shown in Table 5. Every TEDS starts with the length information, continues with a variable number of data blocks and ends with a check sum.

The individual data blocks are described using a Type/Length/Value approach as illustrated in Table 6. The standard provides a number of predefined types. Some of them, e.g. channel identifiers for physical units are mandatory, others are optional. Furthermore, the manufacturer can define additional types. We use this possibility to define additional types to include energy related information as described in the following. It would be beneficial if future standards would natively include such energy related extensions.

Table 6 Description of fields in ISO/IEC/IEEE 21450-2010 TEDS

Field	Description
Type	This code identifies the field in the TEDS that is contained within the value field, e.g. the code 12 in a channel TEDS means that the Physical Units accociated with that channel is described
Length	The number in this field gives the number of octets in the value field. The number of octets in the length field is controlled by an entry in the TEDS Identification TLV.
Value	This field contains the actual information.

Following the discussions in Sect. 2 the data section of an extended datasheet may start with the number of (energy-)modes, as each mode obtains an individual datasheet. The first section for each mode is then the common data sheet, e.g. according to the current standard or including modifications as suggested, e.g. in [44] for PHY TEDS. This is followed by a parametric description of the average power consumption and the combined standard uncertainty for a transducer channel or the latency for the PHY TEDS, again using newly defined types. A possible general form of a parametric model is given by

$$X = \begin{cases} (c_0 + c_p p)^{\lambda}, & \lambda \neq 0 \\ ln(c_0 + c_p p), & \lambda = 0 \end{cases} \quad (3)$$

where c_0, c_p and λ are the coefficients provided in the data sheet using additional types. The parameter p is set using an embedded actuator channel, i.e. an actuator channel that is internally used in the transducer. If this channel is set to zero, then no parameter is used. X may stand for average power, latency, or combined standard uncertainty depending on the data sheet. We believe that (3) provides better fits than a second order polynomial approximation, which would require the same number of coefficients. However, both descriptions could be included using corresponding types.

The section as described above repeats for each mode and thus N-times, where N is the number of modes. The mode is also selected by an embedded transducer channel. If the channel is set to zero, than there is only one mode. The total average power consumption can thus be calculated as the sum of the power consumption of all $N_{submodule}$ modules, i.e. the transducers and the communication unit:

$$P_{total} = \sum_{i=1}^{N_{submodule}} P_i \quad (4)$$

An example for an extended TEDS is given in Table 7.

An energy storage device attached to a TIM is treated as a sensor channel, reporting the state of charge of the device, optionally using a calibration TEDS. Whereas the static information about the nominal capacity of the battery is also stored in the

Table 7 Energy management extension for TransducerChannel TEDS [3]

	Field name	Description	Data type	# octets	
	NbrModes	Number of modes	uint8	1	Extension
Mode 1 (selected by dedicated transducer channel)	Original datafields according to IEEE 21451				
	ModelD	ID of the mode	uint8	1	Extension
	ParameterID	ID of the parameter	uint8	1	
	ParamMin	Minimum value of parameter	float32	4	
	ParamMax	Maximum value of parameter	float32	4	
	Cop	Coefficient for power	float32	4	
	Cpp	Coefficient for power	float32	4	
	λp	Coefficient for power	float32	4	
	Cou	Coefficient for uncertainty	float32	4	
	Cpu	Coefficient for uncertainty	float32	4	
	λpu	Coefficient for uncertainty	float32	4	
Mode 2 (slt. by d. trans. ch.)	Original datafields according to IEEE 21451				
	ModelD	ID of the mode	uint8	1	Extension
	⋮	⋮	⋮	⋮	
⋮	⋮				
Mode N	⋮	⋮	⋮	⋮	

TransducerChannel TEDS (self defined type). Furthermore, a harvesting device is also treated as a sensor channel, reporting actual energy harvesting, optionally using a calibration TEDS. Information on maximum harvesting power is also stored in the corresponding TransducerChannel TEDS. Consequently, only minor extensions of the current standard are needed to implement an energy management information exchange. Please note that the approach to configure the TIM by means of transducer channels also permits the use of virtual TEDS and TEDS caching (compare [13]), as the TEDS information can be read only. However, as explained in Sect. 3, TEDS that are directly stored in the devices offers some distinct advantages.

8 Steps for Using an Autarkic Wireless Sensors Example with IEEE 21450-2010 HTTP API

Accessing a sensor will typically involve the following steps:

- *Find the relevant NCAP*: The wireless device will connect to a NCAP nearby, thus we need to know the address of this NCAP. Usually, this will be known from the installation fo the NCAP to the network.
- *Discovery Request*: Obtain the TIM IDs connected to that NCAP.
- *MetaTEDS Request*: Obtain the META TEDS for the TIMs. The meta TEDS contains the information, how many channels a certain TIM provides.
- *Obtain the channel TEDS*: The channel TEDS is a description of the sensor or actuator channel including, e.g. the physical dimension of the quantity being reported and maximum standard uncertainty linked to this measurement.
- *Obtain additional TEDS (optional)*: More information such as calibration data may be required.
- *Configure Device (optional)*: Use, e.g. embedded actuator channels to configure physical channels.
- *Get the Data*: Read from a sensor channel (Write to an actuator channel).

Considering that we just want to get the data from a sensor, it may appear that above procedure is complex. However, the advantage is that we do not just get numeric values but also interpretation of the data and means to adjust the device to the specific needs in an application. Indeed, several steps can be done in an automated fashion, i.e. getting calibration TEDS and carry out calibration in the background. Thus, the practical application is still simple.

9 Summary

In modern systems, sensors and actuators are frequently used in a network and nowadays the network is often directly connected to the Internet. Consequently, the sensors and actuators become parts of the IoT. Therefore, it will be important to have standardized interfaces on how to connect to the devices. Currently, this field is quite fragmented with many different protocols and interfaces in use. An interface for Internet access to autarkic wireless sensors and actuators has to provide more than just access to sensor data and actuator settings. With the huge number of deployed devices it is important to identify the devices and also get an interpretation of the results. Furthermore, energy management is important and should also be accessible by the interface. By this, the remaining battery lifetime or the sufficiency of energy harvesting of a device can be predicted and safe system states can be achieved before a wireless transducer runs out of power. The ISO/IEC/IEEE 21450-2010 HTTP API allows to access sensors and actuators from the Internet using NCAPs and provides

TEDS that can be extended to include information relevant for energy management. Such extension could also be implemented for other protocols and device descriptions.

References

1. A. Al-Fuqaha, A. Khreishah, M. Guizani, A. Rayes, M. Mohammadi, Toward better horizontal integration among iot services. IEEE Commun. Mag. **53**(9), 72–79 (2015)
2. H. Zhou, *The Internet of Things in the Cloud: A Middleware Perspective*, 1st edn. (CRC Press Inc., Boca Raton, FL, USA, 2012)
3. H. Zangl, M. Zine-Zine, S. Mühlbacher-Karrer, TEDS extensions toward energy management of wireless transducers. IEEE Sens. J. 15(5):2587–2594 (2015), http://ieeexplore.ieee.org/stamp/stamp.jsp?arnumber=7009985
4. B. SIG, *Bluetooth Core Specification 4.1*, Bluetooth SIG, 09 (2013)
5. J. Polastre, J. Hill, D. Culler, Versatile low power media access for wireless sensor networks, in *Proceedings of the 2nd International Conference on Embedded Networked Sensor Systems*, ser. SenSys '04, (ACM, New York, 2004), pp. 95–107. doi:10.1145/1031495.1031508
6. S. Hussain, D. Gurkan, Management and plug and play of sensor networks using snmp. IEEE Trans. Instrum. Meas. 60(5):1830–1837 (2011), http://ieeexplore.ieee.org/stamp/stamp.jsp?arnumber=5740597
7. H. Zangl, K. Hoermaier, Educational aspects of uncertainty calculation with software tools. Measurement (2015), http://www.sciencedirect.com/science/article/pii/S0263224115005916
8. E. Bristol, Swinging door trending: adaptive trend recording, in *ISA National Conference Proceedings* (1990), pp. 749–753
9. J. Lange, F. Iwanitz, T. Burke, *OPC? From Data Access to Unified Architecture*, 4th edn. VDE VERLAG GMBH (2010)
10. G. Monte, V. Huang, P. Liscovsky, D. Marasco, Standard of things, first step: understanding and normalizing sensor signals, in *39th Annual Conference of the IEEE Industrial Electronics Society, IECON 2013* (2013), pp. 118–123, http://ieeexplore.ieee.org/stamp/stamp.jsp?arnumber=6699121
11. S. Mühlbacher-Karrer, L.-M. Faller, R. Hamid, H. Zangl, A wireless steering wheel gripping sensor for hands on/off detection, in *2016 IEEE Sensors Applications Symposium (SAS)*, Apr 2016, pp. 1–5
12. F. Marcelloni, M. Vecchio, A simple algorithm for data compression in wireless sensor networks. IEEE Commun. Lett. **12**(6), 411–413 (2008)
13. ISO/IEC/IEEE information technology—smart transducer interface for sensors and actuators—common functions, communication protocols, and transducer electronic data sheet (TEDS) formats, *ISO/IEC/IEEE 21450:2010(E)*, pp. 1–350, May 2010
14. *RFC 791 Internet Protocol - DARPA Inernet Programm, Protocol Specification*, Internet Engineering Task Force, Sept 1981, http://tools.ietf.org/html/rfc791
15. https://www.google.com/intl/en/ipv6/statistics.html, Aug 2008
16. I. Standardization, Iso/iec 7498–1: 1994 information technology-open systems interconnection-basic reference model: the basic model. Int. Stand. ISOIEC **74981**, 59 (1996)
17. *RFC 2460 Internet Protocol, Version 6 (IPv6) Specification*, Internet Engineering Task Force, Dec 1998, http://tools.ietf.org/html/rfc2460
18. J. Hui, P. Thubert, Compression Format for IPv6 Datagrams over IEEE 802.15.4-Based Networks, RFC 6282 (Proposed Standard), Internet Engineering Task Force, Sept 2011, http://www.ietf.org/rfc/rfc6282.txt

19. IEEE standard for local and metropolitan area networks—Part 15.4: low-rate wireless personal area networks (lr-WPANs), in *IEEE Std 802.15.4-2011 (Revision of IEEE Std 802.15.4-2006)*, pp. 1–314, Sept 2011
20. J. Olson, 6lowpan demystified, Texas Intruments, Technical Report, 2014, http://www.ti.com/lit/wp/swry013/swry013.pdf
21. V. Cerf, R. Kahn, A protocol for packet network intercommunication. IEEE Trans. Commun. **22**(5), 637–648 (1974)
22. J. Postel, User Datagram Protocol, RFC 768 (INTERNET STANDARD), Internet Engineering Task Force, Aug 1980, http://www.ietf.org/rfc/rfc768.txt
23. A. Al-Fuqaha, M. Guizani, M. Mohammadi, M. Aledhari, M. Ayyash, Internet of things: a survey on enabling technologies, protocols, and applications. IEEE Commun. Surv. Tutor. **17**(4), 2347–2376 (2015)
24. R.T. Fielding, Architectural styles and the design of network-based software architectures, Ph.D. Dissertation, 2000, aAI9980887
25. Z. Shelby, K. Hartke, C. Bormann, The Constrained Application Protocol (CoAP), RFC 7252 (Proposed Standard), Internet Engineering Task Force, June 2014, updated by RFC 7959, http://www.ietf.org/rfc/rfc7252.txt
26. A. Banks, R. Gupta (eds.), *MQTT Version 3.1.1*, 10 April 2014, http://docs.oasis-open.org/mqtt/mqtt/v3.1.1/os/mqtt-v3.1.1-os.pdf
27. *An Open Source MQTT v3.1/v3.1.1 Broker*, Aug 2016
28. A. Stanford-Clark, H.L. Truong, Mqtt for sensor networks (mqtt-sn) protocol specification (2013)
29. P. Saint-Andre, Extensible Messaging and Presence Protocol (XMPP): Core, RFC 6120 (Proposed Standard), Internet Engineering Task Force, Mar 2011, updated by RFC 7590, http://www.ietf.org/rfc/rfc6120.txt
30. P. Waher, XEP-0323: Internet of Things - Sensor Data, XMPP Standards Foundation (XSF), 11 2015, http://xmpp.org/extensions/xep-0323.pdf
31. *Data Distribution Servics (DDS)*, Aug 2016, http://portals.omg.org/dds/
32. I. Standard, ISO/IEC 19464:2014 Information technology – Advanced Message Queuing Protocol (AMQP) v1.0 specification, May 2014
33. *STOMP Protocol Specification, Version 1.2*, accessed Aug 2016, https://stomp.github.io/stomp-specification-1.2.html
34. P.M. Barnaghi, W. Wang, C.A. Henson, K. Taylor, Semantics for the internet of things: early progress and back to the future. Int. J. Semant. Web Inf. Syst. 8(1):1–21 (2012), http://dblp.uni-trier.de/db/journals/ijswis/ijswis8.html
35. http://www.sensorml.com, Aug 2016
36. BIPM, JCGM 200:2012 : International Vocabulary of Metrology - Basic and General Concepts and Associated Terms (VIM 3rd edn)
37. BIPM, JCGM 100:2008: Guide to the expression of uncertainty in measurement
38. www.eddl.org, Aug 2016
39. *Semantic Sensor Network Ontology*, Mar 2016, http://www.w3.org/TR/2016/WD-vocab-ssn-20160531
40. *Wolfram Connected Devices Project*, Mar 2016, http://devices.wolfram.com/
41. J. Granjal, E. Monteiro, J.S. Silva, Security for the internet of things: a survey of existing protocols and open research issues. IEEE Commun. Surv. Tutor. **17**(3), 1294–1312 (2015)
42. E. Song, K. Lee, Understanding IEEE 1451—networked smart transducer interface standard—what is a smart transducer? IEEE Instrum. Meas. Mag. 11(2):11–17 (2008), http://ieeexplore.ieee.org/stamp/stamp.jsp?arnumber=4483728
43. R. Matischek, T. Herndl, C. Grimm, J. Haase, Real-time wireless communication in automotive applications, in *Automation and Test in Europe, DATE* (2011), pp. 1036–1041
44. J. Higuera, J. Polo, IEEE 1451 standard in 6LoWPAN sensor networks using a compact physical-layer transducer electronic datasheet. IEEE Trans. Instrum. Meas. **60**(8), 2751–2758 (2011)

Lock-In Amplifier Architectures for Sub-ppm Resolution Measurements

Giacomo Gervasoni, Marco Carminati and Giorgio Ferrari

Abstract Ideally, lock-in amplifiers (LIAs) would be able to measure a minimum signal variation limited by the noise of the front-end amplifier and the filtering bandwidth. On the contrary, a detailed characterization of digital LIAs shows an unforeseen 1/f noise at the instruments demodulated output, proportional to the total signal amplitude. The signal-proportionality and 1/f nature of the measured noise pose a fundamental limit to the LIAs achievable resolution. This limit has been found to be dependent from the instrument maximum operating frequency, from few ppm for LIAs operating up to few hundreds of kHz, to few tens of ppm for LIAs operating up to few MHz or tens of MHz. The additional noise is due to slow gain fluctuations that the signal experiences from the generation stage to the acquisition one. To compensate them, a switched ratiometric technique based on two ADCs alternately acquiring the signal coming from the device under test and the stimulus signal has been conceived. The idea is that both signals should experience the same gain fluctuations, which can be successively removed by means of a division on the outputs of the synchronous demodulation. An FPGA-based LIA working up to 10 MHz and implementing the technique has been realized and results demonstrate a resolution improvement of more than an order of magnitude compared to standard implementations working up to similar frequencies (from tens of ppm down to sub-ppm values).

1 Introduction

Lock-in amplifiers (LIAs) are extensively used for synchronous (phase-sensitive) AC signals detection and measurement in a wide range of scientific fields [1]. Two main reasons justify such a widespread application of LIAs [2]. First, the idea of translating signals in frequency domain (signal modulation), instead of performing DC measurements, arises from the ubiquitous presence of static errors and 1/f noise,

G. Gervasoni · M. Carminati · G. Ferrari (✉)
Politecnico di Milano, Dipartimento di Elettronica, Informazione e Bioingegneria,
Piazza Leonardo da Vinci, 32, 20133 Milano, Italy
e-mail: giorgio.ferrari@polimi.it

B. George et al. (eds.), *Advanced Interfacing Techniques for Sensors*,
Smart Sensors, Measurement and Instrumentation 25,
DOI 10.1007/978-3-319-55369-6_6

given by electronic circuits, devices, sensors or other elements taking part into experiments. The frequency of the measurement is easily selectable using a LIA architecture allowing an operation in the frequency range where the signal-to-noise ratio is maximum. Secondly, the physics of the experiment itself can impose an AC measurement, as in the case of sensors based on impedance measurements, capacitance measurements or resonant phenomena. In both cases, a LIA extracts the amplitude and phase of the AC signal from the background noise and spurious signals with an excellent frequency selectivity, thus combining the advantages of an AC measurement with an optimal signal-to-noise ratio. State-of-the-art lock-in instruments are able to measure voltage signals down to a few nV in a wide range of operating conditions.

For some sensors it is necessary to detect very small variations of an electrical parameter (such as current, resistance or capacitance) with respect to a relatively large baseline value [3–7]. As an example, sensors for counting the number of cells in a liquid [8, 9] or the number of airborne particulate matter [10] can be obtained with an impedance measurement. The fluid under investigation is forced to flow between two electrodes or over them. When no particles are in the volume sensed by the electrodes, the impedance is related to the electrical properties of the fluid and to the geometry of the electrodes. When a particle passes in the sensed volume, the impedance of the sensor changes of a quantity related to the volume of the particle and to its electrical properties. Thus, by monitoring the impedance variations in the time, it is possible to count the number of particles flowing in the fluid. The capability of counting particles with a small (micrometric) size imposes a sensor interface able to sense the correspondingly tiny impedance change superposed to the impedance of the fluid, i.e. a high resolution measurement is required.

Here we discuss the limitations to the best resolution achievable with digital LIAs and how to overcome them. The main noise sources playing a role in measurements with ppm resolution are identified in Sect. 3 and two methods are described to reduce their effects. The first method, described in Sect. 4, is the well-known differential approach where the signal from the sensor is measured with respect to a reference device. By applying to the sensor and to the reference the same stimulus signal and by choosing a reference as similar as possible to the sensor, the baseline signal is effectively reduced, thus relaxing the specification of the LIA in terms of resolution. When a reference path, matching the sensor parameters for all the experimental conditions, is unpractical, it is necessary to design a LIA specifically conceived for high resolution measurements. The second part of the chapter (Sects. 5, 6 and 7) discusses an enhanced lock-in amplifier able to reach sub-ppm resolution in a wide range of operating frequencies without requiring a reference device.

2 Lock-In Amplifier Working Principle

Figure 1 describes a simplified schematic of a lock-in amplifier. A sinusoidal stimulus $A\cos(2\pi f_0 t)$ is generated and applied to a generic Device Under Test (DUT)

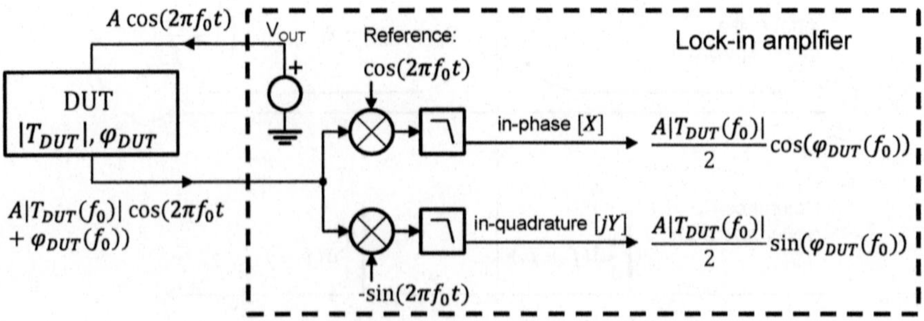

Fig. 1 Simplified scheme of a Lock-In Amplifier

that includes all the elements between the output of the LIA and its input. The sinusoidal stimulus is modified in amplitude and in phase by the transfer function $T_{DUT}(f)$ at the operating frequency f_0. The lock-in amplifier acquires the signal coming from the DUT and multiplies it in parallel by two sinusoidal signals, synchronous with the stimulus signal: $\cos(2\pi f_0 t)$ to obtain the in-phase component and $-\sin(2\pi f_0 t)$ to obtain the in-quadrature one. Both multiplications provide two terms, the first one in DC, while the second one at the double of the working frequency ($2f_0$). The second term is cancelled out through low-pass filtering, while the first terms[1] $X = \frac{1}{2}A|T_{DUT}|\cos(\varphi_{DUT})$ and $Y = \frac{1}{2}A|T_{DUT}|\sin(\varphi_{DUT})$ are respectively proportional to the in-phase and quadrature components of the input signal. From these two outputs, also the signal modulus and phase can be easily obtained: $R = 2\sqrt{X^2 + Y^2} = A|T_{DUT}|$ and $\varphi_{DUT} = \arctan(\frac{Y}{X})$.

For high sensitivity measurements, it is useful to calculate the Signal-to-Noise Ratio (SNR_{OUT}) at the output of the lock-in amplifier. First, from the previous calculations it is possible to define the LIA signal power gain, defined as the ratio between the power of the output DC signal $P(V_{out,DC})$ and the power of the input sinusoidal signal $P(V_{in,AC})$, obtaining:

$$G_{Sign\ Power,X} = \frac{P(V_{out,DC,X})}{P(V_{in,AC})} = \frac{\left(\frac{A|T_{DUT}|\cos(\varphi_{DUT})}{2}\right)^2}{\frac{(A|T_{DUT}|)^2}{2}} = \frac{\cos^2(\varphi_{DUT})}{2} \quad (1)$$

$$G_{Sign\ Power,Y} = \frac{P(V_{out,DC,Y})}{P(V_{in,AC})} = \frac{\left(\frac{A|T_{DUT}|\sin(\varphi_{DUT})}{2}\right)^2}{\frac{(A|T_{DUT}|)^2}{2}} = \frac{\sin^2(\varphi_{DUT})}{2} \quad (2)$$

[1]Here and in the successive calculations $|T_{DUT}|$ and φ_{DUT} are the magnitude and the phase of the transfer function $T_{DUT}(f)$ at the working frequency f_0.

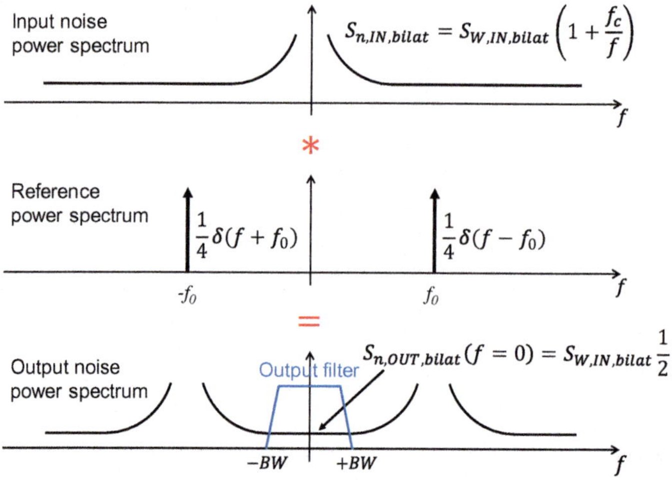

Fig. 2 Frequency shifting and power scaling of the input noise in lock-in amplifiers in the case of $f_0 >> f_c$

In order to compute the LIA output SNR_{OUT}, it is also necessary to calculate how the noise at the input of the LIA is transferred to the output. To do it, an analysis in the frequency domain using the power spectral density of the noise is preferable to a time domain analysis. Looking at the in-phase component extraction chain, the multiplication by the sinusoidal signal $\cos(2\pi f_0 t)$ corresponds, in the frequency domain, to a convolution operation that scales and shifts the power spectral density of the noise at the input around $\pm f_0$ as shown in Fig. 2. The noise power density gain is defined as the ratio between the output noise at $f = 0$ and the input noise at the LIA operating frequency f_0:

$$G_{Noise\ Power\ Density} = \frac{S_{n,OUT,bilat}(0)}{S_{n,IN,bilat}(f_0)} = \frac{S_{n,IN,bilat}(f_0)\frac{1}{2}}{S_{n,IN,bilat}(f_0)} = \frac{1}{2} \tag{3}$$

where $S_{n,IN,bilat}$ and $S_{n,OUT,bilat}$ are the bilateral power spectral density of the input and output noise, respectively. The same result is obtained for the in-quadrature chain.

The power of the noise at the output is obtained by integrating the output power spectral density in the bandwidth of the low-pass filter. Usually the bandwidth is much smaller of the operating frequency f_0 for an effective reduction of the noise and of the spurious harmonics at the output of the demodulator. In this condition the output noise spectral density is commonly assumed white in the bandwidth of the filter irrespective of the frequency dependence of the input spectral density [1, 11, 12]. Consequently, the noise integration in the bandwidth of the low-pass filter is obtained with a multiplication of the spectral density by the equivalent noise bandwidth of the filter. Then, by considering for simplicity the case of the input signal in the in-phase output[2] ($\varphi_{DUT} = 0$), the LIA SNR_{OUT} is:

$$
\begin{aligned}
SNR_{OUT} &= \frac{P_{OUT,Sign}}{P_{OUT,Noise}} = \frac{P_{IN,Sign}\ G_{Sign\ Power,X}}{S_{n,IN,bilat}(f_0)\ G_{Noise\ Power\ Density}\ 2BW} \\
&= \frac{P_{IN,Sign}\ frac12}{S_{n,IN,bilat}(f_0)\ \frac{1}{2}\ 2BW} = \frac{P_{IN,Sign}}{S_{n,IN,mono}(f_0)\ BW}
\end{aligned} \tag{4}
$$

where BW is the equivalent noise bandwidth of the low-pass filter and $S_{n,IN,bilat} = S_{n,IN,mono}/2$ by definition. Equation 4 confirms the LIA efficiency in extracting the information from a sinusoidal signal: SNR_{OUT} is simply given by the ratio between the rms value of the input sinusoidal signal and the input noise density at f_0 integrated in the noise bandwidth BW without a degradation of the SNR at the input of the LIA. This excellent result on the SNR of the measurement is obtained with additional advantages compared to a direct AC measurement of the signal: (i) simplicity in the selection of the operating frequency f_0; (ii) a small bandwidth is feasible independently of f_0 without the implementation of a tuned band-pass filter as required for a standard AC measurement; (iii) both the in-phase and quadrature components are available as DC values at the output of the instrument.

2.1 *Digital Lock-In Amplifiers*

Early LIAs were designed with analog electronics, while digital designs of the low-pass filters emerged in the 1980s. By the early 1990s even the analog demodulators were replaced by high-resolution ADCs and Digital Signal Processors (DSPs). The capabilities of modern DSP-based LIAs in stability, dynamic reserve, and flexibility were revolutionary, making them a mainstay for researchers and engineers. Nowadays, powerful Field Programmable Gate Arrays (FPGAs) are widely used to perform the digital processing required by fast (up to 600 MHz, UHFLI by Zurich Instruments) digital LIAs, whose basic functional scheme is represented in Fig. 3. The FPGA generates the in-phase and in-quadrature reference signals in the digital domain using a Direct Digital Synthesizer (DDS). An output stage based on a digital-to-analog converter (DAC) and a programmable gain amplifier (PGA) generates the output analog signal V_{OUT}. The amplitude of the input signal of the LIA is amplified by a second PGA and low-pass filtered by a wide bandwidth anti-aliasing filter. The digital samples obtained by the analog-to-digital converter (ADC) are demodulated and filtered by the FPGA.

[2]This condition can be easily obtained in a digital LIA by adjusting the phase of the internal reference by a factor $\varphi_{DUT}(f_0)$.

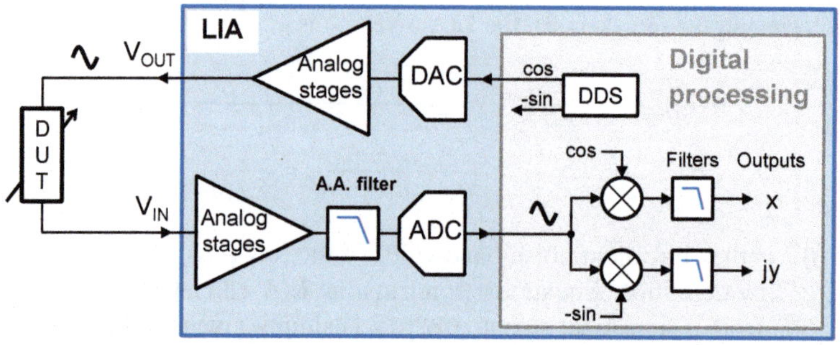

Fig. 3 Basic scheme of a digital lock-in amplifier (LIA) connected to a generic Device Under Test (DUT)

3 Resolution Limit of Digital Lock-In Amplifiers

As shown in Sect. 2, the minimum detectable signal of a lock-in amplifier depends on its input equivalent noise at the working frequency f_0 and the chosen filtering bandwidth BW. State-of-the-art instruments show an input equivalent noise as low as few $\mathrm{nV}/\sqrt{\mathrm{Hz}}$ when used with an input range of few mV.

Ideally, the minimum detectable signal *variation* of a LIA is independent of the signal amplitude given a fixed input range. On the contrary, experimental evidence shows that digital lock-in amplifiers have an unforeseen 1/f output noise proportional to the measured signal [13]. Consequently, when this contribution becomes dominant, the minimum detectable signal variation becomes proportionally dependent to the total signal itself and orders of magnitude greater than the expected minimum detectable signal. Moreover, being the additional noise with a 1/f spectrum, it can not be effectively low-pass filtered, posing a fundamental limit to the lock-in amplifier relative resolution, defined as the ratio between the minimum detectable signal variation and the total signal itself and expressed in ppm. In the following, the relative resolution will be simply called resolution in absence of ambiguity.

Table 1 shows the resolution experimentally measured with different LIAs when the stimulus signal is directly connected to the input of the instrument. The bandwidth was set to 1 Hz and the relative resolution are calculated on a time duration of 100 s.

Results show that the instruments resolution limit can not be improved by increasing the stimulus amplitude or by changing the operating frequency. The LIAs resolution are related to the instrument maximum operating frequency, from few ppm for LIAs operating up to few hundreds of kHz, to few tens of ppm for instruments operating up to few MHz or tens of MHz. The distinctive results obtained with the successively presented Enhanced-LIA (ELIA) are highlighted in red.

Table 1 Experimental comparison of LIA performances. Test conditions are: stimulus signal directly connected to the LIA input, bandwidth of 1 Hz and measurement time of 100 s. The custom LIAs are described in [14, 15]

Model	Maximum frequency [MHz]	Signal amplitude [V]	Measurement frequency [MHz]	Relative resolution [ppm]
Custom LIA [14]	0.1	0.1, 0.3, 1	0.01, 0.05	1
SR830 (Stanford Research Systems)	0.1	0.1, 0.3, 1	0.01, 0.05	12
MCL1-540 (SynkTek)	0.5	1.4	0.1	1.3
SR865 (Stanford Research Systems)	2	0.3	0.5	45
Custom LIA [15]	10	0.03, 0.1, 0.3, 1	0.1, 1	9
HF2LI (Zurich Instruments)	50	0.03, 0.1, 0.3, 1	0.1, 1, 10	39
Enhanced-LIA (Section 5)	10	0.1, 0.3, 1	0.0001-6 6-10	0.6-1 1-3.5

3.1 Lock-In Amplifiers Noise Experimental Characterization

In order to deeply understand the results reported in Table 1, the noise of the state-of-the-art lock-in amplifier HF2LI by Zurich Instruments has been characterized and analyzed. A 1 MHz sinusoidal stimulus ranging from 0 to 1 V has been generated at the instrument output and directly connected to its input (Fig. 4a, top). In all measurements the input and output range have been set at 1.1 V and 1 V respectively. The signal, after synchronous demodulation, is filtered (80 kHz bandwidth) and acquired. The FFT is performed and the results are shown in Fig. 4a.

In the absence of signal (stimulus voltage V_{OUT} set to 0 V) the demodulated signal shows a white noise equal to the noise of the analog front-end at 1 MHz, as expected. However, as the sinusoidal signal amplitude increases, additional 1/f noise proportional to the signal amplitude appears. The same behavior has been observed with all tested LIAs.

As already mentioned, this signal-proportional 1/f noise produces two drawbacks: (i) due to the 1/f spectral distribution, narrowing the filter bandwidth is no more effective in order to reduce the output noise of the instrument; (ii) increasing the signal amplitude does not improve the measurement resolution due to the concomitant noise increase. As a consequence, when dominated by this 1/f noise, the LIA reaches its ultimate resolution limit.

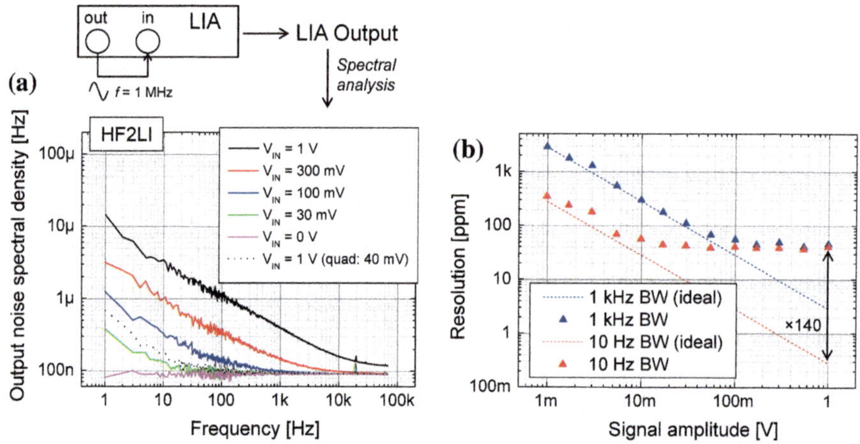

Fig. 4 **a** Noise spectral densities of the demodulated LIA output obtained directly connecting the output terminal to the input of the commercial HF2LI. In addition to the white noise, a 1/f contribution proportional to the signal amplitude is present. The dashed line refers to the noise in the quadrature output with a DC value of 40 mV, **b** resolution obtained with the HF2LI (directly connecting output and input of the instrument and performing measurements 100 s long) for different signal amplitudes and two filtering bandwidths (1 kHz and 10 Hz)

This limitation is confirmed by Fig. 4b, which shows the resolution obtained using the same setup of Fig. 4a and varying the signal amplitude (from 0 to 1 V) and filtering bandwidth (10 Hz and 1 kHz). For comparison, the figure also reports the expected theoretical result considering only the instrument white noise measured for $V_{OUT} = 0$ V. By increasing the signal amplitude, the resolution reaches a plateau of about 40 ppm independently of the measurement bandwidth fixed by the low-pass filter. The resolution degradation due to the additional 1/f noise results remarkable. For example, with a 1 V signal amplitude and 10 Hz low-pass filter, it would be ideally achievable a noise of 280 nV_{rms} and a corresponding resolution of about 0.28 ppm instead of the measured 40 ppm, a factor 140 worse.

3.2 *Flicker Noise Sources Identification*

Benchtop digital LIAs commonly implement a homodyne detection in the digital domain in order to overcome the limitations of analog multipliers in terms of dynamic range, voltage offset and output noise. In particular, a digital demodulator is ideally free from 1/f noise and by operating the digital LIA at frequencies where the noise of the analog circuitry is white, also the noise level at the output of the instrument should be white.

Since the additional 1/f noise is proportional to the signal amplitude, the role of the waveform generator of the LIA has been investigated. The noise added to the

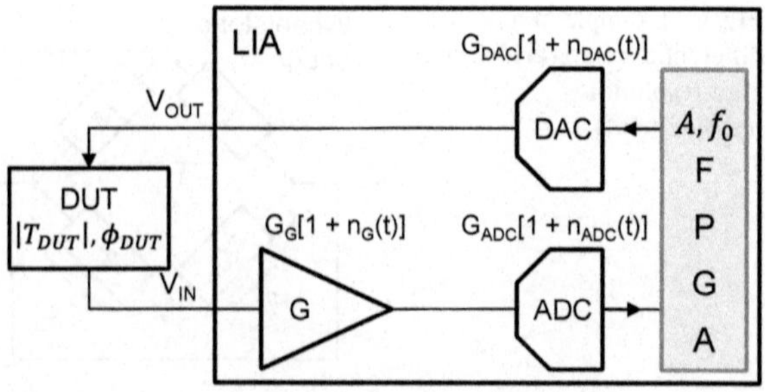

Fig. 5 Visualization of the various multiplicative blocks along the signal path in a digital LIA, affected by slow gain fluctuations which set the resolution limit as detailed in (Eq. 5)

sinusoidal waveform by the output stage of the generator is modulated by the digital demodulator of the LIA giving a non-flicker noise at the output of the instrument, thus cannot explain the obtained results. The phase noise of the signal generator has been discarded as well. In fact, although its contribution is proportional to the signal, it would produce a noise level of the quadrature term higher than the noise of the in-phase term ([1], pp. 71–73). On the contrary, experimental evidence (Fig. 4a) shows a noise of the quadrature term much smaller than the in-phase noise.

A further noise term of a waveform generator is the amplitude noise. A random modulation of the amplitude of the sinusoidal signal is down-converted by the LIA giving an additional noise in the in-phase component (or in the signal amplitude modulus), in agreement with the experimental results. Indeed, every component defining the amplitude of the signal, not only in the generation but also in the acquisition chain, is a source of a random amplitude modulation if its gain fluctuates.

For example, the unavoidable 1/f noise of the reference voltage used by the DAC and ADC results in signal amplitude modulation. In fact, if the DAC reference voltage increases, the corresponding output voltage range results stretched and signal amplitude increases, while if the ADC reference voltage increases, the digital signal processed by the LIA decreases. Similar effects can be originated from internal circuits of the ADC and DAC that define the conversion gains. Regarding the analog stages, the resistors setting the gain of the transfer function are examples of amplitude modulation sources. The intrinsic 1/f noise and temperature fluctuations change the value of resistors and therefore the gain experienced by the signal. Figure 5 shows the effect of these gain fluctuations in a simplified LIA scheme, where the n_{DAC}, n_{ADC} and n_G terms summarize the converters and analog stages different contributions to the fluctuations. The digital processor operates on the digital signal V_{dig} given by

$$V_{dig}(k) = A|T_{DUT}|(1 + n_{DAC})(1 + n_G)(1 + n_{ADC})\cos(2\pi f_0 kT + \varphi_{DUT}) \quad (5)$$

where A is the amplitude of the stimulus voltage and $|T_{DUT}|$, φ_{DUT} are the magnitude and phase at f_0 of the DUT transfer function and T is the sampling period. The amplitude R measured by the LIA is

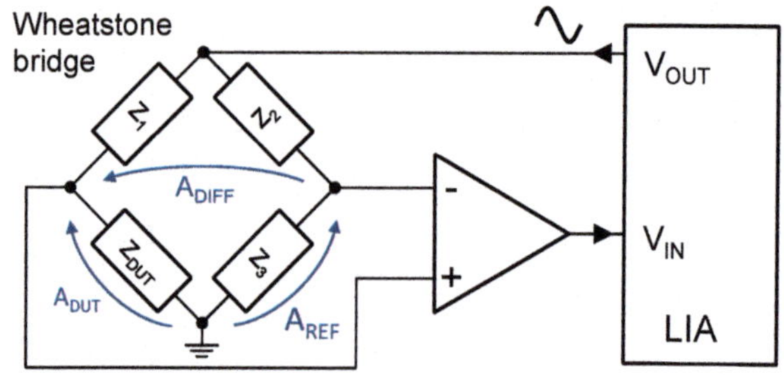

Fig. 6 Example of a differential setup (bridge) for high-resolution measurements

$$R = 2\sqrt{X^2 + Y^2} = A|T_{DUT}|(1 + n_{DAC})(1 + n_G)(1 + n_{ADC}) \tag{6}$$

Thus, the slow gain fluctuations are reflected in slow fluctuations of the amplitude provided by the LIA, setting the resolution limit of the instrument independently of f_0. Whatever signal is measured, it will always be affected by the LIA gain fluctuations, so if these fluctuations are in the order of tens of ppm, it will not be possible to measure the signal with a better resolution.

4 Differential Measurements

The slow gain fluctuations of the LIA add a noise proportional to the signal. This gives a severe limitation of the minimum detectable variation of a relatively large signal. In order to improve the sensitivity in the measurement of a signal variation, one possible approach is to remove the large component of the signal that does not change in the time. This can be performed with differential configuration [13] between the signal to be measured and a reference signal.

An example of a differential scheme for the measurement of an impedance Z_{DUT} is the well-known Wheatstone bridge reported in Fig. 6. The LIA measures the difference A_{DIFF} between the DUT signal and a reference signal. If the bridge is balanced, it gives a null signal A_{DIFF}, making negligible the amplitude-dependent noise produced by the input of the digital LIA. The reference signal is generated starting from the same wave generator applied to the DUT arm, thus its noise is cancelled out by the differential approach as well. The more the reference signal is similar (at any given time) to the DUT signal, the smaller is the differential signal and the less is demanded to the LIA in terms of resolution performance. This widely used technique allows measurements with resolution lower than 1 ppm, as reported for capacitance bridges [16, 17].

The differential solution can also be implemented for current measurements. In Fig. 7 are shown three differential schemes with a TransImpedance Amplifier (TIA) as current sensitive circuit. The difference can be achieved in current mode (i.e. at the virtual ground of the TIA before amplification) or in voltage mode (i.e. by means of a differential INA amplifier connected to the output of the TIA).

Current subtraction can be impl emented in two ways. In Fig. 7a the sensor C_1 is coupled with an identical replica C_2, with the common terminal connected to the virtual ground and the other terminal driven by an inverting buffer in counter-phase with V_{OUT}. If the two impedances are perfectly matched and the inverting buffer is ideal, all the current flowing in C1 flows in C2 and no current is amplified by the TIA, leading to a zero voltage at V_{IN}. Any unbalance between the two arms of the sensor (behaving as a half bridge with differential driving) produces a nonzero current signal that is amplified and demodulated by the LIA. When the technology does not allow the fabrication of a differential sensor, a separate compensation impedance (such as the dummy capacitor C_C in Fig. 7b) could be added of a value close to the expected sensor impedance (C_S in Fig. 7). In this solution, a variable-gain inverting amplifier (G) allows precise tuning of the counter-injected current to obtain perfect cancellation, i.e. accurate matching between the currents in C_C and in C_S. Similarly, solution in Fig. 7c consists in the addition of a full signal chain in parallel to the sensor impedance Z_S, including a dummy impedance Z_C, a second TIA and, if necessary, a phase shifter to be tuned, together with R_{FC}, to adjust the magnitude and phase of this auxiliary signal path. The output voltage of the dummy path is subtracted to the signal path by a differential amplifier, leading to only the residual small signal to be processed by the LIA.

The details of the hardware implementation of solutions (a) and (c) are reported in specific works, focusing on capacitive sensing of dust microparticles in air [10] and on detection of magnetic beads guided by nano-magnetic rails in liquid [18] respectively.

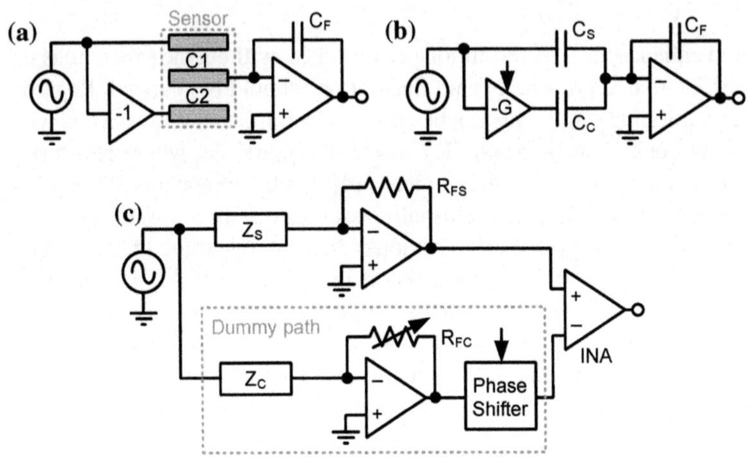

Fig. 7 Differential configurations with a subtraction in current mode: **a** for matched sensors $C_1 - C_2$ and **b** with tunable gain G to match C_C with C_S. **c** Differential subtraction in voltage mode of a compensation signal coming from a parallel dummy path with tunable gain (R_{FC}) and phase (reprinted with permission from [13])

Here the experimental results obtained with these two implementations and using the HF2LI are reported. For solution (a), employing a TIA with capacitive feedback ($C_F = 1$ pF) and a triplet of matched band electrodes, the switch from a single couple of electrodes to a differential configuration allows a reduction of the noise from 30 aF down to 1.1 aF ($V_{OUT} = 1$ V, $f_{AC} = 1$ MHz, $BW = 1$ Hz). The resolution is improved by a factor of 27 reaching ≈ 2 ppm. The performance is still limited by the slow fluctuations of the lock-in amplifier because of the imperfect matching of C1 and C2 (mismatch of 5% limited by lithography of the microelectrodes). For solution (c), the activation of the dummy impedance path and the subtraction at the lock-in input ($V_{OUT} = 50$ mV due to the liquid environment [18], $f_{AC} = 2$ MHz, $BW = 1$ Hz) reduces the conductance noise from 28 nS down to 2.6 nS. In both examples, the reduction of the signal at the input of the lock-in amplifer given by the differential approach has allowed an improvement of the resolution better than an order of magnitude.

Although the differential approach obtains excellent results, the generation of a well-balanced reference signal starting from the same signal source applied to the DUT could be cumbersome. The amplitude and phase of the reference signal has to match the DUT signal for every bias condition, operating temperature and frequency of measurement. Consequently, a calibration process is generally required for every DUT and measurement condition, adding complexity to the system and to the measurement itself.

5 Switched Ratiometric LIA Architectures

In order to enhance the LIA resolution performance, without the restrictions imposed by the differential approach, its gain fluctuations should be reduced. Regarding the analog conditioning stages, this can be obtained by implementing resistors with a low temperature coefficient ($< 5\,\text{ppm/K}$) where necessary, i.e. where resistors set the signal transfer function of the instrument [14]. Such components are less sensitive to temperature variations and are usually associated with a low intrinsic 1/f noise (thus with lower resistance value variations). Instead, regarding the DAC and ADC, which are the major fluctuations contributors in our custom LIAs, there are no studies suggesting a way to reduce the effects of both of them. The simple solution of a common voltage reference would allow the cancellation of the reference effect, but it is not effective for the internal fluctuation sources of ADC and DAC converters such as, for instance, the 1/f noise of a simple buffer at the reference input.

5.1 Digital Switched Ratiometric LIA with a Single ADC

The effect of gain fluctuations added by the DAC and all the generation chain can be reduced using the ratiometric technique, a well-known approach. For example it is a common solution to compensate the light source intensity fluctuations in optical measurements [19]. In LIAs, the acquisition of the stimulus (STIM) signal in addition to the DUT signal (Fig. 8a) and their separate processing with a standard lock-in approach, produces two measured amplitudes (A_{DUT} and A_{STIM}), both proportional to the gain fluctuations of the DAC (and generation chain). Thus, a ratio operation between the two amplitudes gives a value independent of the fluctuations of the generation stage (in Fig. 8 the effect of the analog stages is not considered). The phase can be retrieved as a simple difference between the DUT phase and the STIM phase. Since the phase is less sensitive to a random amplitude modulation, in the following we only analyze the measurement of the amplitude.

This first approach is not effective in many cases because of the gain fluctuations of the two independent ADCs, which are not compensated. In order to also remove their effect, a novel switched ratiometric approach has been conceived. A first single-ADC version, relies on the acquisition of both DUT and the STIM signals with the same ADC, which allows them to be equally affected by the ADC gain fluctuations and consequently to cancel out their effect through a ratio operation. Figure 8b shows the architecture of this switched ratiometric LIA with single ADC. A switch SW periodically changes the input of the ADC alternating the DUT and the STIM signals. The switching frequency is chosen fast enough to assume the same ADC gain for the DUT samples and the STIM samples in a switching period. The digital processor separates the digitized samples in order to reconstruct the two signals in the digital domain, implements the synchronous demodulation of them and finally calculates the ratio of the amplitudes. Although the gain fluctuations are slow (up to few kHz from Fig. 4a), the switching frequency of SW should be chosen fast enough to satisfy the sampling theorem in order to avoid loss of information. Such a condition is difficult to be fulfilled in the case of high frequency digital LIAs, with sampling rate of tens or hundreds of MS/s, in particular for the voltage transients given by the switching of the ADC input.

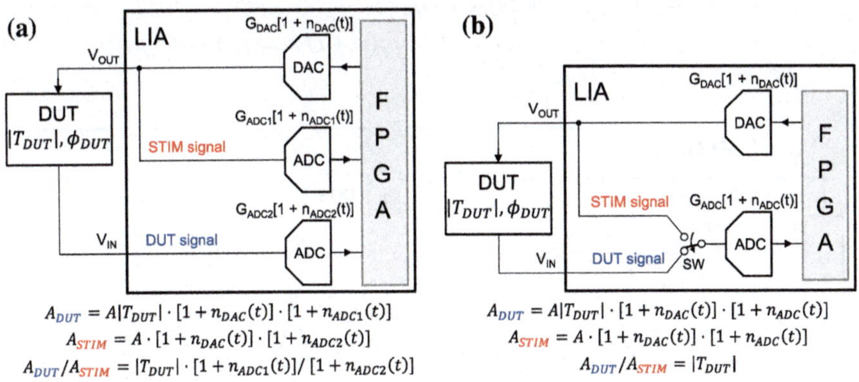

Fig. 8 **a** Simplified scheme implementing a ratiometric measurement. **b** Simplified scheme of a switched ratiometric LIA based on a single ADC

5.2 Digital Switched Ratiometric LIA Based on Two ADCs

In order to reduce the switching frequency required by the scheme of Fig. 8b, an architecture based on two ADCs, as shown in Fig. 9, has been recenty porposed and is patent pending. Two switches SW1 and SW2 are added in front of the ADCs to alternatively acquire the DUT and the STIM signals with each ADC. The signals are reconstructed in real-time using a FPGA to obtain their time evolution with continuity in the digital domain. The following demodulation and averaging are performed as in a standard dual-phase LIA obtaining amplitude and phase of the DUT and STIM signals. The switching frequency f_{SW} of SW1 and SW2 is chosen of few kHz, that is faster than the slow random gain fluctuations of the ADCs. As a consequence, the equivalent gain in a period $1/f_{SW}$ experienced by the two reconstructed signals is the same and is equal to the mean of the two ADC gains. The effect of the gain fluctuations is finally canceled out by performing the ratio between the amplitudes of the DUT and STIM signals. Obviously, the gain fluctuations of the DAC are also included in both signals and therefore removed as well, allowing a high-resolution measurement of the DUT signal.

5.3 Theoretical Analysis

Here the working principle of the presented switched ratiometric scheme is mathematically analyzed. Then, some considerations on the harmonics generated by the switching operation and how to prevent them from causing performance degradation are made.

The two digitally reconstructed signals DUT and STIM (Fig. 9) can be represented by:

$$S_{DUT}(t) = A|T_{DUT}|\cos(2\pi f_0 t + \varphi_{DUT}) \cdot [G_{ADC1}(t)\phi_{SW}(t) + G_{ADC2}(t)(1 - \phi_{SW}(t))] \quad (7)$$

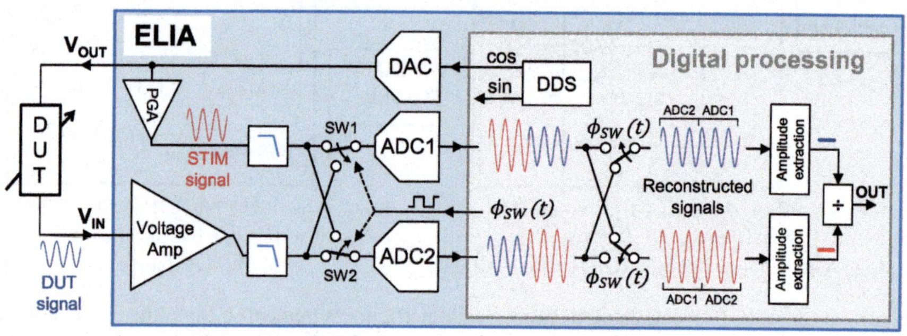

Fig. 9 Architecture of a switched ratiometric LIA based on two ADCs

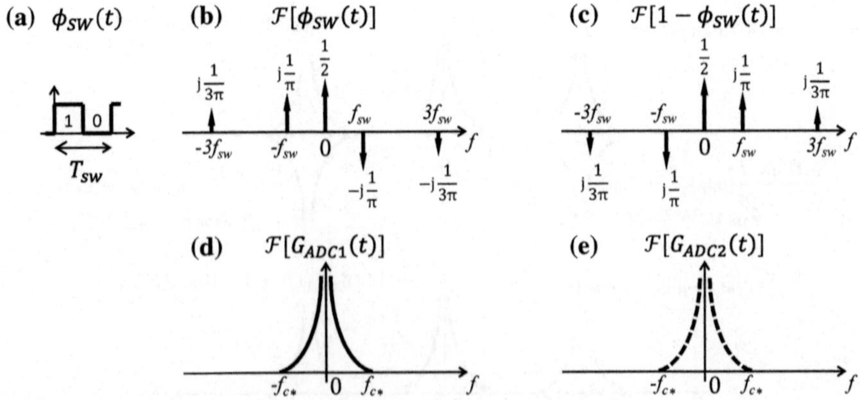

Fig. 10 Fourier transforms of the square wave signals ϕ_{SW} (b) and $1 - \phi_{SW}$ (c) and of the ADCs gains $G_{ADC1}(t)$ (d) and $G_{ADC2}(t)$ (e)

$$S_{STIM}(t) = A\cos(2\pi f_0 t) \cdot [G_{ADC2}(t)\phi_{SW}(t) + G_{ADC1}(t)(1 - \phi_{SW}(t))] \tag{8}$$

where $\phi_{SW}(t)$ is a square wave (0–1) with a duty cycle of 50% and period T_{SW} (as shown in Fig. 10a). $|T_{DUT}|$ is the magnitude at f_0 of the DUT transfer function. For simplicity is assumed $\varphi_{DUT} = 0$ and neglected the noises of DAC, analog stages and quantization.

These reconstructed signals are demodulated in phase and quadrature by the lock-in amplifier. The in-phase component (multiplication by $\cos(2\pi f_0 t)$) is:

$$\begin{aligned} S_{DUT,demod}(t) &= \frac{A}{2}|T_{DUT}|[G_{ADC1}(t)\phi_{SW}(t) + G_{ADC2}(t)(1 - \phi_{SW}(t))] \\ &+ \frac{A}{2}|T_{DUT}|[G_{ADC1}(t)\phi_{SW}(t) + G_{ADC2}(t)(1 - \phi_{SW}(t))]\cos(2\pi 2f_0 t) \end{aligned} \tag{9}$$

$$\begin{aligned} S_{STIM,demod}(t) &= \frac{A}{2}[G_{ADC2}(t)\phi_{SW}(t) + G_{ADC1}(t)(1 - \phi_{SW}(t))] \\ &+ \frac{A}{2}[G_{ADC2}(t)\phi_{SW}(t) + G_{ADC1}(t)(1 - \phi_{SW}(t))]\cos(2\pi 2f_0 t) \end{aligned} \tag{10}$$

These two signals can be studied performing the Fourier transform and exploiting the convolution theorem. Although Eqs. 9 and 10 are characterized by two terms, it is useful to start taking into account only the first one, while the second one, multiplied by $\cos(2\pi 2f_0 t)$, will be treated successively:

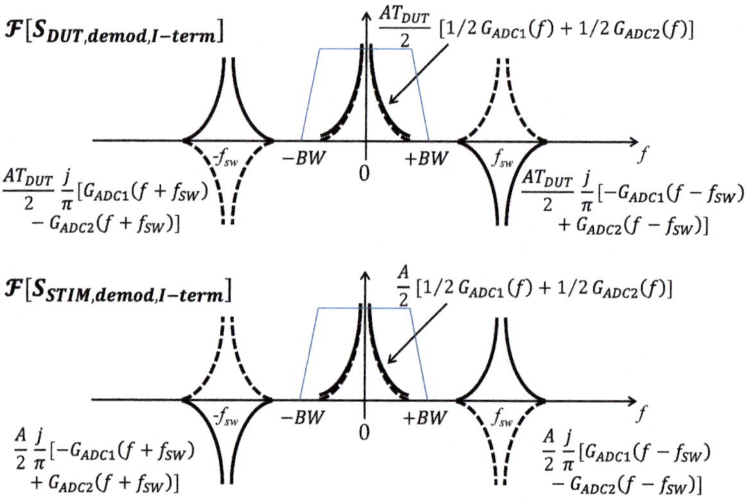

Fig. 11 Spectra obtained from the sum and convolution of the various terms of Eqs. 11 and 12. For graphical clarity the successive harmonics at $\pm 3f_{SW}$, $\pm 5f_{SW}$, etc. are not depicted

$$\mathcal{F}\{S_{DUT,demod,I-term}(t)\} = \frac{A}{2}|T_{DUT}| \cdot \{\mathcal{F}[G_{ADC1}(t)] * \mathcal{F}[\phi_{SW}(t)] + \mathcal{F}[G_{ADC2}(t)] * \mathcal{F}[1-\phi_{SW}(t)]\} \quad (11)$$

$$\mathcal{F}\{S_{STIM,demod,I-term}(t)\} = \frac{A}{2} \cdot \{\mathcal{F}[G_{ADC2}(t) * \mathcal{F}[\phi_{SW}(t)]] + \mathcal{F}[G_{ADC1}(t)] * \mathcal{F}[1-\phi_{SW}(t)]\} \quad (12)$$

In order to compute the Fourier transform of the two signals (Eqs. 11 and 12) it is useful to compute and represent the Fourier transform of their various terms. The Fourier transform of the square wave signals ϕ_{SW} and $1-\phi_{SW}$ are described by a series of Dirac delta functions at frequency 0, $\pm f_{SW}$, $\pm 3f_{SW}$, etc. (as represented in Figs. 10b, c)[3]. The Fourier transforms of $G_{ADC1}(t)$ and $G_{ADC2}(t)$ are sketched in Figs. 10d, e and are characterized by a 1/f noise[4] which becomes negligible starting from a frequency defined f_{c*}, while their white noise is neglected.

In Fig. 11 the spectra obtained from the sum and convolution of the various terms are shown. In DC, both the signals of interest $\frac{A|T_{DUT}|}{2}$ and $\frac{A}{2}$ from the demodulated sig-

[3]Rigorously the Fourier transform should be represented by the real and imaginary part or by module and phase. For simplicity real and imaginary parts are here combined in a single graph.

[4]In Fig. 10 the Fourier transforms of G_{ADC1} and G_{ADC1} are qualitatively sketched with their spectral power density. The analysis in the main text uses the Fourier transform which has the important feature of keeping the phase information, unlike the power spectral density.

nals DUT and STIM respectively, are multiplied by the same factor $\frac{G_{ADC1}(f)+G_{ADC2}(f)}{2}$. This confirms the intuitive idea that the two signals (DUT and STIM) experience an equivalent gain given by the mean of the two ADCs gain. Instead, the terms at $\pm f_{SW}$ and successive harmonics result multiplied by a different factor in DUT and STIM, which in both cases depends on the difference between the two ADC gains.

Finally, the LIA implements a low-pass filter of bandwidth BW. It is evident that it should cut the harmonics at $\pm f_{SW}$ and successive, comprised their tails. The higher harmonics $\pm 3f_{SW}$, $\pm 5f_{SW}$, etc. are not taken into account and represented in Fig. 11 because they are easier to be filtered and smaller respect to the first one. To study the effect of the successive ratio operation, it is useful to work in the time domain, thus the Fourier anti-transform is performed, obtaining:

$$S_{DUT,demod,I-term}(t) = \frac{A}{2}|T_{DUT}| \cdot \{\frac{1}{2}[G_{ADC1}(t) + G_{ADC2}(t)] + \frac{1}{\pi}[G_{ADC1}(t) - G_{ADC2}(t)]\sin(2\pi f_{SW} t)\} \tag{13}$$

$$S_{STIM,demod,I-term}(t) = \frac{A}{2} \cdot \{\frac{1}{2}[G_{ADC1}(t) + G_{ADC2}(t)] + \frac{1}{\pi}[G_{ADC2}(t) - G_{ADC1}(t)]\sin(2\pi f_{SW} t)\} \tag{14}$$

Looking at the terms between brackets we can notice that the first term of both Eqs. 13 and 14 is identical, while the second is not, as expected from the different spectra. Thus, in order to make the ratio operation effective in canceling the effect of $G_{ADC1}(t)$ and $G_{ADC2}(t)$, it is important to filter the harmonics at $\pm f_{SW}$ including their tails. From a different point of view, given a certain filtering bandwidth BW it is important to select f_{SW} in order to make the harmonics (and tails) stay out of the filtering bandwidth BW. Mathematically this necessary condition is:

$$f_{SW} > f_{c*} + BW \tag{15}$$

If this condition is satisfied, it is possible to write:

$$S^{LPF}_{DUT,demod,I-term}(t) = \frac{A}{2}|T_{DUT}| \left[\frac{G^{LPF}_{ADC1}(t) + G^{LPF}_{ADC2}(t)}{2}\right] \tag{16}$$

$$S^{LPF}_{STIM,demod,I-term}(t) = \frac{A}{2} \left[\frac{G^{LPF}_{ADC1}(t) + G^{LPF}_{ADC2}(t)}{2}\right] \tag{17}$$

and performing the ratio between the two signals:

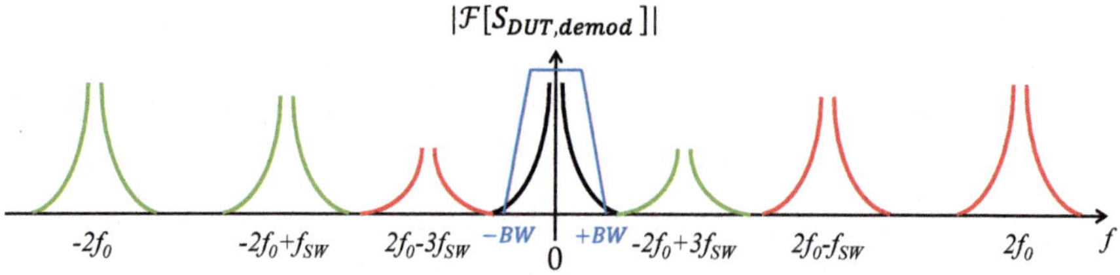

Fig. 12 Useful demodulated DUT signal in DC and harmonics due to the second term (multiplied by $\cos(2\pi f_0 t)$) of Eq. 9 which can fall into the measurement bandwidth $\pm BW$. The same spectral distribution is expected for the demodulated stimulus signal $S_{STIM,demod}$

$$\frac{S^{LPF}_{DUT,demod,I-term}(t)}{S^{LPF}_{STIM,demod,I-term}(t)} = |T_{DUT}| \tag{18}$$

obtaining a measurement of the DUT independent of the two ADCs gain fluctuations.

If the second term of Eqs. 9 and 10 is taken into account, others harmonics appear, in particular at the frequencies $2f_0$, $2f_0 \pm f_{SW}$, $2f_0 \pm 3f_{SW}, \ldots, -2f_0$, $-2f_0 \pm f_{SW}$, $-2f_0 \pm 3f_{SW}$, ..., as shown[5] in Fig. 12. Also in this case it is important that the various harmonics with their tails do not fall into the bandwidth of interest between $\pm BW$. Looking at the bilateral spectrum of Fig. 12, in order to avoid harmonics in bandwidth $\pm BW$ it is necessary to satisfy this condition[6]:

$$BW < 2f_0 - (1 + 2k)f_{SW} - f_{c*} \tag{19}$$

or this other condition:

$$2f_0 - (1 + 2k)f_{SW} + f_{c*} < -BW \tag{20}$$

for each natural number k. At this point it is possible to summarize the conditions in this way:

$$\begin{cases} f_{SW} > f_{c*} + BW \\ \{(1 + 2k)f_{SW} > 2f_0 + f_{c*} + BW \vee (1 + 2k)f_{SW} < 2f_0 - f_{c*} - BW\} \end{cases} \tag{21}$$

In some practical cases, these conditions are considerably simplified. In particular if a narrow bandwidth $\pm BW$ is assumed (much smaller than f_{c*}) and if the switching frequency f_{SW} is much higher than f_{c*}, the conditions becomes:

[5]Only the modulus of the Fourier transform is represented.

[6]The conditions are easily written by considering the components produced by the $2f_0$ term (red harmonics in Fig. 12). The harmonics from $-2f_0$ (in green) are specular and give the same conditions.

$$\begin{cases} BW << f_{c*} \\ f_{SW} >> f_{c*} \\ (1+2k)f_{SW} \neq 2f_0 \end{cases} \quad (22)$$

the last condition must be satisfied with a certain margin (at least the value of f_{c*}) given the performed approximations.

Obviously, working with $f_0 - BW < f_{SW} < f_0 + BW$ is also not a convenient condition. In particular, due to the risk of injecting into the ADCs inputs a feedthrough signal (given by the switches commutations) which exactly falls into the modulated signal bandwidth.

6 Enhanced-LIA Board Realization

An enhanced-LIA (ELIA) instrument based on the switched ratiometric technique has been realized and fully characterized. The prototype comprises a generation channel, two identical acquisition channels and is controlled by a Xilinx Spartan 6 FPGA mounted on a commercial module XEM 6010 by Opal Kelly connected to the board, which also provides an USB interface.

Figure 13 represents the hardware scheme of the realized board. The elements whose gain fluctuations are compensated by the switched ratiometric technique are highlighted in blue. The detailed board design is described here [20].

6.1 Analog Architecture Design Details

Switches Position

The switched ratiometric technique allows compensating the gain fluctuations of the stages following the switches. For this reason, it could be useful to insert the switches as the first stage of the acquisition channels. Nevertheless, in this first implementation, it was decided to implement them just before the ADCs for two reasons: (i) the dominating gain fluctuations source is the ADC; (ii) the samples immediately after the switching of SW1 and SW2 are correctly acquired thanks to the fast differential OpAmp and switches. Figure 14 shows the experimental acquisition and reconstruction of the DUT and STIM signals during the swap between the two ADCs operated by SW1 and SW2. A fast recovery of the correct signals value, after about three acquisition samples (37.5 ns), is shown. With the switches as first stage, the last condition would not be satisfied due to the settling time of the anti-aliasing filter. The noise folding due to the fast differential OpAmp has been taken into account, and its noise results negligible at the end of the acquisition chain.

Not Compensated Gain Fluctuations

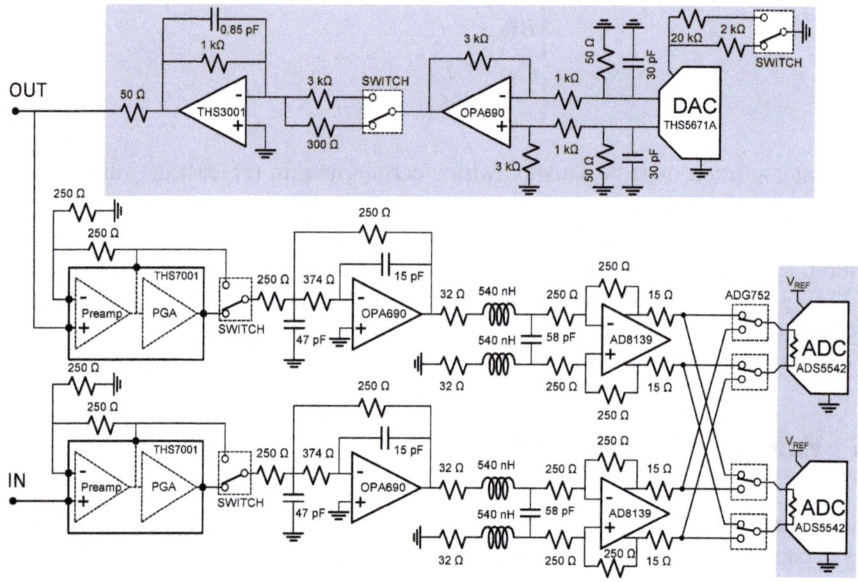

Fig. 13 Detailed hardware scheme of the ELIA instrument. The gain fluctuations of the components in the *shaded boxes* are compensated by the switched ratiometric technique

Fig. 14 Acquisition and reconstruction of the DUT and STIM signals during the swap between the two ADCs. The values are referred to the input of the ADCs, which have a dynamic of ±1.1 V. In this case, after a step of about 300 mV, thanks to the fast settling time of the AD8139, the reconstructed signals reach the correct value (error <1%, i.e. <4 mV) after about three samples (37.5 ns)

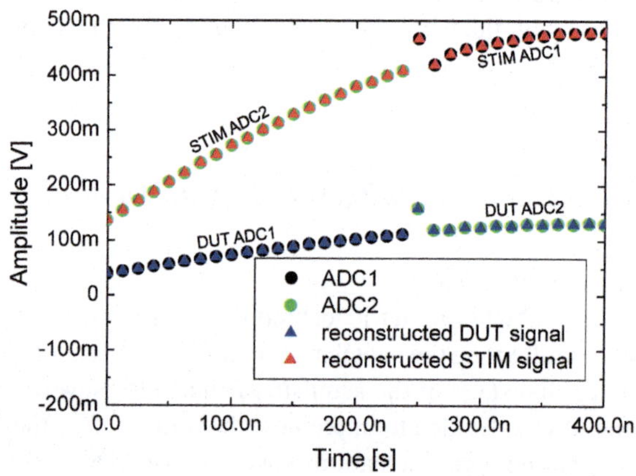

The gain fluctuations of the elements located in the two independent acquisition paths, from the pre-amplifier to the switches, are not compensated by the switched ratiometric technique. In particular, the elements to be considered are: (i) the resistors involved in the channel transfer function; (ii) the PGAs; (iii) the switches before the ADCs.

In order to minimize the resistors effect, non-standard resistors characterized by low thermal coefficient (<5 ppm/K) and low 1/f intrinsic noise, have been employed [14].

Temperature fluctuations also vary the PGAs gain, characterized by a gain temperature coefficient of about 25–35 ppm/K [21]. Experimental results show a small performance degradation when the two PGAs are not by-passed (from 0.6 to 0.85 ppm as shown in the next section). However, it is important to notice that integrated circuits comprising two PGAs with a good channel-to-channel gain temperature coefficient match (few ppm/K) are availabe [22]. This feature will be considered for future implementations.

Regarding the switches network, it is important to consider the voltage divider between the switch resistance and the finite differential input resistance of the ADS5542 ADCs, specified of $R_{ADC} \approx 6.6\,\text{k}\Omega$. A random fluctuation of the ADC input resistance is not an issue itself, because its effect is compensated by the technique. Instead, the resistance of the switches can vary independently each other, causing uncorrelated fluctuations of the two signals acquired by the ADCs. To reduce their effect, low resistance ($R = 15\,\Omega$) switches have been selected. This way, their fluctuations affect the transfer function by a reduced factor $R_{SW}/(R_{SW} + 15\Omega + R_{ADC}) \approx 0.0023$.

6.2 Digital Architecture Additional Modules

Only two additional digital modules, with respect to a standard LIA, are required as shown in Fig. 15. The first is a module to reconstruct in the digital domain the DUT and STIM signals by taking alternatively the samples from the two ADCs coherently with the position of the switches. A delay block allows to set the correct timing for the signals reconstruction considering the acquisition pipeline delay. The second module simply calculates the amplitudes of the DUT and STIM signals and operates the division between them to remove the gain fluctuations.

7 Results

In this chapter the effectiveness of the switched ratiometric technique and the correctness of the previous theoretical analysis are demonstrated. In particular, it is shown that this technique enhances the ELIA instrument resolution limit by more than an order of magnitude (from 9 to 0.6 ppm). After a first technique assessment with a simple resistive DUT, the ELIA instrument has been used to measure a more complex DUT, demonstrating its performance independence from changes in the DUT signal phase and amplitude. Another experiment confirmed the importance of selecting a switching frequency greater than the 1/f noise corner frequency of the single channel as expected.

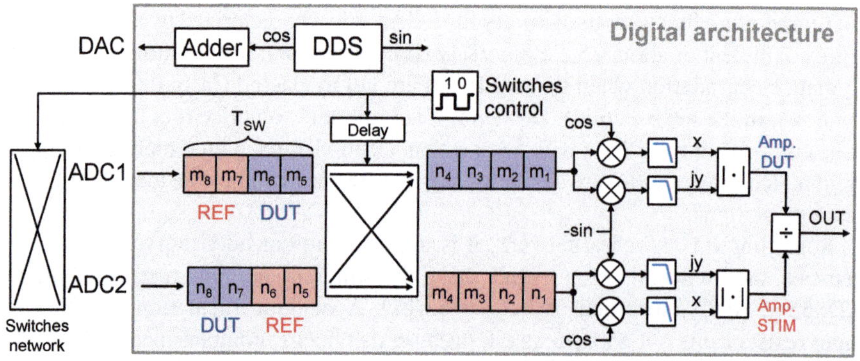

Fig. 15 DUT and STIM signals acquisition and processing chain: after analog-to-digital conversion, the signals are coherently reconstructed taking in consideration the delay between the analog switches and the converted digital words. Successively are processed as in standards LIAs and finally the ratio between the two obtained amplitudes is performed

7.1 Assessment of the Resolution Capability of ELIA

In order to verify the effectiveness of the switched ratiometric technique, a sinusoidal signal of 1 V at a frequency of 1 MHz has been generated and applied to the input of ELIA through a resistive voltage divider as shown in Fig. 16a. In the first experiment, whose results are reported in Fig. 16b, the DUT and STIM signals have been separately acquired with a specific ADC. The demodulated signals are normalized and the fluctuations appear of the same amplitude (about 9 ppm), but not correlated. Although the DAC fluctuations are shared by the two signals, the ones of the two ADCs are indeed uncorrelated, making this simple ratiometric approach ineffective. In a second experiment, the switches have been enabled to perform the switched ratiometric technique. In this case the fluctuations of the reconstructed DUT and STIM signals are clearly correlated (Fig. 16c). Thus, they can be effectively reduced by means of a ratio operation, obtaining a residual uncertainty of only 0.7 ppm in this case (Fig. 16d). Given the obtained results and the fact that the DUT and STIM signals are of different amplitude (0.5 and 1 V respectively), the demodulated output fluctuations cannot be related to an additive noise, but necessarily to gain fluctuations.

As second experiment to evaluate the ELIA performance, Fig. 17a shows the tracking of a time-varying resistance of 250 Ω periodically changed (period of 10 s) of $\Delta R = 1.25$ mΩ, i.e. 5 ppm. The measurement has been performed in three different conditions: (i) using the commercial HF2LI by Zurich Instruments; (ii) using ELIA as a standard lock-in amplifier (i.e. only measuring the DUT signal with a single acquisition channel); (iii) using ELIA with the switched ratiometric technique here proposed (with f_{SW} set to 1 kHz). The signal amplitude applied to the time-varying resistance is 300 mV, the signal frequency is of 3.2 kHz and the filtering bandwidth of 1 Hz. The resolution enhancement of more than an order of magnitude, from 9 ppm to 0.6 ppm, given by the switched ratiometric technique, allows a clear

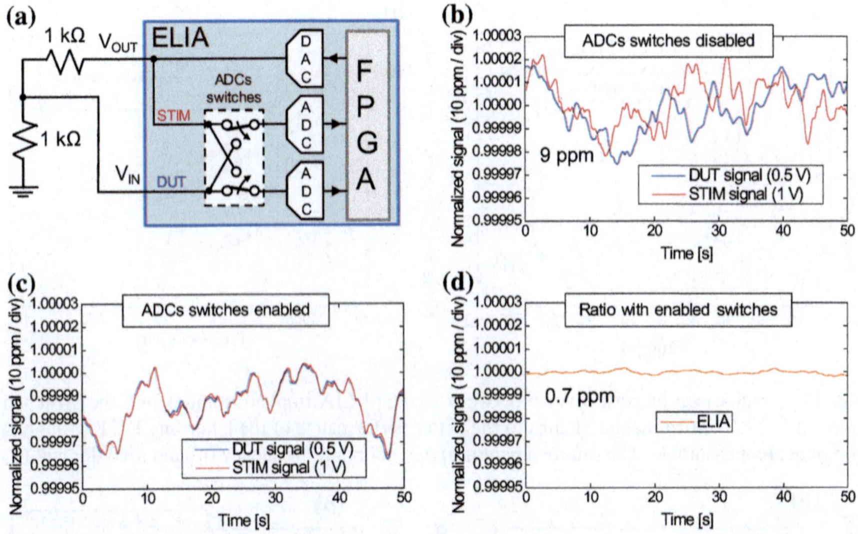

Fig. 16 Demonstration of the technique effectiveness in making the DUT and STIM signals experience the same gain fluctuations in order to allow their cancellation by means of a simple ratio

detection of the tiny (5 ppm) resistance modulation, which is is completely masked by noise in standard LIA implementations, as reported in Fig. 17a. Figure 17b shows the measured noise spectra at the LIAs output in the same three experimental conditions. In order to obtain noise spectra going up to 1 kHz, the LIAs frequency has be increased at 100 kHz, the filtering BW at 1 kHz, selected an internal switches frequency $f_{SW} = 2$ kHz and the modulation of the DUT resistance has been disabled. The spectra clarify that the performance improvement given by the switched ratiometric technique is due to a substantial reduction of the 1/f noise affecting the standard implementations. The technique improves the resolution not only with narrow low-pass filtering (BW $\leq$ 1 Hz), but also with higher filtering bandwidths.

7.2 *Independence from Signal Phase and Amplitude*

Differently from a differential technique, the effectiveness of the switched ratiometric technique is in principle insensitive to the phase and amplitude relationship between the DUT and STIM signals. In order to experimentally prove it, the instrument has been tested with an R-RC network. Figure 18a, b shows the measured transfer function, characterized by a change of a factor 5 of the amplitude and of about $\angle(\pi/4)$ of the phase. In Fig. 18c is shown the measurement resolution at every specific frequency. Sub-ppm performance (< 1 ppm) have been achieved up to about 5 MHz, demonstrating an operation insensitive to the signal phase and amplitude.

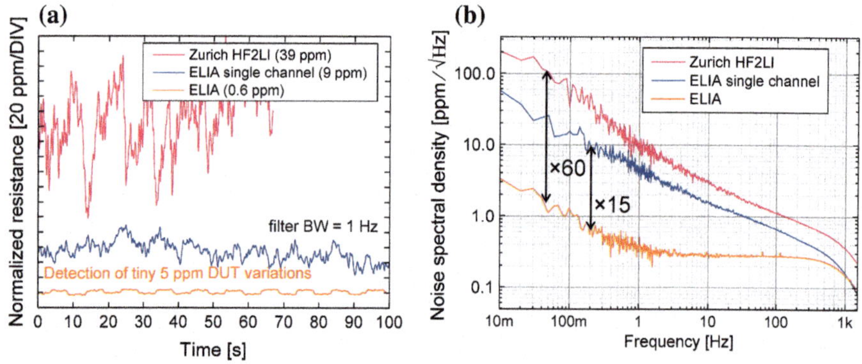

Fig. 17 Comparison of performance between standard LIA implementations and the switched ratiometric ELIA instruments. Thanks to the better performance of the latter tiny DUT variations of 5 ppm are measurable. The improvement is apparent both in frequency (**a**) and time domain (**b**)

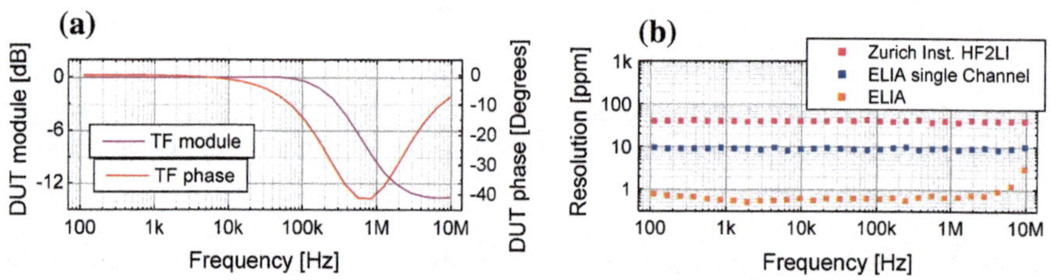

Fig. 18 Measured transfer function TF of the test complex network **a** used to assess the effect of amplitude and phase changes of the DUT signal. The resolution is *below* 1 ppm up to about 5 MHz independently of the DUT signal changes with respect to the STIM one (**b**)

The performance degradation observed for frequencies higher than 5 MHz can be explained by the channel transfer function. At 6 MHz it is decreased by 2% due to anti-aliasing filtering and is starting to rapidly decrease. It means that the capacitors value, characterized by a poor thermal coefficient of 30 ppm/K, becomes increasingly important in defining the transfer function, thus possibly decreasing the resolution performance (about 3 ppm at 10 MHz).

The insensitiveness of the ELIA instrument to the amplitude and phase of the DUT signal with respect to the STIM signal, allows it to operate as a standard LIA, which does not require calibration steps at any new experiment or measurement frequency change, as happen with a differential approach.

7.3 *Resolution and Switching Frequency*

Consistently with the discussion in Sect. 5.3, a relation between the switches frequency f_{SW} and the instrument resolution has been experimentally observed. Figure 19a shows the noise spectrum obtained with the ELIA instrument, but only using a single channel (the DUT channel) as in standard implementations. The stimulus frequency has been set to 500 kHz and the instrument output directly connected to the input. The obtained 1/f noise corner frequency f_c is about 1 kHz. Figure 19b shows the resolution obtained enabling the switched ratiometric technique and varying f_{SW}. The filtering bandwidth was 1 Hz and each measurement was 100 s long. With a switching frequency greater than 1 kHz the resolution flats on sub-ppm values. On the contrary, by lowering f_{SW} the resolution gets worse due to the overlapping with the side harmonics 1/f noise as discussed in Sect. 5.3. The time domain interpretation is that a significant gain fluctuation occurs during the switching period, thus the ADC gain and its fluctuations are not equivalent for the two signals DUT and STIM.

8 Conclusions

The maximum measurement resolution achievable using lock-in amplifiers is limited by an unexpected 1/f noise proportional to the signal to be measured. The measurement resolution cannot be improved narrowing the filtering bandwidth, because the noise is 1/f, neither increasing the signal amplitude, because the noise is signal-proportional, posing a fundamental limit.

The source of this 1/f noise (and of the resolution limit) has been identified in the effect of gain fluctuations of various elements of the generation and acquisition chain, in particular due to the digital-to-analog and analog-to-digital converters.

A differential approach reduces the effects of the gain fluctuations, enabling the measurement of small variations. However, it requires the design of a reference path matched to the signal path for all the experimental conditions. In order to simplify the experimental setup and avoid a calibration step of the reference, a LIA based on two

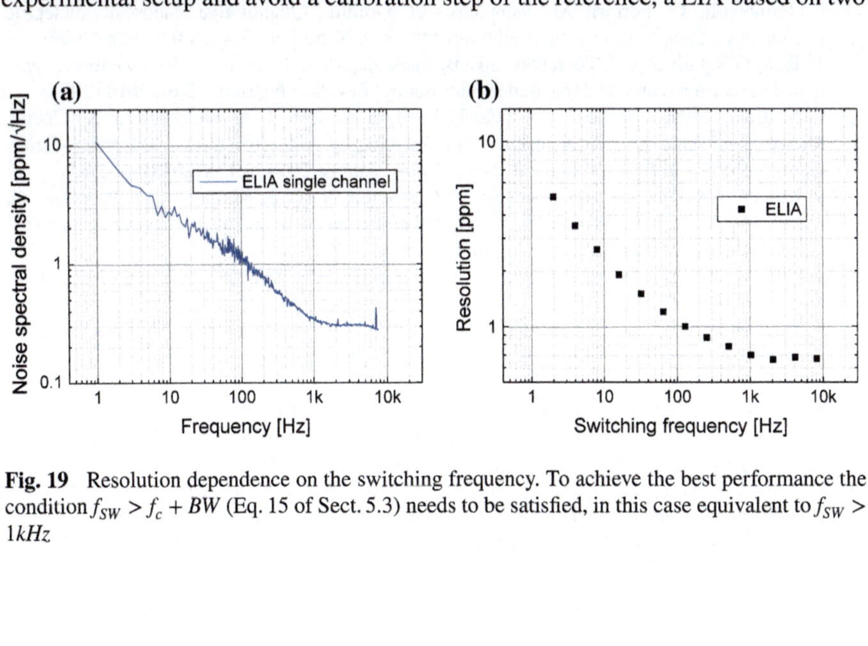

Fig. 19 Resolution dependence on the switching frequency. To achieve the best performance the condition $f_{SW} > f_c + BW$ (Eq. 15 of Sect. 5.3) needs to be satisfied, in this case equivalent to $f_{SW} > 1kHz$

ADCs alternately acquiring the signal coming from the Device Under Test (DUT) and the stimulus (STIM) signal, has been conceived. This enhanced-LIA allows the compensation of the slow gain fluctuations of both DAC and ADC, considerably reducing their effect on the measurements.

Experimental results demonstrate the technique effectiveness. It allows enhancing the instrumentation resolution limit by a factor 15, from 9 to 0.6 ppm, a resolution value which is considerably better than the examined state-of-the-art LIA standard implementations working up to similar frequencies, as shown in Table 1.

The technique does not require additional external elements or accurate case-by-case calibrations, two typical constraints of the alternative differential technique. Instead, it only requires to satisfy some defined conditions when choosing the switching frequency of the internal ADCs switches.

Acknowledgements The authors would like to thank Prof. M. Sampietro for his stimulating interest in this work and for sharing laboratory resources, and Filippo Campoli for the help with the experimental measurements. Financial support by Fondazione Cariplo under Projects Minute and Drinkable and by the European Union under grant agreement no. 688172 (H2020-ICT-STREAMS project) are gratefully acknowledged.

References

1. M.L. Meade, *Lock-in Amplifiers: Principles and Applications* (Peter Peregrinus Ltd., 1983)
2. G. Gervasoni, M. Carminati, G. Ferrari, A general purpose lock-in amplifier enabling sub-ppm resolution, in *Eurosensors 2016, Proceedings* (2016)
3. P. Ciccarella, M. Carminati, M. Sampietro, G. Ferrari, Multichannel 65 zF rms resolution CMOS monolithic capacitive sensor for counting single micrometer-sized airborne particles on chip. IEEE J. Solid-State Circ. **51**(11), 2545–2553 (2016)
4. L. Fumagalli, G. Ferrari, M. Sampietro, G. Gomila, Quantitative nanoscale dielectric microscopy of single-layer supported biomembranes. Nano Lett. **9**(4), 1604–1608 (2009)
5. T. Tran, D.R. Oliver, D.J. Thomson, G.E. Bridges, Zeptofarad (10[sup 21]F) resolution capacitance sensor for scanning capacitance microscopy. Rev. Sci. Instrum. **72**(6), 2618 (2001)
6. C.-S. Liao, M.N. Slipchenko, P. Wang, J. Li, S.-Y. Lee, R.A. Oglesbee, J.-X. Cheng, Microsecond scale vibrational spectroscopic imaging by multiplex stimulated Raman scattering microscopy, in *Light: Science & Applications*, August 2014, vol. 4 (2015), p. e265
7. V. Kumar, N. Coluccelli, M. Cassiniero, M. Celebrano, A. Nunn, M. Levrero, T. Scopigno, G. Cerullo, M. Marangoni, Low-noise, vibrational phase-sensitive chemical imaging by balanced detection RIKE. J. Raman Spectrosc. **46**(1), 109–116 (2014)
8. T. Sun, H. Morgan, Single-cell microfluidic impedance cytometry: a review. Microfluid. Nanofluid. **8**(4), 423–443 (2010)
9. N. Haandbæk, O. With, S.C. Bürgel, F. Heer, A. Hierlemann, Resonance-enhanced microfluidic impedance cytometer for detection of single bacteria, in *Lab on a Chip*, vol. 14, no. 17 (2014), pp. 3313–3324
10. M. Carminati, L. Pedalà, E. Bianchi, F. Nason, G. Dubini, L. Cortelezzi, G. Ferrari, M. Sampietro, Capacitive detection of micrometric airborne particulate matter for solid-state personal air quality monitors. Sens. Actuators A Phys. **219**, 80–87 (2014)
11. M. Carminati, G. Ferrari, M. Sampietro, Attofarad resolution potentiostat for electrochemical measurements on nanoscale biomolecular interfacial systems. Rev. Sci. Instrum. **80**, 12 (2009)
12. SRS, About lock-in amplifiers. Appl. Note **408**, 1–9 (2011)

13. M. Carminati, G. Gervasoni, M. Sampietro, G. Ferrari, Note: differential configurations for the mitigation of slow fluctuations limiting the resolution of digital lock-in amplifiers. Rev. Sci. Instrum. **87**, 2 (2016)
14. G. Gervasoni, M. Carminati, G. Ferrari, M. Sampietro, E. Albisetti, D. Petti, P. Sharma, R. Bertacco, A 12-channel dual-lock-in platform for magneto-resistive DNA detection with ppm resolution, in *IEEE 2014 Biomedical Circuits and Systems Conference, BioCAS 2014—Proceedings* (2014), pp. 316–319
15. M. Carminati, A. Rottigni, D. Alagna, G. Ferrari, M. Sampietro, Compact FPGA-based elaboration platform for wide-bandwidth electrochemical measurements, in *2012 IEEE I2MTC—International Instrumentation and Measurement Technology Conference, Proceedings*, no. 214706 (2012), pp. 264–267
16. M.C. Foote, A.C. Anderson, Capacitance bridge for low-temperature, high-resolution dielectric measurements. Rev. Sci. Instrum. **58**(1), 130 (1987)
17. ANDEEN-HAGERLING, *AH 2700A datasheet*
18. M. Carminati, G. Ferrari, S.U. Kwon, M. Sampietro, M. Monticelli, A. Torti, D. Petti, E. Albisetti, M. Cantoni, R. Bertacco, Towards the impedimetric tracking of single magnetically trailed microparticles, in *2014 11th International Multi-Conference on Systems, Signals Devices (SSD)* (2014), pp. 1–5
19. PerkinElmer, *Low Level Optical Detection using Lock-in Amplifier Techniques* (2000)
20. G. Gervasoni, Novel architecture of digital lock-in amplifier for extremely high resolution measurements. Ph.D. thesis, Politecnico di Milano, 2016
21. Texas Instruments, *THS7001 datasheet*
22. Linear Technologies, *LTC6911-2 datasheet*

Biomedical Sensors and Their Interfacing

Rajarshi Gupta

Abstract Biomedical sensors provide an interface between the electrophysiological activity of different body organs and the instruments those facilitate an in-depth analysis of the pathological conditions. They constitute the most important part of a biomedical health monitoring system. Electrocardiogram (ECG), Photoplethysmogram (PPG), Blood pressure (BP) and respiration are the most fundamental cardiac and respiratory function indicators in a human body. This chapter is aimed to provide a brief overview of these sensors and interfacing techniques for digitized acquisition.

1 Introduction

Many organs in the human body, such as the heart, the brain, muscles, and eyes, have underlying electrophysiological activities and involve generation of electric fields and small biopotentials. Measurements of these and other bio-electric signals can provide information on normal or pathological functions of the organs. The fundamental objective of any medical signal acquisition system is to faithfully capture the information which represents the physiological event or phenomena using a suitable sensor. Over the years, various types of medical sensors have been developed using the principles of materials science, electrical engineering, and electronics science. Historically, the earliest uses of these sensors were confined to clinical use, mainly in hospitals and healthcare clinics. However, with the ever-changing societal demands (busy work schedule of common people) as well as application of medical technology to non-medical domain (space application, sports physiology, military applications, smart home systems to name a few), there were requirements to access medical data remotely, using portable systems (handheld gadgets) and also at the mobile condition of the subject (patient). These demanded

R. Gupta (✉)
Department of Applied Physics, Instrumentation Engineering,
University of Calcutta, Calcutta, India
e-mail: rgaphy@caluniv.ac.in

B. George et al. (eds.), *Advanced Interfacing Techniques for Sensors*,
Smart Sensors, Measurement and Instrumentation 25,
DOI 10.1007/978-3-319-55369-6_7

contribution from embedded technology, information and communication technology, computer science into the broad area of biomedical engineering. The technological challenge in this area in the current century is primarily two fold. First, development of lightweight, miniature and wearable sensors so that medical signals can be acquired from a patient even during his normal daily activity. Second, use of wireless senor networks, embedded technology, and computing tools for continuous monitoring of patients for early detection of pathological events from medical signal(s). Among the various medical signals, cardiovascular signals are considered the most important ones. The frequently used cardiovascular signals, viz., Electrocardiogram (ECG), peripheral pulse, blood pressure (BP) and respiration provide an in-depth assessment of the organs like heart, lungs, and circulatory system. The scope of this chapter is to discuss the basic sensing principles and interfacing of these cardiovascular and respiration sensors for their digitized acquisition in a processor based system. These signals are also functionally and physiologically interrelated in the sense that a pathological symptom obtained in one signal can be sometimes clinically correlated to the other one(s). The layout of the chapter is as follows: Sect. 2 discusses the fundamental bioelectric phenomena underlying the origin of physiological signals and requirements of any medical signal acquisition system. In Sect. 3, basic origin of the cardiovascular and respiratory signals is discussed in brief. Section 4 is dedicated for signal conditioning circuits and interfacing of these signals, followed by a discussion on wireless biomedical instrumentation in Sect. 5. Section 6 highlights two recent developments of modern medical sensors. For the curious and enthusiastic readers, end of chapter references will help to get comprehensive information on the pertinent areas in detail.

2 Basic Aspects of Biomedical Data Acquisition Systems

Biomedical signals, according to their sources and acquisition principles, are divided into two broad classes, viz. (a) endogenous signals and (b) exogenous signals. The endogenous signals arise from natural physiological processes (hence, also called physiological signals) and are measured within or on living creatures. Example of such signals are EEG, ECG, EMG, EOG, blood glucose; etc. The exogenous signals arise from the modulation of an externally applied energy or radiation onto a physiological organ. The external energy or radiation may be of suitable wavelength, X-ray, ultrasound energy, magnetic field etc. Examples of such signals are PPG, SpO_2, blood pressure etc. While endogenous signals can be acquired invasively or non-invasively, exogenous signals are mostly drawn non-invasively.

The sources of nearly all physiological signals may be attributed to the transients in intercellular membrane potentials known as resting membrane potentials (RMP) in excitable living organisms [1, 2]. The electrophysiological basis of generation of such tiny potentials is due to the fact the living cells undergo a reversible change due to an external stimulation. The asymmetric distribution of Na^+, K^+, Ca^{++}, and Cl^- ions which are the principal constituents of intra and extra cellular fluids and semi-permeable nature of the membrane gives rise to the RMP. The RMP of most human living cells is around −20 to −70 mV while measured inside to outside of the cell membrane. The steady state RMP was represented by Goldman–Hodgkin–Katz equation [3]:

$$V_{mo} = -\frac{RT}{F}\left\{\frac{[Na^+]_i P_{Na^+} + [K^+]_i P_{K^+} + [Cl^-]_i P_{Cl^-}}{[Na^+]_e P_{Na^+} + [K^+]_e P_{K^+} + [Cl^-]_E P_{Cl^-}}\right\} \quad (1)$$

where, T is the Kelvin temperature, R is the MKS gas constant (8.314 J/(mol K)), F is the Faraday number, 96,500 Cb/mol, and P_X is the permeability for ion species X.

This 'polarized' behaviour cell of the cell gets changed due to application of short duration external stimulus. An 'ion pump' theory describes the movement of ions through the cell membrane which makes the cell potential rapidly reversed for some time, called 'depolarization' of the cell. After the stimulus is withdrawn, it slowly recovers to the initial polarized state through a slow 'repolarization' and occasional after-potentials. The recording of the membrane potential during these reversible changes is called action potential of the cell, the magnitude, shape and duration of which depends on the type of cell, like nerve, muscle or cardiac [3].

The recording of any physiological signal is the collective representation of these tiny action potentials over space and time at predefined body location(s). The basic aspects of any medical signal acquisition system are [4]:

(a) Sensing of very weak amplitudes, since most biopotentials appear in the order of millivolt or microvolt signal at the body surface. This requires that the sensing electrodes should be sensitive enough to capture these weak biosignals.
(b) Removal or minimization of unwanted signals, collectively named as noise and artifacts, which can be external (interferences from power line, radiofrequency sources, amplifier noise etc.) or internal (physiological signals from other body tissues). If not minimized, these can corrupt the desired medical signal completely, making it either unacceptable for clinical diagnosis or prone to wrong diagnosis.
(c) Safety of the subject (patient) from (i) electrical power supply of medical instrument; (ii) radiation energy imparted on the patient for exogenous signals (for imaging applications), and (iii) any other toxic materials or effects in the process of signal acquisition.
(d) Presentation of the signal in an acceptable format or standard that is suitable for interpretation by the medical experts. This mainly refers to the protocols or standard lead configurations for acquisition of different medical signals, for example, 12-lead ECG for Electrocardiograms.

3 Cardiovascular and Respiratory Signals: Origin and Basic Principle

Cardiovascular and respiratory signals arise from the pumping actions of the heart in conjunction with the circulatory systems that supply the blood to each body cell and, inhalation and exhalation of air to purify the blood through pulmonary circulation, respectively. The signals which represent proper functioning of these two systems are, ECG, peripheral pulse (also known as digital volume pulse or Photoplethysmogram), saturated oxygen in blood (SpO_2), systolic and diastolic BP and respiration.

3.1 Electrocardiogram Signal

The Electrocardiography (EKG or ECG) is the prime investigation tool for preliminary level cardiac functions. It represents time averaged representation of the tiny electrical potentials generated at the sinoatrial (SA) node of the heart and their propagation along specialized conduction fibers over the heart's surface, as recorded at predefined body positions. As shown in Fig. 1, a typical ECG waveform consists of P, QRS and T waves, which represent sequential depolarization and repolarization of the atrial and ventricular myocardium cells [5]. Among these, the P-wave is represented by atrial depolarization during which the electric pulses move from SA node to atrio-ventricular (AV) node, where the conduction is delayed to form the equipotential PQ segment. This is followed by spreading of electric conduction pulses through bundle of His, left bundle branch and right bundle branch along the ventricular region, representing its depolarization to generate QRS complex which suppresses the simultaneous atrial repolarization. A little time pause here gives rise to equipotential ST segment, after which ventricular repolarization

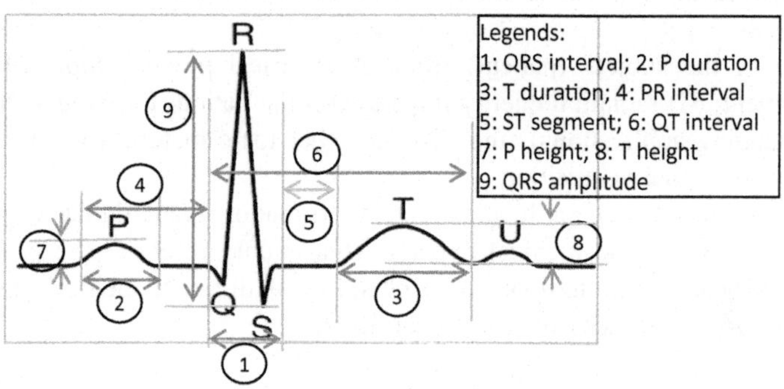

Fig. 1 A typical representative ECG beat in important clinical features (printed from "Hybrid Encoding for Real Time Compressed Electrocardiogram Acquisition", Measurement, Vol. 91, Sept. 2016, (Elsevier Sc.) by P. Bera et al. with permission)

starts with conduction pulses spreading over Purkinjee fibers, with generation of T wave. The important clinical features (marked as 1–9 in Fig. 1) that describe the cardiac function lies in the different time durations and amplitudes of constitutes waves and wave segments. These provide a set of quantitative and qualitative (curvature of a wave, elevation or depression of a segment etc.) metrics which provide a foundation of objective assessment by a cardiologist [6, 7]. A lot of cardiac disorders like arrhythmia, ectopic beats, myocardial infarction, hypertrophy, bundle branch block etc. can be detected by visual analysis or computerized analysis of such features or their suitable combination [8].

'12-lead ECG', an internationally accepted standard for recording of ECG in clinical practice, represents body positions and referencing system to record ECG. Left arm, right arm and left leg are used for recording three standard bipolar leads I, II and III and three unipolar augmented limb leads aVL, aVR, and aVF. Six unipolar chest leads v1, v2, v3, v4, v5 and v6 are attached to different intercostals spaces on the rib-cage. For unipolar leads, the reference is taken as the Wilson Central Terminal (WCT). With the advent of computerized recording and analysis, these 12—lead traces can be simultaneously acquired and analysed to provide preliminary level assessment of cardiac abnormality.

3.2 Peripheral Pulse Signal

Peripheral pulse signal can provide vital diagnostic information on blood circulatory functions through the veins and arteries spread over the whole body. They represent blood volume changes in the micro vascular bed of tissue at the peripheral body parts (like fingertips, earlobes, and forehead) with respect to a cardiac cycle. The most popular technology to acquire such signals is photoelectric plethysmography or, Photoplethysmography (PPG) [9, 10]. Here, an optical source-detector combination attached to the body part is used to illuminate a small portion of skin and sense the change in blood volume information based on difference of absorbance of the light radiation between blood-full and blood-less skin at near infrared (IR) wavelengths. The sensor electronics projects this change in the output voltage. A typical PPG waveform shows a steady (DC) component (representing thermoregulation and autonomic functions), and a pulsatile component (representing blood volume changes). As shown in Fig. 2, the rising part of the pulsatile component is called anacrotic phase, and it represents ventricular systole (contraction), where the supply of blood to the peripheral capillary increases. It is characterized by the systolic peak (point SP), after which the falling part, named catacrotic phase starts, representing ventricular diastole (expansion). For healthy arterial condition, a diastolic peak (point DP) reflects momentary rise in blood supply due to reflection of pulse wave pressure from the closed aorta. Like ECG, PPG also provides a set of objective measurements based on its morphological features, as indicated in Fig. 2. A multitude of cardiovascular parameters like heart rate, cardiac output, respiration, endothelium functions, arterial compliance, and

Notation	Description
amplitude (F-SP)	Systolic amplitude (SA)
duration (h1-h2)	Pulse width (PW)
duration (SP-DP)	Peak to peak time (PPT)
duration (F-SP)	Crest Time (CT)
amplitude (F-DN)	Dicrotic notch amplitude (DNA)
amplitude (F-DP)	Diastolic peak amplitude (DA)
Ratio (SA/DA)	Augmentation index (AI)

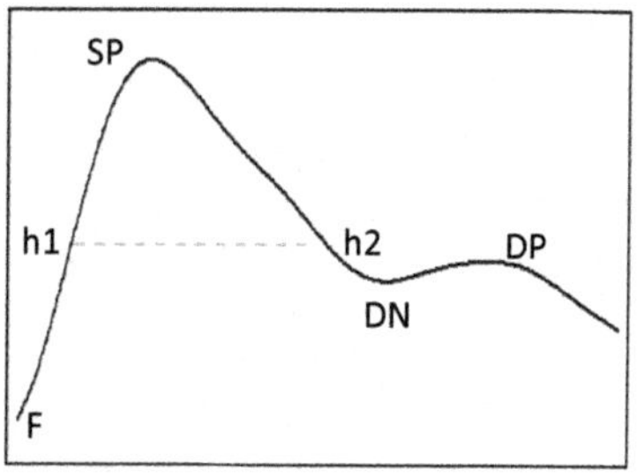

Fig. 2 A finger pulse waveform representing main clinical features

blood pressures can be correlated from the PPG and ECG-PPG combination [11–16]. Over the last two decades, an increase in interest in PPG technique by the biomedical research community has raised its use for various clinical monitoring purposes due to its non-invasive acquisition, low cost instrumentation, single site measurement and favourable sensor configuration [17, 18].

Notation	Description
Amplitude (F-SP)	Systolic amplitude (SA)
Duration (h1-h2)	Pulse width (PW)
Duration (SP-DP)	Peak to peak time (PPT)
Duration (F-SP)	Crest time (CT)
Amplitude (F-DN)	Dicrotic notch amplitude (DNA)
Amplitude (F-DP)	Diastolic peak amplitude (DA)
Ratio (SA/DA)	Augmentation index (AI)

3.3 *Respiration Signal*

The objective assessment of lungs, airways and chest wall functionality can be made by direct or indirect measurement of various variables which are associated with two key events in the respiratory system, viz., transport of the gas from environment and the alveoli via the branching airway tree, and, diffusion of the gases across the physical barrier separating the alveoli from the pulmonary capillaries. The various measurements under the respiration system can be classified under the following heads: (a) gas pressure; (b) gas flow and volume; (c) concentration and partial pressure of gases; (d) pulse oximetry and capnography; (e) lung mechanical function (spirometry and body plethysmography) and, (f) gas exchange [19]. In this chapter, discussion is confined to airflow sensing and breathing rate measurements using plethysmography.

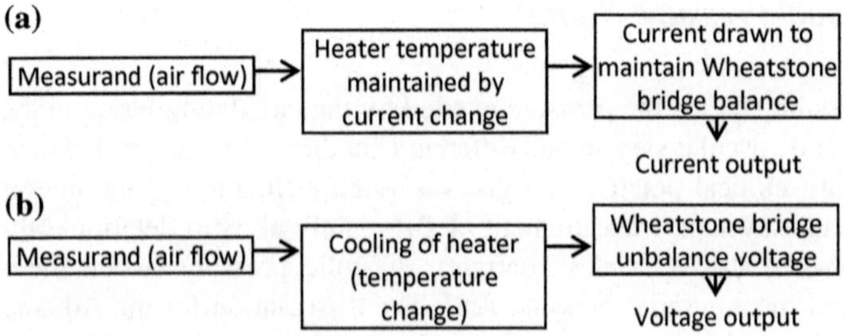

Fig. 3 Basic principle of thermal convection flow meters: **a** Constant temperature mode. **b** Constant current mode

Thermal convection flow meters employing sensing elements like (electrically heated) metal wires, metal films, and thermistors are used to measure gas flow rate through the nasal cavity and provides indirect measurement of respiration in terms of lung volume or its changes [20]. These thermal flow sensors are based on convective heat exchange that takes place when the fluid flow passes over the sensing element (hot body). They are connected in a full Wheatstone bridge and normally designed to operate either in constant temperature mode or, in constant current mode. The basic sensing principle is shown in Fig. 3. The airflow due to breathing activity takes away the heat from the sensing element. In constant temperature mode, the heater element is maintained at constant temperature by adjustment of current from the source, whereas in constant current mode, the heater is supplied with constant current and bridge unbalance provides a measure of air flow rate.

A popular form of respiratory air flow sensing is employing **microwave Doppler radar** [21–23] which avoids sensor attachment, or special clothing, and provides additional advantages like robustness against environmental factors, interference from other sensor signals etc. The sensor operates by transmitting a radio wave (typically in the GHz range) signal and receiving the modulated version of the signal due to quasi periodic motion of the chest wall/abdomen due to inhalation, exhalation and the pause in between.

Plethysmography is a general technique used for measuring respiration, or breathing effort by change in volume (abdominal or chest sectional area). The basic configuration employs change in elastic belt tension (piezo or strain gauge as sensing element), electrical impedance (change in current between two electrode positions) and inductive type (frequency change in a RF signal looped in a chest belt) [24–26]. Among these, chest belt type respiratory inductive plethysmography (RIP) sensors are popular and often taken as standard to calibrate other respiration sensors. In RIP [27], a low current is looped through a coil would around the chest and activated by a high frequency (~kHz) current source. As a result of quasi-periodic motion of chest and abdomen, the body sectional (chest) area changes and so changes the resultant magnetic field. This can be measured by change in frequency in the loop current.

3.4 Blood Pressure Signal

Blood pressure (BP) is the pressure exerted by the circulating blood in the walls of the peripheral vascular system and different chambers of the heart. BP measurement is a standard clinical practice and gives a rough estimation of the integrity of the cardiovascular system. Measurement of BP mainly aims to determine the systolic pressure (when the ventricles contract), diastolic pressure (when the ventricles expand) and mean arterial pressure [28]. The Association for the Advancement of Medical Instrumentation (AAMI) demands a pressure range of –30 to 300 mmHg for a blood pressure transducer. Additionally, it should not be damaged with an overpressure in the range of –400 to 4000 mmHg.

There are several direct and indirect methods for measuring BP. Direct BP measurements involve minor surgical procedure to couple the (body) site pressure to an external sensor, and categorized into extravascular and intravascular sensors. These are applied for pre-and post-surgical patients under hospital set up. Modern electrical type direct BP measurement methods adopt 'catheterization' techniques, which detect deflection of an elastic member by the exerted pressure. The secondary sensing principle may utilize strain gage, piezoelectric, capacitive or fiber optic principles and mostly of intravascular type.

In routine clinical check-ups, however, indirect BP measurements (non-invasive) employ an inflated cuff by air purging for occlusion of a blood vessel (normally brachial artery). In general, the cuff is inflated above the tentative systolic BP and slowly bled off at a rate of 2–3 mm/Hg. With gradual release of cuff pressure, blood leaks through the occluded artery. Based on the sensing principle, the measurement methods are categorized into auscultation, palpation, flush and oscillometric techniques [4, 29, 30]. The auscultation method uses a stethoscope to hear the appearance and disappearance of different phases of Korotkoff sounds with the gradual decreases of cuff pressure. Several improvisations of the auscultation methods facilitate automated detection of systolic and diastolic BPs by employing secondary transducers like piezo crystals, ultrasonic, photoelectric, and electroacoustic transducers. The palpation of the radial artery with gradual release of the cuff pressure can be manually sensed to detect the systolic pressure only. The oscillometric method measures the amplitude of oscillations that appear in the cuff pressure signal created by the arterial wall as the blood leaks through the brachial artery. This method can provide systolic, mean and diastolic (using algorithm) pressures. In the last two decades, several sensor designs and computational methods have been proposed for indirect BP measurement using ECG and PPG [31–33]. These have opened up a new domain of non-invasive [34], continuous and ubiquitous cardiovascular measurements.

4 Sensor Configurations and Interfacing of Cardiovascular and Respiratory Signals

Most of the real world biomedical signals are weak in amplitude (order of millivolt or microvolt) and so, require amplification, scaling, isolation and filtering before being finally acquired in a collecting device. These functions are collectively known as signal conditioning. The biomedical sensors are interfaced with the signal conditioning circuits at the first stage of recording. Among these, filtering is used to pass the clinical bandwidth of the signal and discard any noise from the raw input. Isolation is a special function used for endogenous medical signals (like ECG, EMG). The signal conditioning circuit configuration depends on the biomedical signal being acquired and provides an amplification from $10–10^6$ to the raw input.

4.1 Electrocardiogram Signal

The ECG is measured from the predefined positions of human body. However, for its proper measurement electrodes are required which convert the ionic current (inside the body) to the electronic current in an external circuit, i.e., the measurement electronics. Apart from that, ECG signals being weak in magnitude (±3 mV peak), they are inherently corrupted by other physiological signals (like muscle noise) and external unwanted interferences, collectively known as 'artifacts'. Hence, proper filtering, amplification and isolation are required for its proper recording, presentation and safety of the patient.

The block diagram of a typical ECG signal conditioning is shown in Fig. 4, which can generate a unipolar voltage. The lead selector circuit can be manually operated or controlled by the data acquisition (DAQ) board.

The objective of the isolation circuit is to provide galvanic isolation (1000 MΩ or more) between the patient body and the signal conditioning to prevent any accidental current flowing through the patient body. The total gain of the amplifier is around 2000, distributed over a pre-amplifier and an instrumentation amplifier (INA). The first stage, pre-amplifier provides a low gain, typically 10–20. The filter circuit mainly discards high frequency noise and passes the clinical bandwidth of 0.05–100 Hz to the output. The INA provides a gain of 100–200. After INA, the

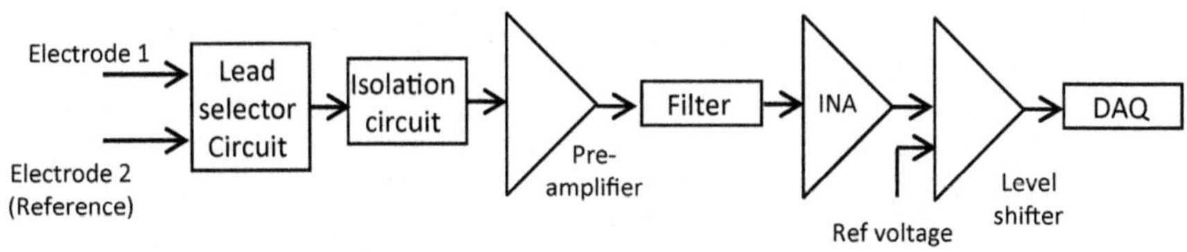

Fig. 4 Block schematic of a ECG amplifier circuit

final level shifter is used to get a unipolar voltage 0–5 V. In the following sections, the components of the ECG signal conditioning are elaborated.

4.1.1 Electrodes

Electrodes facilitate conversion of the ionic current within the body tissue to an external current using a metal-electrolyte interface. Electrodes are categorized into polarized and non-polarized according to their electrical behaviour. A detailed analysis of the biopotential electrodes are available in [2, 4]. Silver-Silver Chloride electrodes provide the most stable performance. They are made of silver disks coated electrolytically by silver chloride. The electrolyte (gel) additionally provides protection against slippage of the electrodes. Commercially, two types of electrodes are used in ECG recording, viz., reusable and disposable.

Figure 5 shows common electrodes used for ECG recording. Here, (a), (b) and (c) are metal reusable electrodes used in common clinical settings. Materials used for these types of electrodes include German silver (a nickel-silver alloy), silver, gold, and platinum. Among these, (a) is a suction cup electrode for precordial or chest leads. The (b) type is used as limb electrode. The type (c) is a general metal disc electrode which is normally used in long-term ECG recording. A lead wire is soldered into the back surface for electrical connection to recorder. A thin insulating material epoxy or polyvinylchloride protects the conducting part from lead wire. (d) Shows another reusable Silver-Silver Chloride electrode attached to the skin by non-allergic adhesive tape, and provides high quality measurement. (e) and (f) are disposable electrodes used in research due to their improved design. Configuration (e) provides a secure arrangement by attaching a large foam pad to the electrode

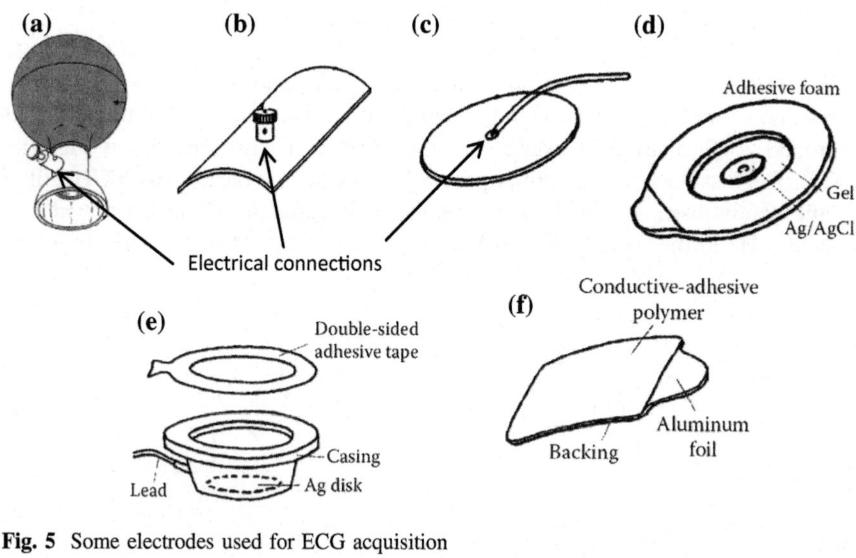

Fig. 5 Some electrodes used for ECG acquisition

body with adhesive coating on one side. The gel is trapped in the pocket between the electrode contact surface and skin. A double sided adhesive tape fixes the electrode on one side, and the body surface on the other. Such electrodes are particularly suited in ambulatory recordings for long term use. Configuration (f) shows special conductive polymer electrodes developed in recent times, with the advantage of conducting and adhesive properties simultaneously. The polymer is attached to a metallic backing made of silver or aluminum foil, allowing electric contact to external instrumentation.

4.1.2 Isolation Circuits

According to guidelines from AAMI, all medical amplifiers must comply with certain safety standards. Among them, one important is IEC 60601-1 standard, which prescribes a limiting patient auxiliary current (current that can flow between two separate leads connected to the patient body) up to 100 μA at 0.1 Hz. The isolation between the patient and the electrical recorder is achieved in two ways. First, complete galvanic isolation using optical, capacitive or magnetic coupling between the recording leads and amplifier inputs. Second, using surge protection arising from electrosurgical equipments or defibrillators [35].

In magnetic isolation, a transformer is used to achieve nearly 7 kV of isolation between the primary and secondary. The low frequency ECG signal modulates a high frequency carrier (around 500 kHz) signal, and transferred to secondary through magnetic coupling. At the output, a demodulator (low pass filter) extracts the ECG. This is schematically shown in Fig. 6a.

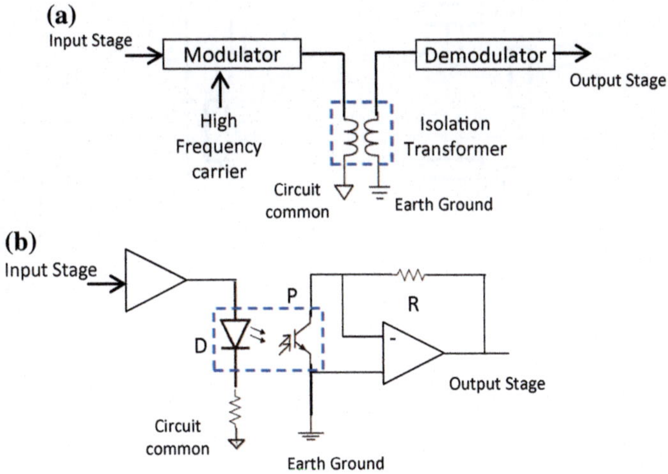

Fig. 6 Different types of isolation strategy in ECG amplifiers: **a** magnetic isolation. **b** Optical isolation

An optical isolation amplifier uses a linear opto-coupler using infrared LED and a phototransistor to separate the input and output (Fig. 6b).

4.1.3 Filtering Circuits

Among the major artifacts which corrupt the ECG signal are electromyography noise, baseline wander, motion artifacts, power line interference (PLI), electrode pop or contact noise, baseline wander, electrosurgical noise, and amplifier noise [8]. Some of these artifacts have overlapping spectra with the ECG, and hence only filtering in analog domain may not be sufficient for their complete removal. However, for practical purposes, a band pass filter in the frequency range 0.05–70 Hz can remove most of the artifacts, except PLI. Two filtering circuits at the first stage (pre-amplifier) are shown in Fig. 7. In (a), small ferrite beads (T1) and inductors at the lead wires, along with a low pass filter using R_0 and C_0 can substantially reduce the high frequency electromagnetic and radio frequency interference. A band pass filter using the R_1, C_1 and R_2, C_2 combination can select the clinical bandwidth of ECG using band pass filter, with the pass band given by:

$$f = \frac{1}{2\pi}\left(\frac{1}{R_1 C_1} - \frac{1}{R_2 C_2}\right) \tag{2}$$

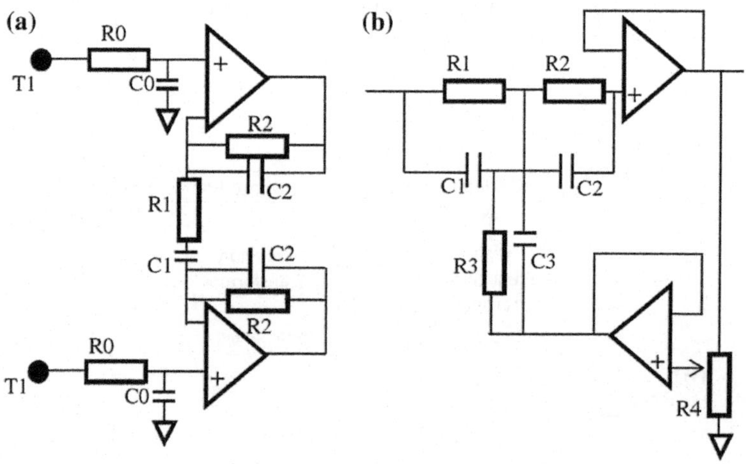

Fig. 7 Filtering circuit for noise reduction: **a** band pass filtering; **b** power line interference (printed from "ECG Acquisition and automated remote processing", (Springer, 2014) by R. Gupta et al. with permission)

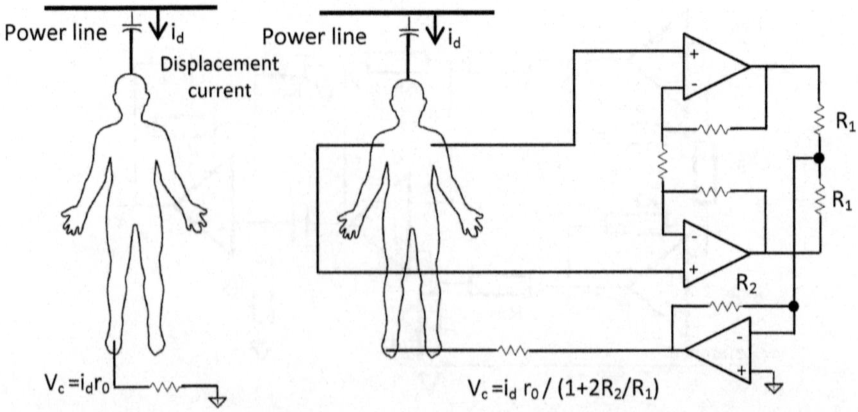

Fig. 8 Power line interference reduction using driven *right* leg circuit (printed from "ECG Acquisition and Automated Remote Processing", (Springer, 2014) by R. Gupta et al. with permission)

A twin T notch filter, shown in (b) can minimize the PLI. The rejected frequency is given as:

$$f = \frac{1}{2\pi R_1 C_1} \tag{3}$$

where, $R_1 = R_2 = 2R_3$; $C_1 = C_2 = C_3/2$.

Another improvisation to minimize power line noise is to use driven right leg circuit, frequently used in commercial ECG recorders and research equipments. The right leg of the subject is used as reference for unipolar lead ECG measurements. A feedback circuit from the first stage (pre-amplifier) output using a voltage divider can reduce the PLI, as shown in Fig. 8. Stray electrical capacitances are formed with the patient body, lead wires and the power line. The induced displacement current i_d creates a common mode voltage ($V_0 = i_d \times r_0$) at the lead wires (having resistance r_0) which, at power line frequency can be as high as 20 mV. The DRL circuit uses a feedback amplifier to reduce this common mode voltage to:

$$V_c = \frac{i_d r_o}{1 + \frac{2R_2}{R_1}} \tag{4}$$

Here, by proper choice of R_2 the PLI effect can be minimized.

4.1.4 Instrumentation Amplifier (INA)

An ECG INA should have some basic characteristics for faithful amplification of the low amplitude signal. The most essential of them are: high common mode

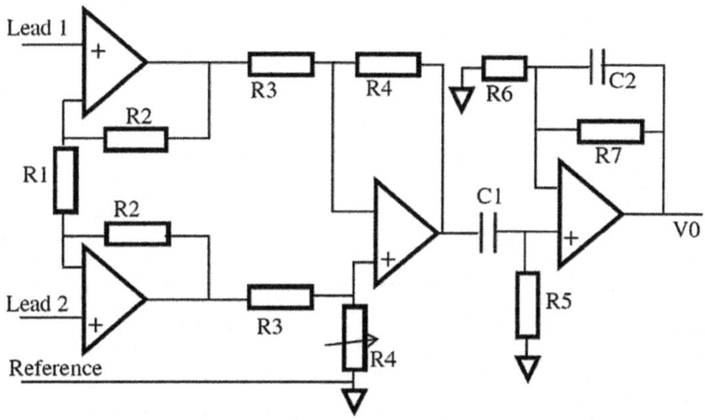

Fig. 9 Three OPAMP classical ECG amplifier configuration (printed from "ECG Acquisition and automated remote processing", (Springer, 2014) by R. Gupta et al. with permission)

rejection (CMR) and high input impedance. The CMR at the first stage (pre-amplifier) is one prime factor to minimize the noise generated from PLI as well as potential arising out of electrode-skin interface, which can saturate the output of the final amplifier. A typical CMR ratio of 250–300 dB and input impedance of 100 MΩ at 60 Hz can provide good results. Apart from these, the amplifier should have a good 'recovery' property which enables it to track the input after saturation due to sudden movement of the patient during ECG procedure. A modified configuration of classical three-OPAMP circuit is shown in Fig. 9 [8], where the total gain is distributed among the three stages, given by:

$$G = (1 + \frac{2R_2}{R_1})(\frac{R_4}{R_3})(1 + \frac{R_7}{R_6}) \tag{5}$$

This design also offers a band pass filter in the final stage:

$$\Delta f = \frac{1}{2\pi}(\frac{1}{R_7 C_2} - \frac{1}{R_5 C_1}) \tag{6}$$

Some low power monolithic INA designs are available in [36, 37]. AD620 from Analog Instruments is used in some ECG amplifier designs in recent times. This IC has some attractive features, like, using a single external register (R_G) to achieve a gain in the range 1 to 10^3, very good CMRR of 100 dB at gain of 10, low operating voltage 2.3 V, which outperforms a classical three-OPAMP IA circuit for ECG amplifiers.

A practical ECG amplifier (without the isolation stage) with AD620 is shown in Fig. 10. The circuit consists of four stages. In the first stage AD620 provides low gain, typically 10–15 and high CMRR. The filter at this output can prevent the saturation due to electrode offset voltage. The second stage (CA3140) implements a

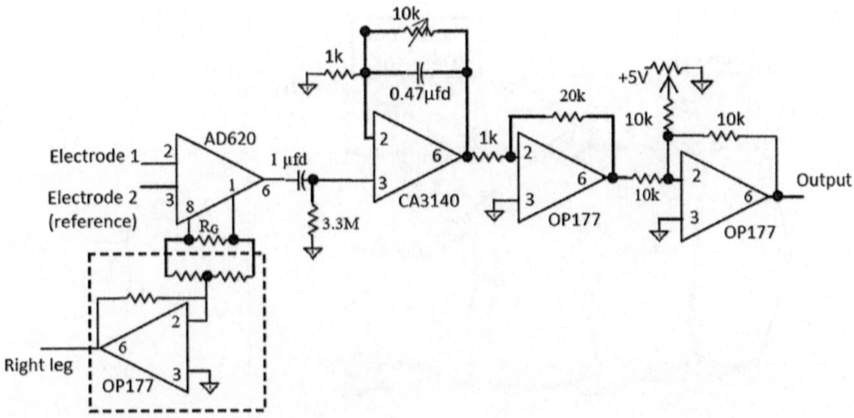

Fig. 10 Practical ECG amplifier using AD620

low pass filter of 70 Hz cutoff and a gain of 10. The third (OP177) stage, a simple inverting amplifier, provides a gain of 20. The final OP1-177 provides an adjustable bias to get unipolar final output. This design also implements driven right leg circuit.

4.2 Peripheral Pulse Signal (Photoplethysmogram)

There are two configurations of PPG sensors, viz., transmission mode, and, reflectance mode. The transmission type is more common in use and can be applied in peripheral body parts. In this mode a matched pair of LED and photodiode is attached to opposite surface of body extremity. In the 0.8–1 μm wavelength (named optical water window), the light is least absorbed by the water content of the tissue. A pulse of blood in the illuminated capillary increases the optical density, and there is a decrease in received light intensity at photodiode. The absorption is quantitatively expressed by Beer Lambert's law, which states that the intensity (I) decays from the original (I_O) according to following equation:

$$I = I_0 e^{-alc} \tag{7}$$

where, α: absorption coefficient of the material; l: length of travel; c: concentration of the medium. However, there are other factors like scattering, reflection and refraction of light from the bones and skin tissues causing reduction of intensity those are not covered in the law. By convention, the waveform is inverted so that it correlates positively with blood volume [38].

In reflection mode, the source-detector combination is placed on the same side of tissue. In the recent time, green LED operating in 0.5–0.6 μm wavelength has

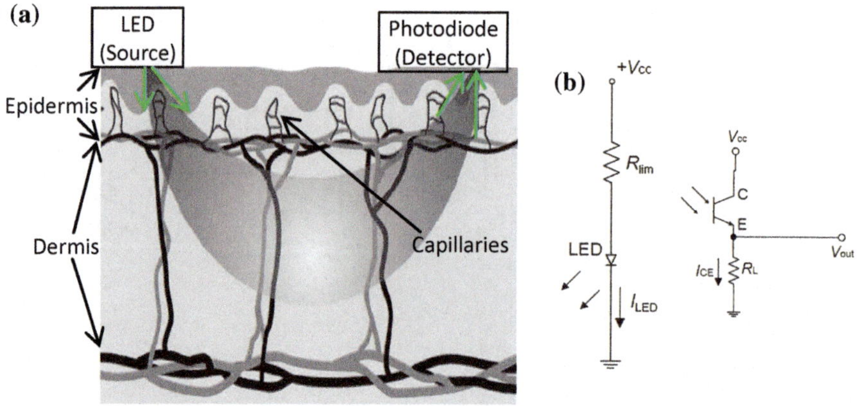

Fig. 11 Photoplethysmography: **a** tissue penetration and backscattering; **b** source and detector circuit configuration

become popular. In this mode, the emitted photons follow banana-shaped path from photodiode to the detector, and barely penetrate deep into the skin [39]. Figure 11a shows the basic principle of reflective PPG sensing.

A typical PPG sensor electronics contains a forward biased LED, producing a light intensity proportional to conductive current with a voltage drop of nearly 2 V across the LED. In Fig. 11b, a current limiting resistance R_{lim} is so chosen that the current (I_{LED}) is sufficient to drive the LED. In the detecting circuit, a load resistor in the phototransistor is used to convert the output current to a voltage as per the following equation:

$$V_L = I_{CE} R_L \tag{8}$$

A typical PPG signal conditioning circuit is shown in Fig. 12 [40]. The total circuit consists of two stages, viz., stage 1 amplifier and filter, and stage 2 amplifiers and filter. The stage 1 and stage 2 are similar in configuration, and use a passive high pass filter of 0.5 Hz and active low pass filter at 3.4 Hz to select the clinical bandwidth of the PPG, which is around 1.5 Hz. The potentiometer P provides an adjustable gain to the final analog output. The reference voltage is kept fixed at 2.0 V. In recent years several wearable PPG sensors have been proposed [41–43].

4.3 Respiration Signal

This section describes the sensor and their interfacing principles for measuring respiration signal using most common methods, viz., differential pressure, thermal convection, microwave Doppler radar, ultrasonic and inductive plethysmography approaches.

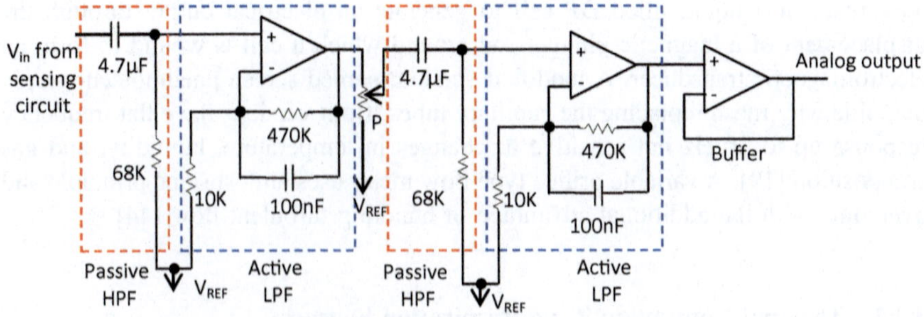

Fig. 12 Signal conditioning of PPG sensor

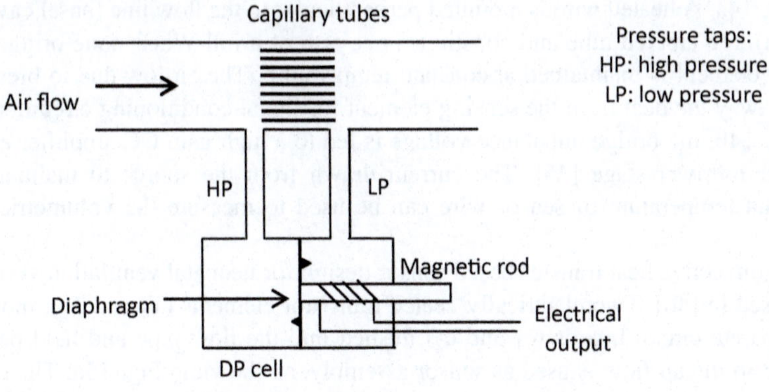

Fig. 13 Schematic of Fleisch pneumotachograph

4.3.1 Differential Pressure (DP) Respiration Sensors

Differential pressure flow meters are one of the oldest and widely used for air flow rate and volume measurement devices. The principle is based on measuring the differential pressure across an element of known resistance, popularly known as pneumotachography. The older version offered by Fleisch is schematically shown in Fig. 13. It provides small capillary tubes as restrictors parallel to the flow line inside the flow conduit. At laminar flow rates, the pressure drop (ΔP) across the restriction is given as:

$$\Delta P = \frac{12\mu l}{\pi N d^4} V \tag{9}$$

Where, d: diameter of each capillary tube, N: Number of tubes; μ: viscosity of the air medium; l: gap between pressure taps; V: flow rate.

The pressure taps (high pressure and low pressure) activate a diaphragm movement in a liquid filled DP cell to generate an electrical output through the displacement of a magnetic plunger rod around which a coil is wound to form an electromagnetic transducer. A modified version, named screen pneumotachograph, uses fine wire mesh replacing the capillary tubes. Both versions have flat frequency response up to 20 Hz but sensitive to changes in temperature, humidity, and gas composition [19]. A variable orifice type flow meter uses almost same principle and overcome with the additional advantage of handling turbulent flow [44].

4.3.2 Thermal Convection Type Respiration Sensors

Hot wire anemometers are the most popular type air flow measurement systems used in commercial systems under clinical setting. A typical configuration is shown in Fig. 14a. A heated wire is mounted perpendicular to the flow line (nasal cavity or mouth) in a tapered tube and constitutes one arm of a full Wheatstone bridge. The heater element is maintained at constant temperature. The airflow due to breathing takes away the heat from the sensing element. In signal conditioning circuit, shown in Fig. 14b, the bridge unbalance voltage is fed to a high gain DC amplifier and an emitter follower stage [45]. The current drawn from the source to maintain the constant temperature of sensor wire can be used to measure the volumetric flow rate.

A convective heat transfer based sensor design for neonatal ventilation system is proposed in [46]. Two electrically heated transistor elements (T_1 and T_2), mounted on separate circuit boards (c_1 and c_2) flushed into the flow pipe and held perpendicular to the air flow is used as sensor assembly, as shown in Fig. 15a. The circuit boards are held at a small gap, with sensing elements mounted on opposite faces, thus thermally insulating them. Each transistor is used in common-emitter configuration and supplied with constant current. Thus, the heat taken away by the cold (er) air will be different for each sensor in bidirectional flow (inspiration and

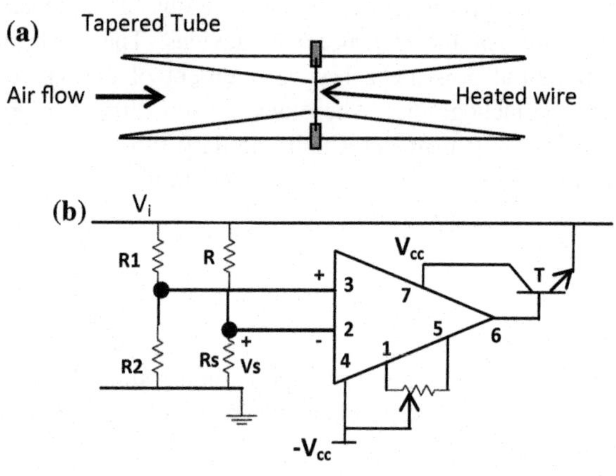

Fig. 14 Hot wire anemometer: **a** Sensor configuration and **b** interfacing circuit

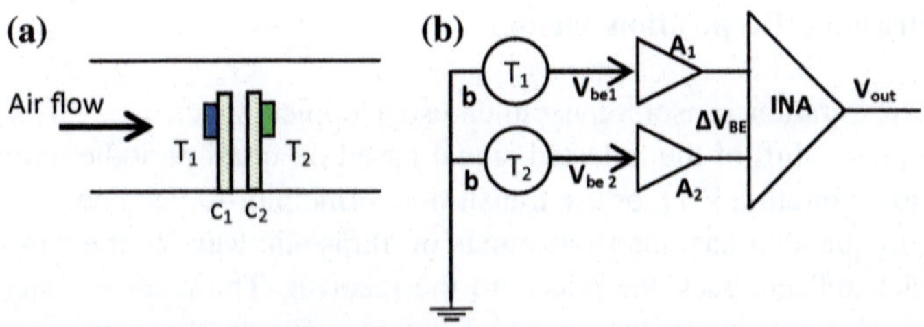

Fig. 15 Thermal convection type respiration sensing using differential heating of transistors: **a** Sensor mounting. **b** Signal conditioning schematic

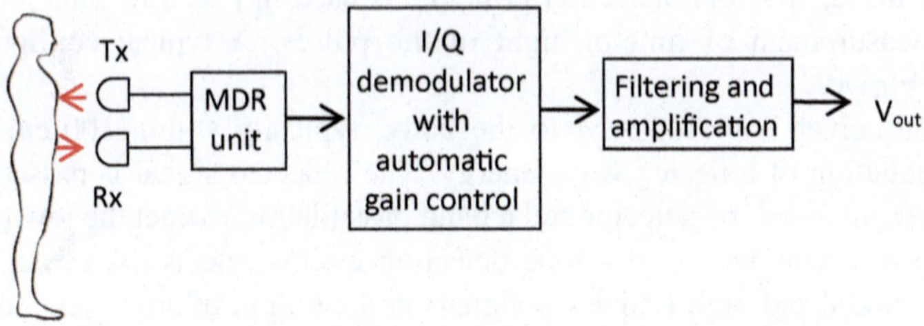

Fig. 16 Typical test set up of a MDR system for respiration measurement

expiration cycle). The signal conditioning circuit as shown in Fig. 15b, utilizes the difference in base-emitter voltage of two transistors through first stage and INAs to get the final analog output (V_{out}). Another silicon based thermal convection type respiration measurement unit is described in [47].

4.3.3 Microwave Doppler Radar (MDR) Type Respiration Sensors

A typical MDR system employs a pair of transmit and receive antenna mounted close to the body of the subject and captures the frequency shift of the reflected wave from the target (chest or abdomen) based on its quasi periodic motion due to respiration. The theoretical analysis [23] shows that the inhalation, exhalation and the pause in between them can be represented as a phase modulated signal, with the phase shift directly proportional to the object's chest movement.

A typical test-set up with interfacing schematic is shown in Fig. 16. A pair of Tx-Rx antenna transmits and receives the frequency modulated CW microwave radiation in the GHz range (based on antenna size), controlled by the MDR unit. The received signal is passed through a direct quadrature demodulator with RF and baseband automatic gain control, facilitating quadrature demodulation into direct baseband frequencies [48]. The final output is low pass filtered (with amplifiers) to extract the respiration signal V_{out} [49, 50].

4.3.4 Ultrasonic Respiration Meters

Ultrasonic respiration sensors can either use Doppler effect, i.e., change in frequency or phase shift of the reflected signal based on quasi-periodic motion of the chest due to respiration [51], or use transit time principle. A piezo crystal operating at 40 kHz frequency transmits short bursts of ultrasonic wave to the target (human chest), which reflects back the pulses to the receiver. The receiver calculates the time of travel, considering the speed of sound wave in the medium (air). The resolution in measurement can be improved with higher frequency. However, direct time of flight measurements suffer from the inaccuracies that occur in measurement of the flight time (~μs or less), and inertial delay problem in piezo sensor. To counteract these, envelope detection principle is used in [52, 53]. This provides an indirect measurement of time-of-flight of the pulses. A typical configuration is shown in Fig. 17.

The transceiver is placed near to the body, typically within 100 cm to avoid heavy attenuation of reflected wave energy. The reflected signal is passed through an amplifier, an envelope detector and a band pass filter to extract the low frequency respiration information. An envelope detection circuit detects the peaks from the amplitude modulated high frequency signals at the output of amplifier. Finally, the analog output is digitized and collected in a personal computer.

4.3.5 Respiratory Inductive Plethysmography (RIP) Sensors

The common RIP sensor consists of an elastic chest and abdomen belt (around 1 inch wide) with a zigzagging wire sewn into it and worn around the chest (rib cage under the armpits) and abdomen. A small current (some 10–20 mA) is passed through the coil from an oscillator. During inspiration, the abdomen (rib cage) area decreases (increases) with the lung volume. A typical RIP configuration is shown in Fig. 18. The outputs of the sensing units are separately amplified, and filtered to extract the respiration information before being added. Since the amplitudes of abdomen and chest sensors are unequal, the outputs of filters are normalized before addition.

An improved version of RIP sensor [54] uses pulsed current excitation (15 mA at 400 kHz for 100 μs) in time division multiplexing (TDM) to activate multiple

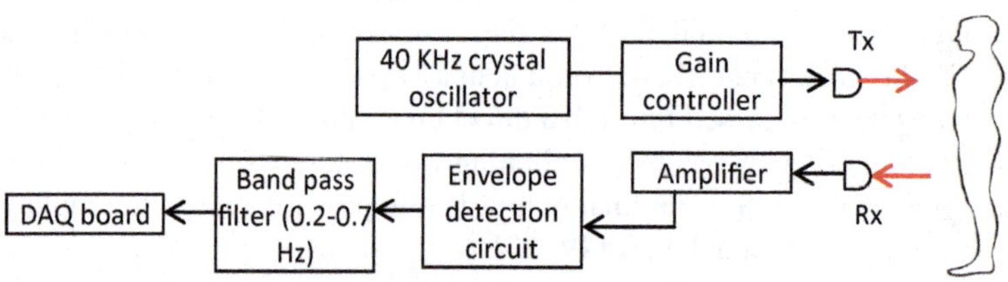

Fig. 17 Ultrasonic respiration measurement schematic

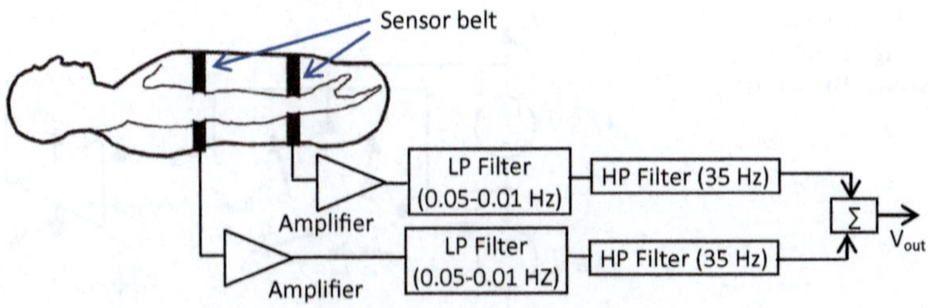

Fig. 18 Basic RIP sensor configuration and signal conditioning

sensors controlled by CMOS switching circuit. For sensing, a parallel LC circuit is used with a fixed C and L being the loop wire inductance. This combination virtually acts as a band pass filter with variable central frequency, which changes the output amplitude as per change in inductance due to breathing. Thus voltage change across the sensing unit can be detected to reflect the respiration rate.

4.4 *Blood Pressure Signal*

This section briefly describes the sensor interfacing for catheter type electrical blood pressure sensors. The most common configuration involves an elastic member (diaphragm) to be deflected by the blood pressure. The different secondary sensing principles can be categorized into the following.

4.4.1 Capacitive Blood Pressure Sensors

The displacement of the elastic diaphragm can be utilized to form a variable capacitance transducer, using the diaphragm with a thin conductive coating as movable plate. For a circular diaphragm fixed at the periphery, the approximate deflection (y) at the centre is given as:

$$y = \frac{3PR^4\left[\left(\frac{1}{\sigma}\right)^4 - 1\right]}{16E\left(\frac{1}{\sigma}\right)^2\delta^3} \tag{10}$$

where, E: Young's modulus of the material, σ: Poisson's ratio, P: applied pressure, δ: thickness of the diaphragm, R: radius of the diaphragm.

The sensor structure and lead connectors are encapsulated in a small volume by macro machining technique to achieve a sensor dia of 500 μm, diaphragm thickness and plate gap of 1 μm each, giving a capacitance value of 3 pF [55]. The problems encountered with this type of sensors are higher inherent electrical noise,

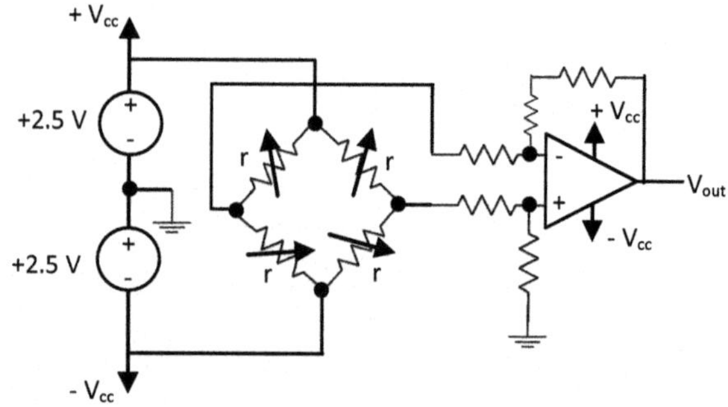

Fig. 19 Typical signal conditioning circuit of piezoresistive BP sensor

requirement of complex signal processing, internal oscillator and demodulator, which also increases the overall cost.

In another configuration [56], the variable capacitance is realized in a FET structure, so that, with the movement of the diaphragm reduces the gate and source resistance and source-drain current is altered.

4.4.2 Piezoresistive Blood Pressure Sensors

A four arm Wheatstone bridge is used with four active resistive elements each connected to one arm; with two in compression mode and the rest in expansion mode. Common piezoresistive transducers are ion-implanted into a thin silicon monocrystalline membrane [56, 57]. Typical values are 100–3 kΩ and powered by 3 V. The output (v_O) of the four-arm Wheatstone bridge excited by supply v_s is obtained as,

$$v_O = v_s \frac{\Delta r}{r} \tag{11}$$

where, Δr/r is the fractional change in resistance, assumed to be equal in each piezoresistive arms. The schematic circuit diagram is shown in Fig. 19. The bridge output is fed to an INA, followed by a successive approximation type AD converter.

4.4.3 Optical Blood Pressure Sensors

Their most advantageous feature is use of an optical beam, thus the risk of inducing current to the patient body is avoided, along with decoupling of associated electronics at the sensor site. The basic principle of sensing can be divided into the following categories: (a) Intensity reflective fiber optic sensor; (b) light coupling between two fiber optic sensors, (c) microbending fiber optic sensor, and (d) fiber optic sensing based on interferometer principles. The most straightforward

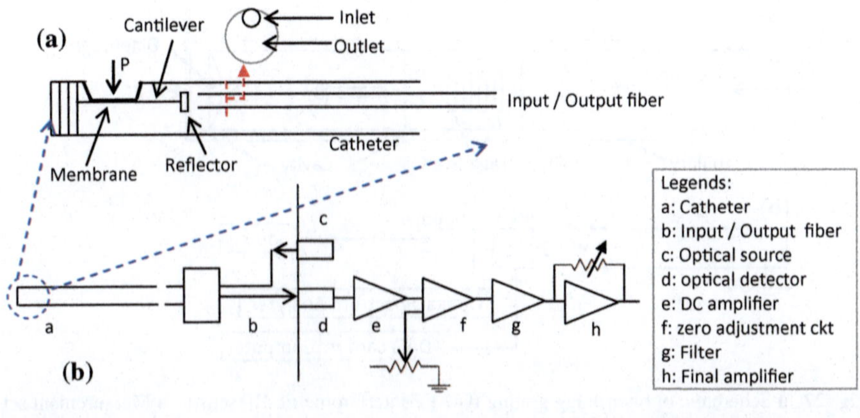

Fig. 20 **a** Schematic configuration and **b** signal conditioning of FO cantilever pressure catheter sensor

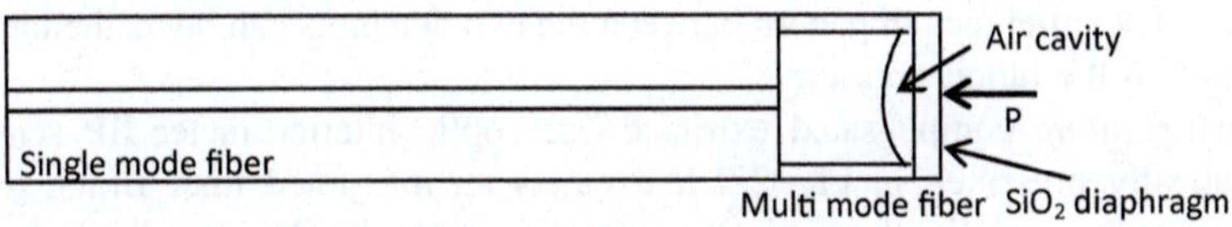

Fig. 21 Schematic blood pressure sensor using FP interferometry

configuration uses passing or reflecting light beam through the deflection system (elastic member) [58, 59] and measures the modulation by an electronic circuit. A classical configuration of a cantilever fiber optic catheter probe is shown in Fig. 20a. A cantilever arrangement accepts the pressure on a membrane coupled to a reflector, which faces the single Fiber cable acting as inlet and outlet and guides the light pulses to the catheter tip. A change in pressure alters the amount of reflected light into the output fiber. The modulation in light intensity is sensed by a photo electric detector and the final output is filtered, amplified to get a scaled output. The signal conditioning schematic is shown in Fig. 20b.

In another configuration [60] a Fabry-Perot (FP) interferometry is described to measure aortic arch and right coronary artery pressure. As shown schematically in Fig. 21, it consists of a single mode fiber, a multi-mode fiber, and a SiO_2 diaphragm. The SiO_2 diaphragm couples the pressure (P) at the catheter tip. An FP cavity is formed by multi-mode fiber—air cavity interface and air cavity—diaphragm interface. The single mode fiber guides the light exciting on the FP cavity and collects the reflected light. The multiple reflections of light in the FP cavity form an interference pattern.

The change in pressure on the diaphragm alters the FP cavity length and this is reflected in the phase change of the interference pattern. This in turn brings sinusoidal changes the intensity of the final reflected optical beam. The reflected light can be split into two fibers using an optical splitter, one directly connected to an

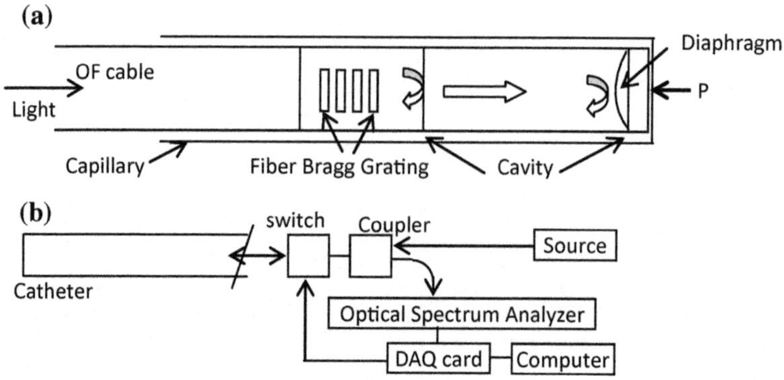

Fig. 22 **a** Schematic of fiber bragg grating type FP interferometric BP sensor. **b** Measurement set up

optical detector (broadband channel) and the other via a tunable filter (narrowband channel). The difference in pattern between the two detectors can be mathematically correlated to the blood pressure.

A temperature compensated extrinsic fiber optic interferometer BP sensor is schematically presented in Fig. 22. It contains an integrated fiber Bragg grating (FBG) core by periodic change in the refractive index inside a single mode fiber. The fiber is drawn into a glass capillary, at the end of which a diaphragm is sealed by splicing a multimode, which accepts the blood pressure. A FP cavity is formed between the diaphragm and the FBG which facilitates reflection of light. The outside pressure can change the cavity length (ΔL) through deformation of the diaphragm, calculated as in [61]:

$$\Delta L = \frac{3}{16}\frac{(1-\mu^2)}{E}\frac{r^4}{h^3}\Delta P \tag{12}$$

The measurement set-up consists of an optical source, coupler with switch (operating in time multiplexing mode and controlled by DAQ card), an optical spectrum analyzer (OSA), data acquisition card and computer. The coupler directs the light pulses from sources and redirects the reflected light into the receiver fiber. The OSA converts the received radiation into intensity modulated electrical pulses, which are finally analysed in the computer.

4.4.4 Oscillometric Blood Pressure Sensor

Most of the automatic BP monitors use the principle of oscillometry, where the oscillations in occluded cuff pressure with gradual release of cuff pressure is sensed by peizoresistive sensors connected in a Wheatstone bridge. The bridge output is filtered, amplified and processed by a microcontroller to compute the systolic,

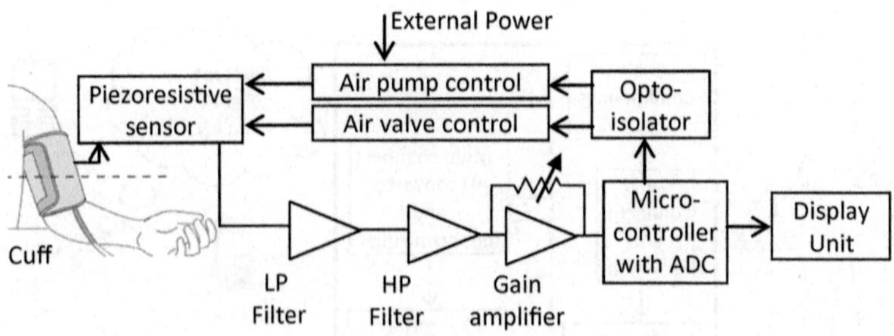

Fig. 23 Block schematic of an automatic BP monitor using oscillometry

diastolic and mean arterial pressures (MAP). Figure 23 describes a schematic layout of such a system. The air cuff is provided with a controlled air supply and air valve control system, both controlled from a central microcontroller. The sensor output is typically filtered by a series of low pass (10 Hz) and high passes (2.2 Hz) filters and a gain control stage to generate an analog output that is fed to the microcontroller unit. The microcontroller determines the MAP by taking the cuff pressure when the pulse with the largest amplitude appears. Typically, in some commercial units [62] calculates the systolic pressure (diastolic pressure) equal to the measurement taken in the cuff with a pulse with 70% (50%) of the amplitude of the MAP while the cuff pressure is above (below) the MAP value.

5 Wireless Biomedical Instrumentation

With the advent of low power microelectronics, embedded systems, sensing technology and wireless communication techniques, the area of medical instrumentation has undergone sea change in the last two decades. Using miniature wearable sensors with digital intelligence, a patient's medical condition can be remotely monitored under his normal activity. Development of low power hungry digital microcontrollers has enabled fast real-time analysis of the acquired signals for conditioning monitoring of critical patients. Hence, modern biomedical instrumentation systems are not only confined to measurement, and acquisition of data, but include significant computing tasks for analyzing the acquired signals [63]. Over the last decade, low power industrial, scientific and medical (ISM) band protocols (ZigBee, WLAN and Bluetooth) have been widely used in short range health monitoring applications [64–66].

A typical configuration of multi (two)-channel wireless biomedical instrumentation system is shown in Fig. 24. The hardware acquisition module (HAM) is a small circuit encompassing the signal conditioning functions and digital interface to transfer the signals through a wireless media to handheld gadgets, or remote servers. The HAM includes a low power microcontroller which controls the

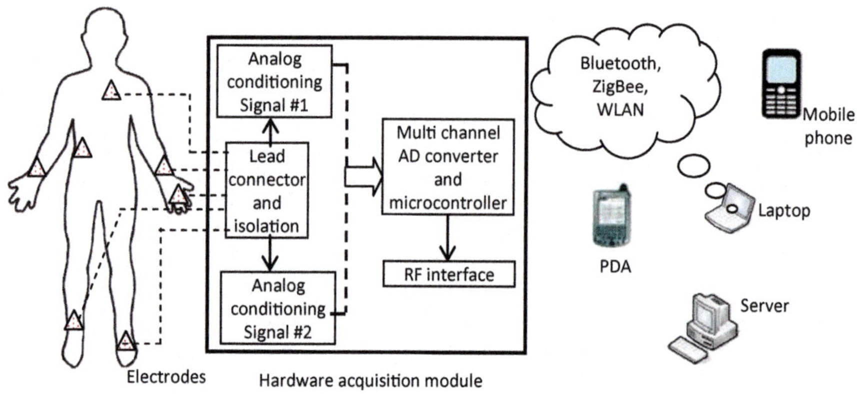

Fig. 24 Block schematic of a typical multi-channel wireless biomedical signal acquisition system

multiplexing of the conditioned signals and transfers the data using a serial protocol to the on-board RF interfaces. For short distance monitoring, these RF modules utilize license free ISM bands and comply with either ZigBee or Bluetooth or WLAN protocol. For remote-end connectivity, all the sensors' data is routed through a gateway to a public communication network (PSTN, internet or ISDN etc.). Now a days, a variety of low power RF modules are commercially available in the consumer market, which can be used to develop custom make health monitoring applications [67]. Body area network (BAN), a specialized application of sensor network has enabled wireless networking among various sensors attached to different parts of the body using dedicated wireless protocols (medical implant communication service, MICS, at 402–405 MHz and wireless medical telemetry services, WMTS at 608–614 MHz).

6 Trends in Biomedical Sensor Technology

Considering the huge potential of biomedical science and healthcare in the coming days, the following developments are expected to play key roles in medical instrumentation domain.

6.1 Use of Wearable Integrated Biomedical Sensors

With the increasing demand from the patients to stay in normal life while being monitored for their physiological activity, wearable biomedical sensors can play an

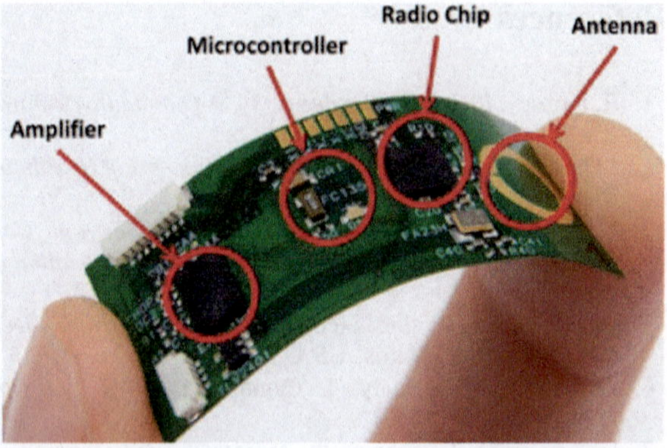

Fig. 25 Flexible integrated ECG sensor from IMEC, Netherlands

important role. Towards this objective, new adhesive tape type wearable sensors enable integration of sensors, signal conditioning, digitization, controller, energy source (collector) and low power RF module in a small area (less than 2 sq. inch) and fixed on human body [68, 69]. Figure 25 shows the diagram of a flexible ECG sensor which contains three electrodes, an amplifier, microcontroller and RF chip and transceiver units on small flexible polymer strip.

The design challenges for these units are low power consumption, ease of wearing, and energy harvesting for continuous data supply.

6.2 Context Aware Smart Biomedical Sensors

For continuous monitoring of patients in their normal activity, an important requirement is real-time recognition of abnormal 'events' (fall, unconsciousness etc.) which can be communicated to a caregiver. Autonomous sensor systems, as described in last section lack in computational capability due to small size of the embedded processor. However, such smart capability can be achieved in conjunction with a device with higher computational capability, typically a mobile phone or PDA which the patient often carries. The software in the device can be intelligent enough to discriminate a normal and abnormal condition, however, in such systems additional sensors (like accelerometers, temperature sensors) may be deployed.

Acknowledgements The author would like to Mrs. Priyanka Bera, research scholar at Department of Applied Physics, University of Calcutta, India for the help in copyediting and proof reading of the manuscript.

References

1. R. Plonsey, R.C. Barr, *Bioelectricity: A Quantitative Approach*, 3rd edn. (Springer, NY, USA, 2007)
2. J.D. Bronzino (ed.), *The Biomedical Engineering Handbook*, 2nd edn. (CRC Press LLC, Boca Raton, USA, 2000)
3. L.A. Geddes, L.E. Baker, *Biopotential and Electrophysiology Measurement*
4. J.G. Webster (ed.), *Medical Instrumentation: Application and Design*, 4th edn. (Wiley, USA, 2010)
5. A. Natale, F. Marchlinski (ed.), *Handbook of Cardiac Electrophysiology* (Informa Healthcare, CRC Press, Boca Raton, USA, 2007)
6. F. Morris, W.J. Brady, J. Camm (eds.), *ABC of Clinical Electrocardiography* (BMJ Publishing, USA, 2008)
7. L. Schamroth, C. Schamroth, *An introduction to Electrocardiography*, 7th edn. (Blackwell Science, France, 2008)
8. R. Gupta, M. Mitra, J.N. Bera, *ECG Acquisition and Automated Remote Processing* (Springer, India, 2014)
9. A.B. Hertzman, The blood supply of various skin areas as estimated by the photoelectric plethysmograph. Am. J. Physiol. **124**, 328–340 (1938)
10. A.A.R. Kamal, J.B. Harness, G. Irving, A.J. Mearns, Skin photoplethysmography—a review. Comput. Methods Programs Biomed. **28**(4), 257–269 (1989)
11. J. Allen, Photoplethysmography and its application in clinical physiological measurement. Physiol. Meas. **28**(2007), R1–39 (2007)
12. S.C. Millasseau, R.P. Kelly, J.M. Ritter, P.J. Chowienczyk, Determination of age-related increases in large artery stiffness by digital pulse contour analysis. Clin. Sci. 371–377 (2002)
13. E. Tam et al., Correction of cardiac output obtained by modelflow® from finger pulse pressure profiles with a respiratory method in humans. Clin. Sci. 371–376 (2014)
14. P. Wang, W. Zuo, D. Zhang, A compound pressure signal acquisition system for multichannel wrist pulse signal analysis. IEEE Trans. Instrum. Meas. **63**(6), 1556–1565 (2014)
15. Y Kurylyak, K. Barbe, F. Lamonaca, D. Grimaldi, W. Van Moer, Photoplethysmogram-based blood pressure evaluation using Kalman filtering and NN, in *Proceedings of International Symposium on Medical Measurements and Applications (MeMeA)*, May 3–4 (2013), pp. 170–174
16. L. Nilsson, A. Johansson, S. Kalman, Respiratory variations in the reflection mode photoplethysmographic signal: relationships to peripheral venous pressure. Med. Biol. Eng. Comput. **41**(3), 249–254 (2003)
17. R. Erts, E. Kviesis-Kipge, J. Zaharans, E. Zaharans, J. Spigulis, Wireless photoplethysmography finger sensor probe, in *Proceedings of 12th Biennial Baltic Electronics Conference (BEC2010)*, Tallinn, Estonia, October 4–6 (2010), pp. 283–284
18. E.M. Lee, J.Y. Shin, J.H. Hong, E.J. Cha, T.S. Lee, Glass-type wireless PPG measuring system, in *32nd. Annual International Conference of the IEEE EMBS*, August 31–September 4 (2010), pp. 1433–1436
19. J.G. Webster (ed.), *The Physiological Measurement Handbook* (CRC Press, USA, 2015)
20. S. Silvestri, E. Schena, Micromachined flow sensors in biomedical applications. Micromachines **3**, 225–243 (2012)
21. Y.S. Lee, P.N. Pathirana, C.L. Steinfort, T. Caelli, Monitoring and analysis of respiratory patterns using microwave doppler radar. IEEE J. Trans. Eng. Health Med. **2** (2014)
22. C. Gu, C. Li, Assessment of human respiration patterns via noncontact sensing using doppler multi-radar system. Sensors **15**(3), 6383–6398 (2015)
23. Y.S. Lee, P.N. Pathirana, R.J. Evans, C.L. Steinfort, Noncontact detection and analysis of respiratory function using microwave doppler radar. J. Sens. 1–13 (2015)
24. G.D. Garguilo et al., A wearable contactless sensor suitable for continuous simultaneous monitoring of respiration and cardiac activity. J. Sens. 1–6 (2015)

25. S.T.A. Hamdani, A. Fernando, The application of a peizo-resistive cardiorespiratory sensor system in an automobile safety belt. Sensors **15**(4), 7742–7753 (2015)
26. Z. Zhang et al., Development of respiratory inductive plethysmography module supporting multiple sensors for wearable systems. Sensors **12**(10), 13167–13184 (2012)
27. K. Konno, J. Mead, Measurement of the separate volume changes of rib cage and abdomen during breathing. J. Appl. Physiol. **22**(3), 407–422 (1967)
28. L.A. Geddes, *Handbook of Blood Pressure Measurement* (Springer Science Business Media, NY, 1991)
29. J.G. Webster (ed.), *The Measurement, Instrumentation and Sensors Handbook* (CRC Press LLC, USA, 1999)
30. J.S. Kim, Y.J. Chee, J.W. Park, J.W. Choi, K.S. Park, A new approach for non-intrusive monitoring of blood pressure on a toilet seat. Physiol. Meas. **27**(2), 203–211 (2006)
31. M. Nitzan, A. Patron, Z. Glik, A.T. Weiss, Automatic noninvasive measurement of systolic blood pressure using PPG. Biomed. Eng. Online 1–8 (2009)
32. J.H. Shin, K.M. Lee, K.S, Park, Non-constrained monitoring of systolic blood pressure on a weighing scale. Physiol. Meas. **30**, 679–693 (2009)
33. M. Mukkamala et al., Toward ubiquitous blood pressure monitoring using pulse transit time theory and practice. IEEE Trans. Biomed. Eng. **62**(8), 1879–1901 (2015)
34. D. Buxi, J.M. Redoute, M.R. Yuce, A survey on signals and systems in ambulatory blood pressure monitoring using pulse transit time. Physiol. Meas. **36**(3), R1–R26 (2015)
35. R.B. Northorp, *Analysis and Application of Analog Electronic Circuits to Biomedical Instrumentation*, 2nd edn. (CRC Press, Boca Raton, USA, 2012)
36. M. Degrauwe, I. Verbauwhede, A micropower CMOS-instrumentation amplifier. IEEE J. Solid State Circ. **Cs-20**(3), 805–807 (1985)
37. E.M. Spinelli, R.P. Areny, M.A. Mayosky, AC-coupled front-end for biopotential measurements. IEEE Trans. Biomed. Eng. **50**(3), 391–395 (2003)
38. A. Reisner, P.A. Shaltis, D. McCombie, H.H. Asada, Utility of the photoplethysmogram in circulatory monitoring. Anesthesiology **108**, 950–958 (2008)
39. T. Tamura, Y. Maeda, M. Sekine, M. Yoshida, Wearable photoplethysmographic sensors—past and present. Electronics **3**(2), 282–302 (2014)
40. Easy Pulse PPG sensor from Embedded Lab, http://www.embedded-lab.com
41. S. Borik, I. Cap, Implementation of wireless data transfer to photoplethysmographic measurements. Proc. ElEKTRO **21–22**, 407–410 (2012)
42. G. Angius, L. Raffo, Cardiovascular disease and sleep apnoea: a wearable device for PPG acquisition and research aims, in *Proceedings of Computing in Cardiology*, 9–12 September (2012), pp. 513–516
43. K. Davoudi, M. Shayegannia, B. Kaminska, Vital signs monitoring using a new flexible polymer integrated PPG sensor. Proc. Comput. Cardiol. **22–25**, 265–268 (2013)
44. G. Tardi, C. Massaroni, P. Saccomandi, E. Schena, Experimental assessment of a variable orifice flowmeter for respiratory monitoring. J. Sens. 1–8 (2015)
45. G.A.L. Araujo, R.C.S. Freire, J.F. Silva, A. Oliveira, E.F. Jaguaribe, Breathing flow measurement with constant temperature hot wire anemometer for forced oscillations technique, *in Proceedings of instrumentation and measurement technology conference*, Italy, Vol. 1, May (2004), pp. 730–733
46. M. Giorgino et al., Design and characterization of a bidirectional low cost flowmeter for neonatal ventilation. IEEE Sens. J. **14**(12), 4354–4360 (2014)
47. M.J.A.M. van Putten, M.H.P.M. van Putten, A.F.P. van Putten, J.C. Pompe, H.A. Bruining, A silicon bidirectional flow sensor for measuring respiratory flow. IEEE Trans. Biomed. Eng. **44** (2), 205–208 (1997)
48. AD8347 Direct Conversion Quadrature Demodulator Datasheet, http://www.analog.com
49. O. Postolache, P.S. Girao, E. Lunca, P. Bicleaknu, Unobtrusive cardio-respiratory monitoring based on microwave doppler radar, in *Proceedings of 2012 International Conference and Exposition on Electrical and Power Engineering (EPE 2012)*, 25–27 October, Iasi, Romania (2012), pp. 597–600

50. M. Zakrzewski, A. Vehkaoja, A.S. Joutsen, K.T. Palovuori, J.J. Vanhala, Noncontact respiration monitoring during sleep with microwave doppler radar. IEEE Sens. J. **15**(10), 5683–5693 (2015)
51. P. Arlotto, M. Grimaldi, R. Naeck, J.M. Ginoux, An ultrasonic contactless sensor for breathing monitoring. Sensors **14**, 15371–15386 (2014)
52. Yusuke Yamana, et al., A sensor for monitoring pulse rate, respiration rhythm, and body movement in bed, in *Proceedings of 33rd Annual International Conference of the IEEE EMBS*, Boston, Massachusetts USA, August 30–September 3 (2011), pp. 5323–5326
53. S.D. Min, Noncontact respiration rate measurement system using an ultrasonic proximity sensor. IEEE Sens. J. **10**(11), 1732–1739 (2010)
54. Z. Zhang et al., Development of a respiratory inductive plethysmography module supporting multiple sensors for wearable systems. Sensors **12**, 13167–13184 (2012)
55. K.E. Babbitt, L. Fuller, B. Keller, A surface micromachined capacitive pressure sensor for biomedical applications, in *Proceedings of the Twelfth Biennial University/Government/ Industry Microelectronics Symposium*, 20–23 July (1997), pp. 150–153.
56. Samaun, K.D. Wise, J.B. Angell, An IC Piezoresistive Pressure Sensor for Biomedical Instrumentation. IEEE Trans. Biomed. Eng. **BME-20**(2), 101–109 (1973)
57. M. Esashi et al., Fabrication of catheter-tip and sidewall miniature pressure sensors. IEEE Trans. Electron Devices **-ED-39**(1), 57–63 (1982)
58. H. Matsumoto, M. Saegusa, K. Saito, K. Mizoi, The development of a fibre optic catheter tip pressure transducer. J. Med. Eng. Technol. **2**(5), 239–242 (1978)
59. A. Lekholm, L. Lindström, Optoelectronic transducer for intravascular measurements of pressure variations. Med. Biol. Eng. **7**, 333–335 (1969)
60. N. Wu et al., A miniature fiber optic blood pressure sensor and its application in vivo blood pressure measurements of a swine model. Sens. Actuators B: Chem. **181**, 172–178 (2013)
61. S. Poeggel, D. Tosi, G. Leen, E. Lewis, Low drift and high resolution miniature optical fiber combined pressure—and temperature sensor for cardio-vascular and other medical applications, in *Proceedings of the 2013 IEEE Sensors*, Baltimore, MD, USA, 3–6 November (2013), pp. 1–4
62. S. Lopez, *Blood Pressure Monitor Fundamentals and Design* (Freescale Semiconductor Application Note)
63. H. Eren, J.G. Webster (eds.), *Telemedicine and Electronic Medicine* (CRC Press, Boca Raton USA, 2015)
64. S.K. Chen et al., A reliable transmission protocol for zigbee-based wireless patient monitoring. IEEE Trans. Inf Technol. Biomed. **16**(1), 6–16 (2012)
65. H.J. Lee et al., Ubiquitous health care service using zigbee and mobile phone for elderly patients. Int. J. Med. Inf. **78**(3), 193–198 (2009)
66. R. Gupta, J.N. Bera, M. Mitra, An intelligent telecardiology system for offline wireless transmission and remote analysis of ECG. J. Med. Eng. Technol. (Informa Healthcare) **36**(7), 358–365 (2012)
67. S. Roy, R. Gupta, Short range centralized cardiac health monitoring system based on zigbee communication, in *2014 IEEE Global Humanitarian Technology Conference (GHTC)-South Asia Satellite*, September 26–27, 2014, Kerala, India (2014), pp. 177–182
68. S. Patel, H. Park, P. Bonato, L. Chan, M. Rodgers, A review of wearable sensors and systems with application in rehabilitation. J. NeuroEng. Rehabil. **9**(21), 1–17 (2012)
69. L.G. Villanueva, S. Cagnoni, L. Ascari, Design of a wearable sensing system for human motion monitoring in physical rehabilitation. Sensors **13**, 7735–7755 (2013)

Interfacing and Pre-processing Techniques with Olfactory and Taste Sensors

Rajib Bandyopadhyay and Anil Kumar Bag

Abstract In human beings, analytical conclusion about any object is governed by the brain from the sensual influences coming from visual, auditory, olfactory, taste and touch organs. However, these conclusions are purely subjective in nature and target specific and may widely vary due to different human factors due to variability in mood and health conditions. Scientists and engineers have been trying to mimic the sensory organs of the humans in order to provide sufficiently reliable and selective analysis on the object, and imitation of these sensual organs, nowadays, has advanced to a great extent due to the immense development in sensor technology and intelligent smart electronics. In this regard, the electronic gadgets for visual, audio, and touch system have traversed a long journey in terms of accuracy and resolution and these three sensual organs now can be faithfully reproduced with reasonable accuracy. However, the electronic gadgets like electronic nose for olfactory system and electronic tongue for taste system, till now, are under research to find a profile to be commercialized widely. There are many challenges in this regard; firstly the selection of right sensor or sensor array for the application in hand. Secondly, design of reliable signal conditioning and pre-processing. In this chapter, we present the different types of sensors used for electronic nose and electronic tongue systems and associated interfacing circuits and preprocessing techniques.

R. Bandyopadhyay (✉)
Department of Instrumentation and Electronics Engineering,
Jadavpur University, Kolkata, India
e-mail: rb@iee.jusl.ac.in

A.K. Bag
Department of Applied Electronics and Instrumentation Engineering,
Heritage Institute of Technology, Kolkata, India
e-mail: anilkumarbag@gmail.com

B. George et al. (eds.), *Advanced Interfacing Techniques for Sensors*,
Smart Sensors, Measurement and Instrumentation 25,
DOI 10.1007/978-3-319-55369-6_8

1 Electronic Nose and Tongue

Smell and taste sensations in mammals are due to multidimensional effects of specific and nonspecific molecular recognition by the olfactory receptors or gustatory receptors. But factors like individual variability, decrease in sensitivity due to prolonged exposure, infection, adverse mental state affect the sensual responses and thus, mimicking the mammalian olfactory or taste systems is difficult and challenging. So far, the electronic nose and electronic tongue systems have been developed to mimic these two sensual systems for specific applications only and universal gadgets for the mammalian olfactory and taste system are still far from reality. These application specific electronic gadgets comprise of sensor array, signal conditioning unit, data acquisition system, signal processing unit and a pattern recognition system.

1.1 Electronic Nose

"An instrument which comprises an array of electronic chemical sensors with partial specificity and an appropriate pattern recognition system, capable of recognizing simple or complex odor" [1] is an accepted definition of an electronic nose. The purpose of electronic nose is to produce rapid, inexpensive analysis of volatile organic compounds without any assistance of specialist technicians [2]. The block diagram of the conventional electronic nose is shown in Fig. 1, where the odour handling and sample delivery module ensures delivery of the odour molecules to the surface of the sensors in the array. This is usually done using delivery or suction pumps. Once exposed to the chemicals, the sensors in the array interact with the molecules to produce sensual responses/patterns. The signals obtained from these sensors are conditioned and passed through the data acquisition and signal processing block to make the dataset appropriate for pattern recognition. The pattern recognition

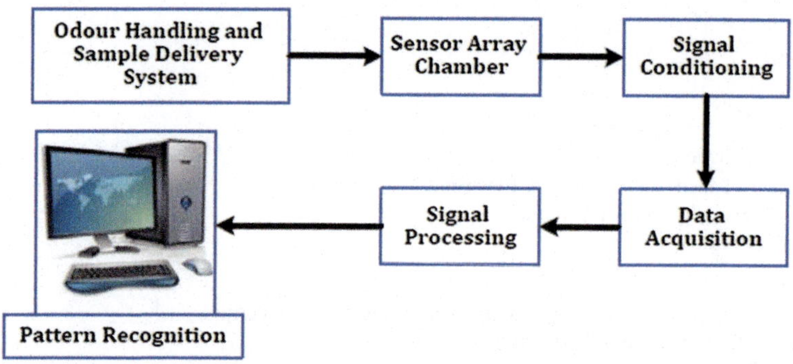

Fig. 1 Block diagram of conventional electronic nose

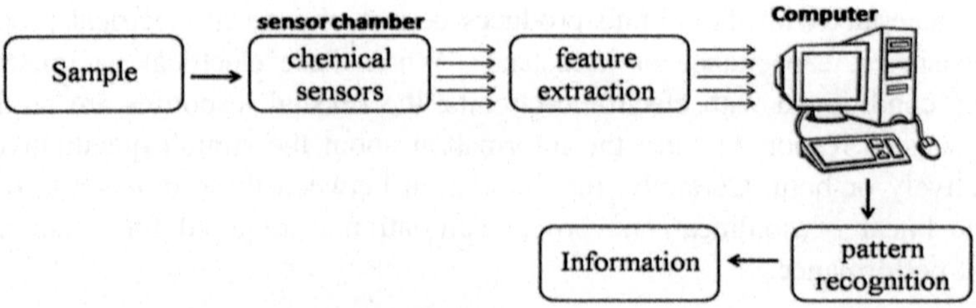

Fig. 2 Block diagram of electronic tongue [36]

system is analogous to the human brain which provides the result of classification or the final result depending upon the application for which the electronic nose is used. Out of all the blocks in the diagram, the most critical ones are the sensor array and pattern recognition. The other modules also play important roles and their performance is crucial for the desired output from an electronic nose.

1.2 Electronic Tongue

An electronic tongue (e-tongue) is an analytical instrument that artificially reproduces the sensation of taste. It typically consists of an array of chemical sensors coupled to chemometric processing used to characterize complex liquid samples [3] and the block diagram of the e-tongue is shown in Fig. 2. According to IUPAC technical report it is defined as "a multisensor system, which consists of a number of low selective sensors and uses advanced mathematical procedures for signal processing based on the pattern recognition (PARC) and/or multivariate data analysis" [4]. The chemical sensor array produces electrical signals (characteristic pattern) which are not necessarily specific to a particular species, but do form a pattern that can be correlated to certain features or qualities of the sample [5, 6]. The sensual pattern thus produced needs to be processed using suitable preprocessing techniques and then is interpreted by multivariate statistics including principle component analysis (PCA), linear discriminant analysis (LDA), discriminant function analysis (DFA), soft independent modelling of class analogy (SIMCA) and partial least squares (PLS). For non-linear responses, artificial neural network (ANN) can be used for data modelling [7].

2 Sensors/Transducers for Electronic Nose and Tongue

The devices which produce some form of response to any stimulus are termed as sensors. When these sensual responses are converted into electrical form with the help of suitable signal conditioning unit (SCU), the sensor along with its SCU is termed as a transducer. The physical, chemical or bio-chemical interaction of the sensor materials with the stimuli produces certain change in electrical properties like resistance, capacitance or inductance. When these electrical parameters are suitably conditioned with electronic circuits, the sensual responses are produced. These sensual responses carry the information about the stimuli quantitatively or qualitatively or both. Certainly, the correlation between these inputs and outputs may be linear or nonlinear and proper calibration is required for obtaining the desired performance.

2.1 Electronic Nose Sensors

2.1.1 MOS Sensor

A metal oxide semiconductor (MOS) sensor uses metal oxide based sensing thick films deposited on to a Si-micro-machined substrate. The electrodes attached with the sensor measure the resistance of the sensing layer. When oxygen in the atmosphere is adsorbed on the surface of the n-type semiconductor, the free electrons are trapped from the semiconductor and consequently a high resistive layer is formed in the vicinity of the surface. The reaction between the reducing gas and oxygen generates electrons and increases the conductivity of the resistive layer as a result of increase in carrier concentration. The detection principle of the MOS sensor is sketched as shown in Fig. 3.

This mechanism follows the reaction as written below:

$$e + \frac{1}{2}O_2 \rightarrow O(s)^-$$

$$R(g) + O(s)^- \rightarrow RO(g) + e$$

where, ***e*** is the electron, ***R(g)*** is reducible gas, ***s*** and ***g*** imply surface and gas, respectively.

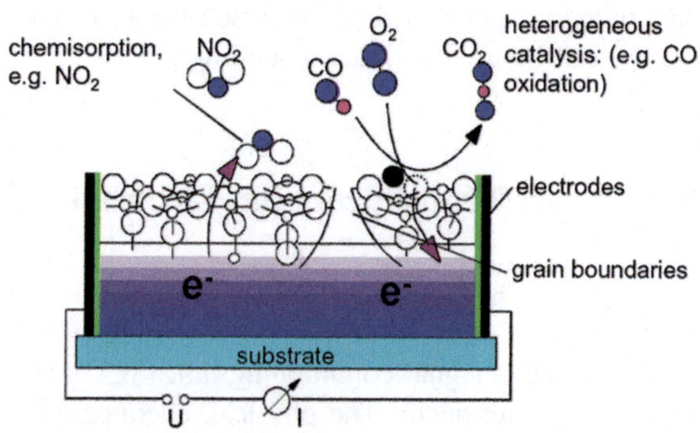

Fig. 3 Sketch of a MOS sensor illustrating the detection principle [37]

MOS sensors are responsive to many reducible gases in the sensing environment and undergo change in resistance upon redox reaction between the gas and oxygen [8, 9]. The sensor resistance value depends on the concentration of the reducible gases and their adsorption and desorption of gas to the sensor surface. The temperature of the sensor decides the level, type and rate of adsorption and desorption of the gas and oxygen. At moderate concentrations, above the noise floor and below the saturation level of the sensor, resistance is related to the gas concentration as follows:

$$R_S = \frac{1}{AC^{\alpha}}$$

where, R_s is the resistance of the metal oxide sensor, C is the concentration of the gas, A and α are constants. The constants A and α depend on the type of the reducing gas and the sensor temperature. The noise in the sensor heaters and the surface chemistry variation put limitations on this relationship at the lower end. At the upper end, increase in sensor surface temperature induced by exothermic reactions between the gas and the sensor surface produces changes in the above relationship. In a controlled environment, this relationship fits well to the gas concentration between these high and low limits, although effects of diffusion, turbulent air flow and similar other factors are not considered in the above equation. Nature of sensor response significantly depends on operating temperature and the amount of oxygen available on the sensor surface to be desorbed in the presence of reducible gases. Sometimes the presence of impurities and catalytic metals like platinum or palladium increases the sensitivity and selectivity. Amount and type of impurity material decide sensitivity and selectivity of the sensors respectively.

Most commonly used semiconductor material as a gas sensor is SnO_2 doped with small amount of impurities and catalytic metal additives. Operating temperature of the sensor is application specific and varies from 200 to 500 °C.

2.1.2 Organic Conducting Polymer

Organic polymers are also a type of chemoresistors [10]. They undergo change in conductance upon interaction with the reducible or oxidizable gases. Adsorption and desorption of the chemical substances from the metal surface produce reversible changes in conductivity of the sensor. Organic CPs show response to a wide range of polar molecules at temperature as low as room temperature and they have high sensitivity down to 0.1 ppm. This is why the organic CP materials find applications in odour sensing and electronic nose technology. Popularity of the organic CP sensors is due to the following facts:

1. A wide range of materials is available for the sensor fabrication,
2. They are relatively low cost materials,
3. Sensors are sensitive to various kinds of organic vapors,
4. Operating temperature is low.

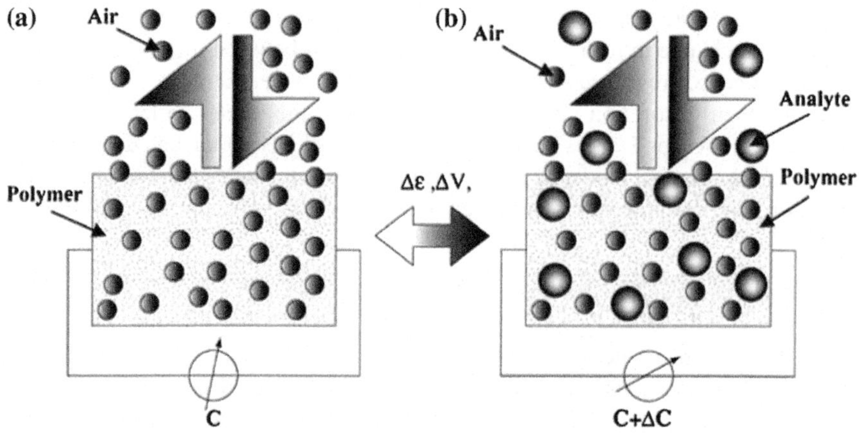

Fig. 4 Chemocapacitor based on capacitance measurement of sensitive layers [11]

2.1.3 Chemocapacitors

Certain polymers adsorb gaseous analyte molecules and show changes in capacitance of the substrate [11]. Change in dielectric property as well as physical property (volume) of the polymer matrix occurs upon adsorption of the analyte molecules. Capacitance under air environment gives the baseline signal and that in sample environment gives the analyte signal. The detection mechanism of the chemocapacitor is shown in Fig. 4.

2.1.4 Potentiometric Odour Sensors

Change in work function of the Schottky diodes in presence of the chemical species on their surface is the principle of working of this type of odour sensors. In another type, the MOSFET can be used [12], where the gate is made of a gas sensitive metal. Concentration of the analyte gas in the sample environment determines the threshold voltage. Use of different gate material makes the sensor to be responsive to different analyte gases.

2.1.5 Thermal (Calorimetric) Sensors

These sensors are formed from the basic thermal sensors [13] like pyroelectric [14] or thermopiles which use coatings that adsorb the analyte of interest. Heat is generated in the coating as the analyte gas is absorbed by the coating and provides the quantitative information about the gas.

Another type of thermal sensor is the catalytic sensor along with the platinum resistance thermometer. At about 500 °C, the hydrocarbon in the analyte sample undergoes catalytic oxidation and liberates heat of combustion. This heat, in turn increases the temperature of the platinum wire and so also the wire resistance. Higher the concentration of the hydrocarbon, larger is the wire resistance. These types of sensors were used in the very early versions of the electronic noses.

2.1.6 Gravimetric Odour Sensors

Recent research on electronic nose sensors is mostly concentrated on these types of sensors [15]. Sensors are of piezoelectric type and detect the effect of sorbed molecules on the propagation of acoustic waves. Two basic types of gravimetric sensors, quartz crystal microbalance (QCM) and surface acoustic wave (SAW), have been used as the odour sensor. Sorbent materials are coated on the sensor substrate and when sorbs the vapor molecules from the sample environment exhibit change in mass of the sensor and hence change in resonant frequency of the sensor. This frequency shift from the baseline provides qualitative as well as quantitative information about the sorbent molecules. Selectivity of both the sensors depends upon the type of coating material used on the sensor surface to produce sensing membrane.

QCM odor sensors [16] are fabricated from single crystal of quartz and have thin film gold electrode on the both surfaces of the crystal. Schematic of the QCM is shown in Fig. 5. Fundamental frequency of the sensor depends on the thickness of the sensor following the equation $\lambda/2 = t$. The Sauerbrey equation states the change in resonant frequency of the QCM sensor is related to the change of the mass of QCM loading by

$$\Delta f = -2f_0^2 m_f / A\left(\rho_q \mu_q\right)^{1/2}$$

where,

Δf change in resonant frequency,
f_0 resonant frequency,
m_f mass change due to adsorption of gases,
A electrode area,
q_q density of quartz and
μ_q shear modulus

QCM sensors are generally operated at 10 MHz and are sensitive to about 1 ng of mass change to give a change in frequency of 1 Hz. Measurement model and mechanism of QCM are as in Fig. 6.

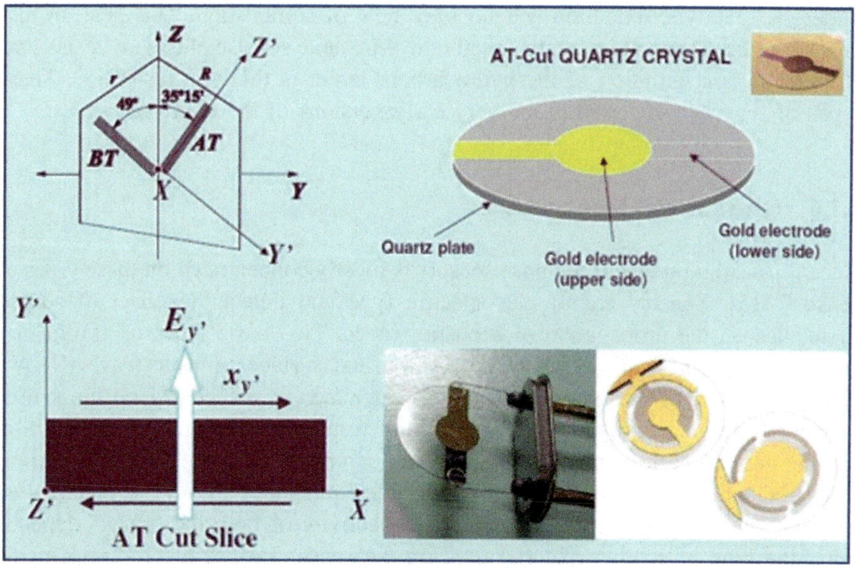

Fig. 5 Quartz crystal microbalance (QCM) [38]

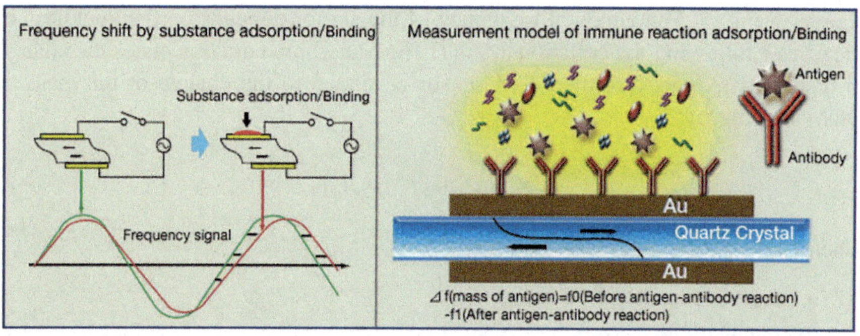

Fig. 6 Measurement model and mechanism of QCM [38]

SAW sensors [17] are fabricated from relatively thick piezoelectric material of ST quartz crystal. Propagation of the acoustic wave can be trapped or otherwise be modified near the surface of the sensor and is affected by changes in the properties of the piezoelectric crystal surface. Basic structure of SAW sensor is as given in Fig. 7.

There are two interdigital transducers (IDT) on the surface, one of which is used as the transmitter of acoustic wave by means of application of ac signal to it and the other is to receive the modified signal in terms of output voltage. Application of ac signal on IDT produces a strain pattern (acoustic wave) on the surface induced by inverse piezoelectric effect. This strain pattern when received by the receiver IDT

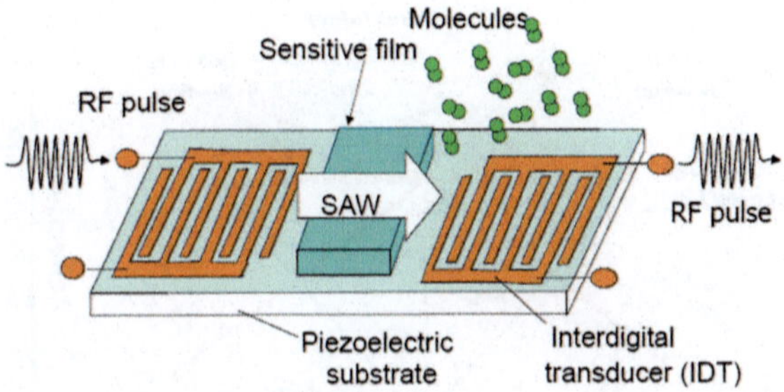

Fig. 7 Basic structure of a SAW device [39]

produces voltage following piezoelectric effect. The SAW devices are usually operated in one of two configurations such as a delay line and a resonator. The change in frequency of the SAW with sorption of vapor, Δf_v for a simple mass loading effect, is given by

$$\Delta f_v = \Delta f_p c_v K_p / \rho_p$$

where,

Δf_v is the change in frequency caused by polymer membrane itself,
c_v vapor concentration,
K_p partition coefficient and
q_p is the density of the polymer membrane used

Operating frequency of the SAW sensor is of the order of GHz compared to QCM sensor (several MHz).

2.1.7 Optical Odour Sensors

Surface plasmon resonance (SPR) sensors utilize the evanescent field of a special mode of electromagnetic field propagating at a metal/dielectric interface. The surface plasmon measures changes in the refractive index of the dielectric in the proximity of the interface [18]. The principle of analysis is based on the measurement of light reflected from the sensor interface and includes methods like angle modulation [19], wavelength modulation [20], intensity modulation [21], and phase modulation [22]. The sensor module consists of a glass slide with a thin gold coating mounted on a prism as shown in Fig. 8. Light passes through the prism and slide, reflects off the gold and passes back through the prism to a detector.

Changes in reflectivity versus angle or wavelength give a signal that is proportional to the volume of biopolymer bound near the surface. The resonance angle at which the minimum reflection intensity occurs is due to coupling of energy between the incident light and the surface plasmon waves. A flow cell allows solutions above the gold surface to be rapidly changed.

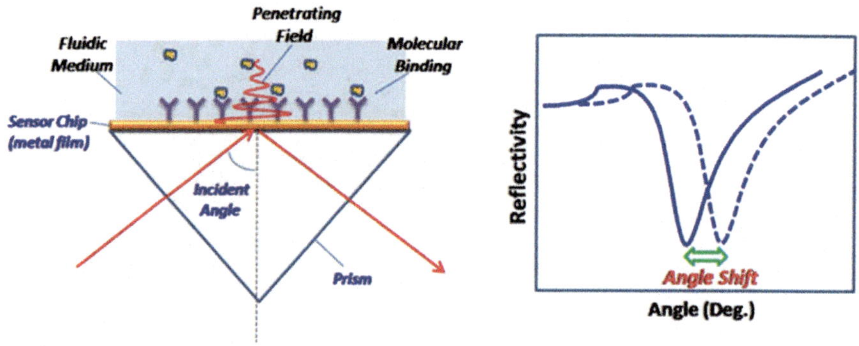

Fig. 8 Scheme of an optical system of an SPR sensor [18]

2.1.8 Fluorescent Odour Sensors

This type of chemosensor consists of optical fibers deposited with fluorescent indicator Nile Red dye in polymer matrices of varying polarity, hydrophobicity, pore size, elasticity, and swelling tendency to create unique sensing regions that interact differently with vapor molecules [23]. In a fiber-optic chemosensing system, the optical sensing element is typically composed of a reagent phase immobilized at the fiber tip by either physical entrapment or chemical binding. The chemical indicator in reagent phase experiences some changes in optical properties, such as wavelength shift in fluorescence, intensity change, lifetime change and spectrum change, upon interaction with analyte gases or vapors. The responses depend upon the nature of the organic vapor and the strength of its interaction with the different polymer systems used.

2.1.9 Amperometric Sensors

Principle of working of these sensors [24] involves the measurement of the current in the electrochemical cell between the working and counter electrodes as a function of the analyte concentration. The sensor uses a potentiostat circuit and produces current or signal when exposed to a gas/vapor containing an electroactive analyte. The analyte diffuses into the electrochemical cell and participates in a redox reaction with the working electrode surface. The cell current is directly related to the rate of reaction taking place at the electrode surface as described by Faraday's Law, relating the mass, W, of a substance of molecular mass M (grams/mol) as:

$$W = \frac{QM}{Fn}$$

where,

Q the charge per unit electrode area,
F Faraday's constant in coulombs/ equivalent, and
N the number of electron equivalents per mole of the reacting analyte

The observed current, $i\ (dQ/dt)$ is directly proportional to the amount of analyte, W, that is supplied to the working electrode and, this in turn can be related to the gaseous analyte concentration.

Commercially available electronic nose systems mostly use the MOS and CP sensors. With the advancement of the fabrication technology and other associated techniques, QCM and SAW sensors have attracted considerable interest among the researchers for use in electronic nose. Besides the quantitative and qualitative analysis of different products based on their odour, now a days, more complex and pertinent problems regarding contamination, purity, hazards, etc. are being targeted using sensor array. As a result the application area of an electronic nose has become more broadened.

2.2 *Electronic Tongue Sensors*

2.2.1 Potentiometric Sensors

Any type of charge transfer involves generation of potential and been represented by means of the Nernst equation. Potentiometry is a direct application of the Nernst equation to measure the potential between two non-polarized electrodes under the conditions of no current flow (Fig. 9). The concentration and nature of ions present in the solution as well as on the medium and the type of electrodes employed [25] determines the strength of potential to be developed. The developed potential in the

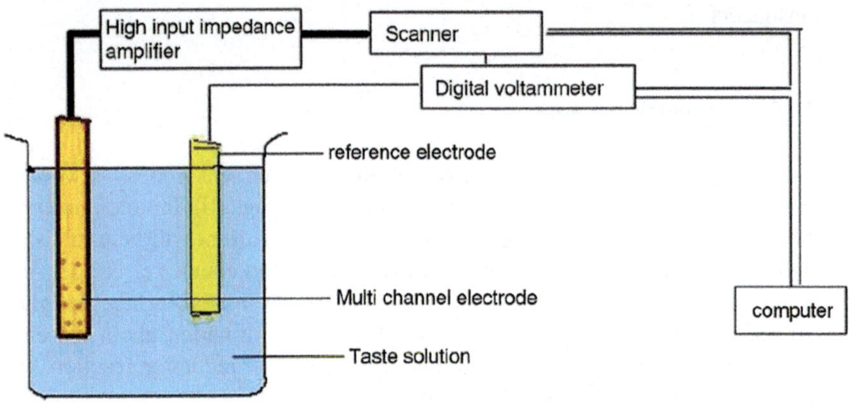

Fig. 9 Potentiometric multichannel taste sensor system [40]

electrochemical cell is the result of the free energy change that would occur if the chemical phenomena were to proceed until the equilibrium condition has been satisfied. Therefore, the potential thus developed carries the analytical information about the components of the solution under test.

Potentiometry based electronic nose comprises of an ion selective electrode, a reference electrode and a voltage measuring device. The silver/silver chloride (Ag/AgCl) electrode is commonly used as reference electrode. The electrode consists of a silver wire coated with silver chloride placed into the chloride ions solution. A porous plug serves as a salt bridge to the outer solution. On the other hand, an ion-selective electrode uses an ion selective membrane instead of the salt bridge. The differential potential developed at the two sides of the membrane carry the information about the analyte. The membrane should be non-porous, water insoluble and mechanically stable and should have an affinity for the selected ion [26].

The underlying measurement principle is potentiometric and sensor response are recorded as mV values. According to Nernst equation, the electrode potential depends on activity of the substance and is expressed as in equation below.

$$E = E_0 + \frac{RT}{zF} \ln a_i$$

where, $\boldsymbol{E}$ is the electrode potential; $\boldsymbol{E_o}$ is standard electrode potential; $\boldsymbol{R}$ is universal gas constant; $\boldsymbol{T}$ is temperature ($\boldsymbol{K}$); z is ionic valence of the substance; $\boldsymbol{F}$ is Faraday constant; and $a_i(=f_i c_i)$ is activity of the substance. c_i is the concentration of the substance; f_i is the activity coefficient of the substances.

Toko and co-workers [27] presented the first electronic tongue based on potentiometry. Several kinds of lipid/polymer membranes were used for the transformation of taste quality information into an electric signal. The output responses carry the information about the taste quality and intensity, because different output electric patterns are obtained for chemical substances producing different taste qualities. Legin et al. [28] introduces another type of potentiometric electronic tongue using several non-specific sensors based on chalcogenide glasses as transducers.

2.2.2 Voltammetric Sensors

A voltammetric electronic tongue consists of three different electrodes—a working electrode, a counter electrode and a reference electrode (Fig. 10). In voltammetry, a potential is applied on the working electrode and thus a current will be established through the electro-chemical cell between the working and counter electrode. As a result, an electrochemical redox reaction occurs at the electrodes' surface and gives rise to the measured current. The current carries the information about the electroactive species as they are participating in the oxidation or reduction reaction. The current in a voltammetric cell is equal to

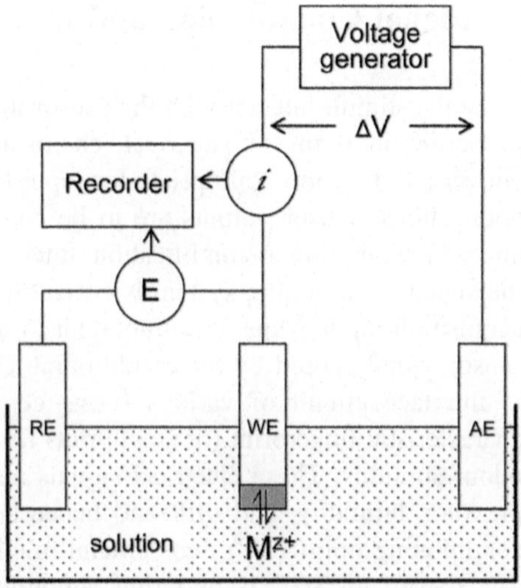

Fig. 10 Voltammetry setup [41]

$$i = \frac{nFAD(C_{bulk} - C_{x=0})}{\delta}$$

where $\boldsymbol{n}$ is the number of electrons in the redox reaction, $\boldsymbol{F}$ is Faraday's constant, $\boldsymbol{A}$ is the area of the electrode, $\boldsymbol{D}$ is the diffusion coefficient for the species reacting at the electrode, $\boldsymbol{C_{bulk}}$ and $\boldsymbol{C_{x=0}}$ are its concentrations in bulk solution and at the electrode surface, and $\boldsymbol{\delta}$ is the thickness of the diffusion layer [29].

In practice, multiple working electrodes are used in an electronic tongue and they are switched one by one to get more information regarding the analyte. An example of a voltammetric electronic tongue is the developed by Winquist et al. [5], where multiple working electrodes made of inert materials such as gold, platinum, palladium, iridium and rhodium are used. The excitation voltage may be different and based on the variation of applied excitation voltage, the voltammetry may be of cyclic, pulse, square-wave type.

2.2.3 Impedance Spectroscopy based Sensors

Electrochemical impedance spectroscopy (EIS) has now being used for qualitative analysis using electronic tongue [30]. Riul et al. [31] first introduce the electronic tongue based on impedance spectroscopy. The sensors were constructed from pure and composite nano-structured films of conducting polymers. Impedance measurements are advantageous because of their simplicity and shorter response times [30] compared with potentiometry or with voltammetry.

3 Signal Conditioning and Interfacing Circuits

Once the stimuli interact with the sensor to produce certain changes in the material properties in terms of electrical, chemical or of any form, it now needs to be converted to a form appropriate for input into the data acquisition system. In most applications, sensor outputs are to be converted into the voltage form (if it isn't already) in addition to amplification, filtering, digitization etc. The correct design of the signal conditioning system is critical for mapping the sensor output to the data acquisition input. Thus, it is important to note the changes in the properties of the sensor signal caused by the conditioning circuitry.

Interface circuit of various forms generates an electrical (analog or digital) characteristic fingerprint (or smellprint) from the sensor array corresponding to the odour stimulus. These electrical signals may be severely affected by temperature, pressure, humidity, and drift and hence become noisy. In this regard, the signal conditioning unit (SCU) finds crucial role in association with the interface circuit. The role of an SCU is buffering, amplification, filtering and some other special functions. The function of buffer is to isolate different electronic stages and provide impedance matching, which benefits to maximize energy transfer between circuits or systems. The voltage buffer and the current buffers are the two types, purposes of each is same that is to isolate circuits to avoid loading the input circuit or source from the output stage. Amplification of the sensor signal is necessary to suit the dynamic range of the analog to digital converter or some other dedicated data acquisition (DAQ) card used to interface with computer. Different types of filtering circuits are used to remove unwanted frequency components from the sensor signals. Sometimes various special functions like linearization, integration, differentiation, logarithmic and antilogarithmic conversion, peak-to-peak and phase detection, and temperature compensation are included to enhance the sensor signal quality in terms of compensation for cross-sensitivity, nonlinearity and deficiencies. The sensor array signal after proper signal conditioning is digitized and either analysed online or stored for offline analysis.

The chemoresistor output is in change in resistance value and a voltage divider with this chemoresistor or a Wheatstone bridge circuit forms the basic interface circuit as shown in Fig. 11. Care should be taken to design these circuits to have enhanced sensitivity. Standard condition for figure of merit is adopted in the bridge circuit. For improving selectivity of the sensor, the interface circuit often adopts the AC impedance spectroscopy where a small amplitude AC voltage with a sweep frequency of preset range is impressed across the sensor and the corresponding current values are measured. The current actually shows the contributions made by the sensors for different molecular structures of the vapours/gases. Plots of the resistance to reactance ratio for different vapours at different concentrations show sharp peaks at different frequencies [32].

Interfacing circuits of QCM or SAW devices are required to measure shift in the resonance frequency and frequency counters are employed as shown in Fig. 12. The crystals which are employed for this purpose have typical resonant frequencies

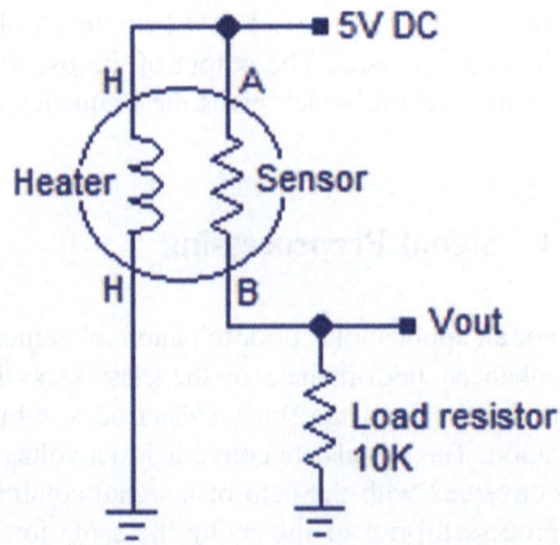

Fig. 11 Signal conditioning circuit for the MOS sensor

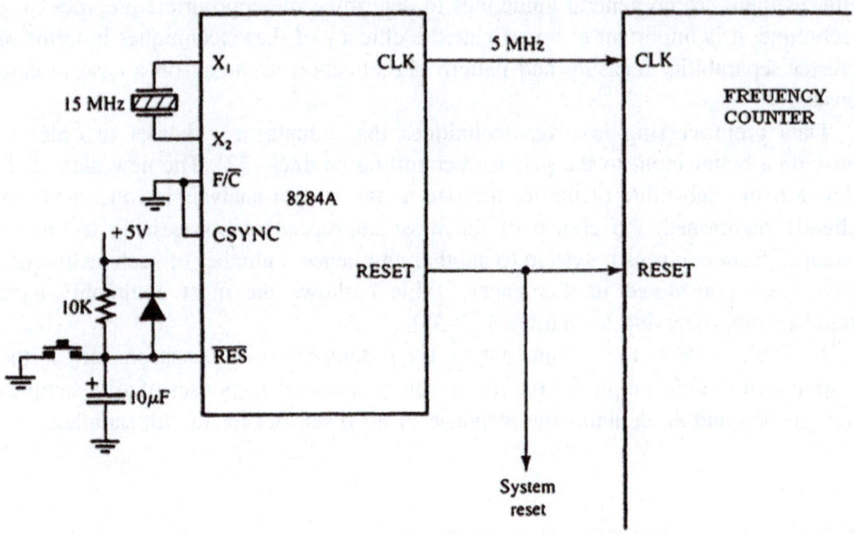

Fig. 12 Interfacing circuit (counter based) with QCM

of 10 MHz or 20 MHz. The frequency deviation is of order of a few hundred Hertz or even fraction of a hundred Hertz and the counter circuits can measure this deviation accurately.

A crystal oscillator circuit is commonly used as an interfacing circuit for QCM sensor. It uses a piezoelectric crystal (commonly a quartz crystal) as a frequency sensitive element. The crystal mechanically vibrates as a resonator when excited by voltage and the resonance frequency is modified upon addition or removal of a small mass due to film deposition at the surface of the resonator. The output frequency of the oscillator circuit is the resonance frequency of vibration.

The circuit as shown in Fig. 12 employs a clock generator circuit using IC 8284 and the quartz crystal. The output of the oscillator circuit is connected to a frequency counter circuit, which gives the frequency of the circuit.

4 Signal Preprocessing

For all applications, prior to pattern classification, data preprocessing is essential for enhanced performance of the classifiers. The signals collected are the transient responses from the sensors/electrodes, which constitute the raw data for computation. The signals are converted to a voltage level suitable for the analog-to-digital converters with the help of a signal conditioning circuit. These raw data are then processed prior to subjecting the same for pattern classification algorithms. There are a number of mathematical techniques for preprocessing of multivariate data. But, as there are no general guidelines to determine the appropriate preprocessing technique, it is important to investigate the efficacy of these techniques in terms of cluster separability measure and pattern classification accuracy on a case to case basis.

Data preprocessing involves techniques that transform a dataset in order to provide a better input to the pattern recognition engines [33]. The new data set is thus a better substitute of the former one as far as data analysis is concerned. As already mentioned, the choice of the most appropriate preprocessing technique changes from one sensor system to another and hence a number of such techniques have been considered in this paper. Table 1 shows the most commonly used standard preprocessing techniques [33–35].

In Table 1, M is the feature matrix for n samples from p sensors, M_{ij} is the response of the ith sample for the jth sensor, m_j contains responses of all n samples for sensor j and m_i contains the response of all p sensors for the ith sample.

Table 1 Standard data preprocessing techniques

Preprocessing technique	Mathematical expression
Relative scale$_1$	$M_{ij} = M_{ij}/\max(M)$
Relative scale$_2$	$M_{ij} = M_{ij}/\max(m_j)$
Relative scale$_3$	$M_{ij} = M_{ij}/\max(A_r), A_r$ is a $1 \times n$ reference response
Relative scale$_4$	$M_{ij} = M_{ij}/\lVert m_i \rVert$
Background subtraction	$m_i = m_i - a_b, a_b$ is the $1 \times p$ blank response
Range scale$_1$	$M_{ij} = (M_{ij} - \min(m_j))/(\max(m_j) - \min(m_j))$
Range scale$_2$	$M_{ij} = \left(2\left(\frac{(M_{ij} - \min(m_j))}{(\max(m_j) - \min(m_j))}\right)\right) - 1$
Baseline subtraction	$M_{ij} = M_{ij} - M_{1j}$
Standard normal variate transform	$M_{ij} = (M_{ij} - mean(M))/std(M)$
Auto scale	$M_{ij} = (M_{ij} - mean(m_j))/std(m_j)$

Different preprocessing techniques bear different implications [35]. Relative scaling is used to reduce the effect of concentration of the sample. While in relative $scale_1$, global compression with a maximum value of 1 is realized, relative $scale_2$ compresses values per feature with a maximum value of 1. In relative $scale_3$, a reference reading is taken, and the compression is relative to that response. The reference reading in our experiment is the response of the electrode array with only distilled water as the sample. Relative $scale_4$ compresses data with respect to the Euclidean norm. Background subtraction attempts noise correction by removing a blank response of the sensors. Range $scale_1$ and range $scale_2$ set the limits at (0, 1) and (−1, 1) respectively. A baseline subtraction removes the base reading of a sensor prior to applying the pulse waveform, and is often used in temporal data collection. Mean-centre sets the mean of the data to the origin. Standard normal variate transform reduces within-class scattering. Auto scale uses the mean-centre and sets the variance within the data to 1. This method is often used when responses are on different magnitude scales.

5 Conclusion

Electronic nose and tongue are relatively new entrants in the measurement and instrumentation domain and because of uncertain inputs, the interfacing circuits, sample delivery methods and preprocessing techniques are difficult to design and unique in nature. Research in this area is being carried out by the scientists and technologists all over the world. There are many variants, and in this chapter a broad overview of the techniques are mentioned. As the sensors used in electronic nose and tongues are different from other conventional sensors, a brief outline on the basic principle of the sensing mechanism is also illustrated. In both the instruments, the multiple sensors are simultaneously used to extract more information and the interfacing and preprocessing techniques presented in this chapter may be applicable in other areas as well.

References

1. J.W. Gardner, P.N. Bartlett, *Electronic Noses: Principles and Applications*, vol. 233 (Oxford University Press, New York, 1999)
2. C.D. Natale, A. Macagnano, A. Mantini, F. Davide, A. D'Amnico, R. Paolesse T. Boschi, M. Faccio, G. Ferri, Advances in food analysis by electronic nose, in *Proceedings of the IEEE International Symposium on Industrial Electronics*, vol. 1 (1997), pp. 122–127
3. L. Escuder-Gilabert, M. Peris, Review: highlights in recent applications of electronic tongues in food analysis. Anal. Chim. Acta **665**(1), 15–25 (2010)
4. Y. Vlasov, A. Legin, A. Rudnitskaya, C. Di Natale, A. D'amico, Nonspecific sensor arrays ("electronic tongue") for chemical analysis of liquids (IUPAC Technical Report). Pure Appl. Chem. **77**(11), 83–1965 (2005)

5. F. Winquist, P. Wide, I. Lundström, An electronic tongue based on voltammetry. Anal. Chim. Acta **357**, 21–31 (1997)
6. F. Winquist, C. Krantz-Rülcker, I. Lundström, Electronic tongues and combinations of artificial senses. Sens. Update **11**(1), 279–306 (2002)
7. E.A. Baldwin, J. Bai, A. Plotto, S. Dea, Electronic noses and tongues: applications for the food and pharmaceutical industries. Sensors **11**, 4744–4766 (2011)
8. D.M. Wilson, K. Dunman, T. Roppel, R. Kalim, Rank extraction in tin-oxide sensor arrays. Sens. Actuators B: Chem. **62**, 199–210 (2000)
9. J.R. Stetter, J. Colloid Int. Sci. **65**(3) (1978)
10. H.V. Shurmer, P. Corcoran, J.W. Gardner, Integrated arrays of gas sensors using conducting polymers with molecular-sieves. Sens. Actuator B: Chem. **4**, 29–33 (1991)
11. C. Cornila, A. Hierlemann, R. Lenggenhager, P. Malcovati, H. Baltes, G. Noetzel, U. Weimar, W. Gopel, Capacitive sensors in CMOS technology with polymer coating. Sens. Actuator B: Chem. **25**, 357–361 (1995)
12. I. Lundstrom, M.S. Shivaraman, C.S. Svenson, L. Lundkvist, Hydrogen sensitive MOS field-effect transistor. Appl. Phys. Lett. **26**, 55–57 (1975)
13. W.J. Buttner, J.R. Stetter, G.J. Maclay, An integrated amperometric microsensor. Sens. Mater. **2**, 99–106 (1990)
14. J.N. Zemel, Future directions for thermal information sensors. Sens. Actuators A **56**(1–2), 57–62 (1996)
15. K.A. Marx, Quartz crystal microbalance: a useful tool for studying thin polymer films and complex biomolecular systems at the solution-surface interface. Biomacromol. Am. Chem. Soc. **4**(5), 1099–1120 (2003)
16. M. Slater, E.J. Watt, Examination of ammonia poly(pyrrole) interactions by piezoelectric and conductivity measurements. Analyst **116**, 1125–1130 (1991)
17. http://www.bureau.tohoku.ac.jp/kohyo/kokusai/06Feb3research-overview.htm
18. M. Piliarik, J. Homola, Surface plasmon resonance (SPR) sensors: approaching their limits? Opt. Express **17**(19), 16505–16517 (2009)
19. E. Kretschmann, The determination of the optical constants of metals by excitation of surface plasmons. Z. Phys. **241**, 313–324 (1971)
20. K.S. Johnston, S.R. Karlson, C. Jung, S.S. Yee, New analytical technique for characterization of thin films using surface Plasmon resonance. Mater. Chem. Phys. **42**, 242–246 (1995)
21. B. Chadwick, M. Gal, An optical temperature sensor using surface plasmons. Jpn. J. Appl. Phys. Part I 32, 2716–2717 (1993)
22. S.G. Nelson, K.S. Johnston, S.S. Yee, High sensitivity surface plasmon resonance sensor based on phase detection. Sens. Actuators B: Chem. **35–36**(1–3), 187–191 (1996)
23. J. White, J.S. Kauer, T.A. Dickinson, D.R. Walt, Rapid analyte recognition in a device based on optical sensors and the olfactory system. Anal. Chem. **68**, 2191–2202 (1996)
24. G.J. Maclay, W.J. Buttner, J.R. Stetter, Microfabricatedamperometric gas sensors. IEEE Trans. Electron Devices **35**, 793–799 (1988)
25. G. Pioggia, F. Di Francesco, A. Marchetti, M. Ferro, A. Ahluwalia, A composite sensor array im-pedentiometric electronic tongue Part I. Characterization. Biosens. Bioelectr. **22**, 2618–2623 (2007)
26. F. Winquist, C. Krantz-Rülcker, I. Lundström, Electronic tongues and combinations of artificial senses. Sens. Update **11**(1), 279–306 (2002)
27. K. Hayashi, M. Yamanaka, K. Toko, K. Hamafuji, Multichannel taste sensor using lipid membranes. Sens. Actuators B: Chem. **2**(3), 205–213 (1990)
28. A.V. Legin, Y.G. Vlasov, A.M. Rudnitskaya, E.A. Bychkov, Cross-sensitivity of chalcogenide glass sensors in solutions of heavy metal ions. Sens. Actuators B **34**, 456–461 (1996)
29. D. Harvey, *Modern Analytical Chemistry* (McGraw-Hill, 2000)
30. M. Cortina-Puig, X. Muñoz-Berbel, M.A. Alonso-Lo-millo, F.J. Muñoz-Pascual, M. del Valle, EIS multianalyte sensing with an automated SIA system—an electronic tongue employing the impedimetric signal. Talanta **72**, 774–779 (2007)

31. A. Riul, R.R. Malmegrim, F.J. Fonseca, L.H.C. Mattoso, An artificial taste sensor based on conducting polymers. Biosens. Bioelectron. **18**, 1365–1369 (2003)
32. U. Weimar, W. Gopel, AC measurements on tin oxide sensors to improve selectivities and sensitivities. Sens. Actuators B **26**, 13–18 (1995)
33. P.C. Jurs, G.A. Bakken, H.E. McClelland, Chem. Rev. **100**, 2649–2678 (2000)
34. J.W. Gardner, P.N. Barlett, *Electronic noses: principles and Applications* (Oxford University Press, 1999)
35. S.M. Scott, D. James, Z. Ali, Data analysis for electronic nose systems. MicrochimActa **156**, 183–207 (2007)
36. Wadehra, P.S. Patil, Application of electronic tongues in food processing. Anal. Methods **8** (3), 474–480 (2016)
37. http://www.tehnomagazin.com/Sensors/Air-quality-sensor.htm
38. http://www.ecvv.com/product/3060721.html
39. J.W. Grate, S.J. Martin, R.M. White, Acoustic-wave microsensors. 1. Anal. Chem. **65**, 940–948 (1993)
40. R.S. Latha, P.K. Lakshmi, Electronic tongue: an analytical gustatory tool. J. Adv. Pharm. Technol. Res. **3**(1), 3–8 (2012)
41. C.B. Braungardt, "Evaluation of Analytical Instrumentation", Part XXVI: Instrumentation for Voltammetry. Anal. Methods **7**, 1249–1260 (2015)

Harnessing Vision and Touch for Compliant Robotic Interaction with Soft or Rigid Objects

Ana-Maria Cretu and Pierre Payeur

Abstract The chapter discusses recent research achievements related to sensing issues and interfacing techniques to enable safe interaction of commercial-grade robot manipulators with objects exhibiting rigid or soft surfaces. The main challenges are described, including the identification of proper combinations of vision and touch sensor technologies, and their placement and trajectory with respect to the objects of interest to enable safe navigation and close interaction. Various selective data acquisition procedures are also examined to ensure fast and sufficient monitoring of the interaction behaviour of the object under forces imposed by a robotic manipulator or a multi-finger gripper. Issues related to sensor calibration and data fusion are detailed. Potential solutions are presented in the context of various interaction tasks, including adaptive surface and contour following, object characteristics identification, and dexterous robot hand manipulation of soft objects using the Barrett hand. Numerous experiments demonstrate the validity of the proposed solutions.

1 Introduction

Modern robotic systems to be employed in industrial, security and space applications require the development of a new generation of autonomous robot manipulators able to intelligently perform sophisticated manipulation tasks [1] in environments that are often unknown, variable or unstructured. Over the past

---

A.-M. Cretu (✉)
Computer Science and Engineering Department,
Université du Québec en Outaouais, Gatineau, Canada
e-mail: ana-maria.cretu@uqo.ca

P. Payeur
School of Electrical Engineering and Computer Science,
University of Ottawa, Ottawa, Canada

B. George et al. (eds.), *Advanced Interfacing Techniques for Sensors*,
Smart Sensors, Measurement and Instrumentation 25,
DOI 10.1007/978-3-319-55369-6_9

decades, a huge research effort was invested in the design and development of robot systems able to sense and react intelligently to their environment and safely handle rigid and deformable objects. Building such systems that can interact autonomously with unknown objects is a complex task, requiring a combination of sensing technologies, control systems, knowledge of computer and mechanical engineering, as well as an understanding of human abilities that could be mimicked in order to produce more flexible, general and intelligent solutions.

Human vision-touch experience shows the ability of vision in assisting grasping, handling and manipulation tasks. In a similar manner, autonomous robot manipulators can count on the coordination of these two sensory capabilities to adapt to unpredictable situations and work efficiently in unknown environments. Vision sensing, as provided by stereo cameras, RGB-D sensors or range scanners provides rich information on the geometry and topology of the objects to be manipulated. Along with advanced image processing techniques, it can also enable the monitoring or tracking of soft object deformations under forces exerted by the manipulator. Visual feedback can improve the grasping and manipulation process by guiding the robot manipulator and assist in the estimation of the relationship between the object and the end-effector. Integrating visual feedback with touch (contact, force) sensing also compensates for the inaccuracy of vision systems alone due to occlusions and the inability of vision sensors to provide force measurements. Moreover, the use of vision sensing in the system can guide the touch probing towards areas of relevant features in order to shorten the exploration time which can be long, as the manipulator must execute multiple complex motions to collect tactile data.

Most of current research effort in the robotics literature is focused on manipulation and grasping of rigid objects. Relatively few researchers yet dedicated their interest to the interaction with deformable objects, while in fact numerous real-world objects are mostly unsymmetrical, compliant, and exhibit alterable shapes. The robotic manipulation of deformable objects still offers an important challenge to the robotics community and makes it a subject of significance for the development of future generations of autonomous robots. This chapter discusses challenges in rigid and deformable object grasping and manipulation based on a combination of vision and touch sensor technologies. The authors' research groups investigated their placement and trajectory with respect to the objects of interest to enable safe navigation and close interaction, the various selective data acquisition procedures to ensure fast and relatively complete monitoring of the interaction behaviour of the object under forces imposed by a robotic manipulator, as well as object modelling techniques. These issues and some potential solutions are exemplified in the context of various interaction and manipulation tasks, including adaptive surface and contour following, surface characteristics identification, and dexterous robotic hand manipulation of soft objects using the Barrett hand.

2 Challenges of Robotic Interaction with Soft or Rigid Objects

While the interaction with objects exhibiting soft or rigid surfaces is one of the fundamental capabilities of autonomous robot systems, the design and development of comprehensive autonomous robotic systems able to interact with surfaces and manipulate objects, in particular soft deformable objects, without human intervention remains a challenging task. As briefly highlighted in the introduction, such complex interaction can only take place with the assistance of multisensory data acquisition systems that combine vision and touch (tactile, force-torque) measurements. Such sensors allow for the recuperation of crucial information on the interaction, including the location (pose) of the object in the environment, the occurrence of a contact between the manipulator and the object, the size and shape of the object, its material properties, the magnitude and position of forces exerted by the manipulator or the detection of slippage of the object from the manipulator. A coordinated fusion of this information opens door to dexterous manipulation. However, there are several issues that complicate the automation of this information acquisition and fusion.

In the case of large rigid objects, if multiple vision sensors are involved, a calibration process is required prior to their use. An object model is indispensable to represent the geometry of the object and to enable its close inspection or the interaction with it. Ideally, this model needs to be compact to support the robot's operation in real time. Moreover, because local tactile probing is time consuming, intelligent selective algorithms should be employed to only select areas of interest, such as areas where the local geometry changes, for enhancing the sensing procedure. Path planning algorithms have to be employed accordingly to guide the interaction with the object.

In contrast to the manipulation of rigid objects which has been extensively studied in the literature and for which well-established procedures exist, the investigation of the manipulation of soft deformable objects represents a more recent undertaking. While several 1D and 2D solutions tackle the issue of grasping and manipulation of soft objects [2, 3], few researchers have addressed the manipulation and grasping of 3D objects [4–7]. This is due to its complexity and to the fact that a majority of researchers hope to tackle simpler 1D and 2D modelling problems before generalizing to a 3D solution. One of the most critical issues is the difficulty to estimate or predict in real-time the deformation properties of the object [8]. These properties tend to vary greatly among various objects. Their understanding and their prediction, ideally without making assumption on the material (such as linearity, homogeneity and isotropy), is necessary to coordinate the motion of the manipulator and its interaction with the object. If touch sensors are involved as well (i.e. force-torque sensors), a synchronization of visual and force (or tactile) data is required as different sensing technologies work at different sampling rates. As well, in this case, as for rigid objects, the probing should be restricted only over areas of interest. Furthermore, monitoring the coupling between the contact forces and the object deformation is also necessary to study the impact of the position, the magnitude and the angle of forces applied to the object over the various stages of the shape deformation. In order to implement and evaluate the interaction tasks of a soft object with a robot manipulator, a soft object model is required to represent the deformation characteristics during the physical interaction. In classical models, the deformation characterization generally implies the approximate identification of elastic parameters of the model, generally a mass-spring model [5, 9] or a finite element representation [10–12], by comparing the real and simulated object subject to interaction and aiming to minimize the differences. However, these approaches work only by making assumptions on the object material, such as linearity, homogeneity or isotropy, which do not transpose well to multiple materials such as foam or rubber. These justify our interest in the development of methods that do not make assumptions on the material of the object, but rather directly employ experimental data to make decisions on the properties of the object and capture implicitly the deformation behaviour.

Once the model is developed, a control scheme has to be proposed to ensure smooth interaction with the object. For rigid object exploration or contour following, a path planning algorithm is required to guide the motion of the manipulator to achieve the desired task. In case of dexterous grasping and manipulation of soft objects, this operation needs to be performed robustly in spite of possible uncertainties in the robot environment in which a deformable object is neither located at a precise position, nor modelled with high accuracy. For such dexterous manipulation, it is important to consider the difference between the ways of handling a rigid or a deformable object, in particular the major distinction between the definitions of grasping and manipulation respectively [13]. The manipulation of a rigid object requires only the control of its location and therefore grasping and manipulation can be performed independently. Grasping of a rigid object requires the control of grasping forces only, while manipulation of a freely moving rigid object results in the change of its position and orientation. On the other hand, grasping and manipulation interfere with each other in the manipulation of deformable objects. Handling of a deformable object requires controlling both the location of the object and its deformation. But grasping forces yield the deformation of a deformable object, which may change the shape and location of the object. Hence contact between fingers and the object may be lost and grasping may be compromised due to the deformation at the fingertips. Therefore, in the handling of deformable objects, grasping and manipulation must be performed in a collaborative way.

These issues will be exemplified in the context of practical interaction (manipulation) applications in Sect. 4, after the following section describes some of the relevant work on robotic interaction with rigid or soft objects, respectively.

3 Related Work on Robotic Interaction with Rigid or Soft Objects

The literature reports on relatively few research that has been performed in the context of 3D soft object modelling and on robotic interaction with 3D rigid and deformable objects. In [7], an approach is proposed for the in-hand modelling of 3D rigid objects using RGB-D data. An estimate of the position of a robot manipulator, the object and the Kinect sensor is produced at each frame by a Kalman filter based on depth and visual information. These estimates enable the segmentation of the object and its model is built using a series of surfels. Also using surfel models, the authors of [14] propose a registration method on multi-resolution surfel maps that provides a dense displacement field between deformable object shapes monitored in RGB-D images. Petit et al. [11] explore the issue of real-time tracking of 3D elastic objects in RGB-D data. Assuming that a prior segmentation of the object of interest is available, the object is tracked using a graph-cut approach. The iterative closest point (ICP) method is then applied on the resulting point-cloud to estimate a rigid transformation from the point-cloud to a linear tetrahedral finite element model (FEM) representing the object. Linear elastic forces exerted on vertices are computed from the point cloud to the mesh based on closest point correspondence and the mechanical equations are solved numerically to simulate the deformed mesh. A linear isotropic 3D deformable object in interaction with a three-fingered robot hand is modelled by Zaidi et al. [12] as a mass-spring system based on a tetrahedral mesh. The object deformations and the contact points estimation is based on tracking the node positions by solving the dynamic equations of Newton's second law. The authors of [15] measure the stiffness of a 3D planar elastic object by the curvature of surface points from the object geometry and describe the local deformation in terms of a level curve set. In the same line of research, the authors of [16] inscribe markers on the surface of a paper to track its folding in the visual data input. The paper in interaction with a robot hand is represented as a 2D grid of nodes connected by links that specify the bending constraints, namely a resting distance between two nodes and the stiffness coefficient that are tuned manually. Choi et al. [9] propose to tune elasticity parameters of moving deformable balls, painted red against a blue background, by tracking their global position in a video stream and optimizing the differences between the real object captured and its mass-spring representation. In [17], models are acquired and tracked via a webcam. While visual features alone work correctly for some objects, many objects lack sufficient texture for this type of tracking. Sparse sets of oriented 3D points along contours of objects manipulated by a robotic manipulator are monitored in Kraft et al. [18] using a stereo camera, and then predicted based on the motion induced by the robot. Schulman et al. [19] track deformable objects from a sequence of point-clouds by identifying the correspondence between the point-cloud and a model of the object composed of a collection of linked rigid particles, governed by dynamical equations. An expectation-minimization algorithm aims at finding the most probable node positions for the model given the measurements. Tests are performed in a controlled environment, against a green background that limits its applicability to normal conditions. A solution for robot manipulation of elastic objects that allows to control simultaneously the object's final position (i.e. points of interest over the object and its centroid) and its deformations (i.e. compression distance between points of interest, folding angle and normalized curvature of the object, as estimated by the curve passing through 3 points of interest) is proposed in [8]. In Hur et al. [20], a 3D deformable spatial pyramid model is introduced to find the dense 3D motion flow of deformable objects in RGB-D data without assuming a prior model or template for the object. The point-cloud is corrected with a depth hole-filling algorithm and treated with a Gaussian filtering prior to the computation of a series of perspectively normalized descriptors. The 3D deformable spatial pyramid finds dense correspondences between instances of a deformed object by optimizing an objective function, in form of an energy corresponding to a Markov random field that takes into consideration the translation, the rotation, the warping costs and the descriptors matching costs.

4 Vision and Touch Sensing Systems for Soft Object Interaction

The main challenge in developing autonomous robotic systems able to handle deformable objects originates from the fact that a series of interconnected problems have to be solved, starting with data acquisition, data fusion, data modelling, simulation and validation of objects properties, and the definition and tuning of a control scheme to safely handle the manipulation or the interaction with an object. Figure 1 illustrates these issues in the context of a combined use of vision and

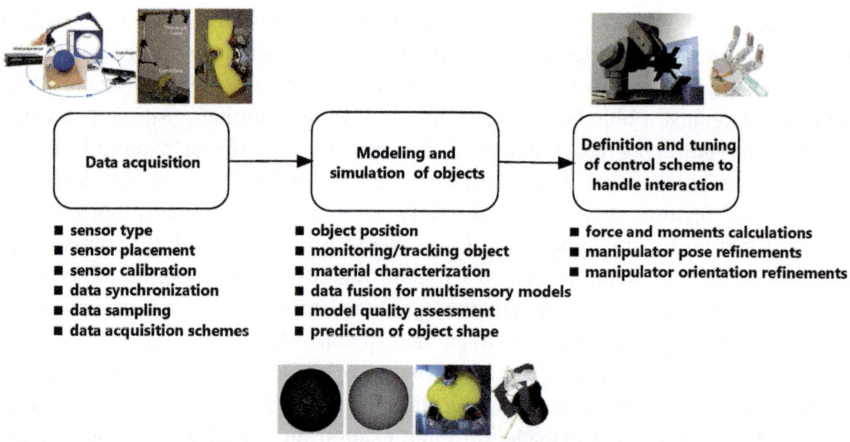

Fig. 1 The series of interconnected problems in rigid or soft object interaction with a manipulator

tactile sensing to enable an autonomous interaction and/or manipulation of rigid or soft objects.

These issues and some possible solutions are discussed in the following subsections.

4.1 Data Acquisition

In the data acquisition process, there are multiple aspects to be taken into account in practical applications, including: the sensor technologies to be used, the placement of the sensors in the environment, the calibration between multiple sensors and sensor technologies, data sampling strategies and selective acquisition schemes to allow for the collection of only relevant data and acceleration of sensing.

4.1.1 Sensor Type and Placement

Realistic, plausible models for objects require the acquisition of experimental measurements using physical interaction with the object in order to capture its complex behavior when subject to various forces. Tests can be carried out based on the results of instrumented indentation tests and usually involve the monitoring of the evolution of the force (e.g. its magnitude, direction, and location) using a force-feedback sensor (i.e. force-torque) (Fig. 2a) or measuring forces applied by the fingers of a robot hand (Fig. 2b, c) accompanied by a visual capture of the deformed object surface to collect geometry data.

In order to collect 3D geometry data, a classical solution offering high precision, are laser scanners. However, they are expensive and the acquisition is often lengthy. In this case, algorithms should be employed to only collect relevant data (Sect. 4.1.3). Stereo systems (Fig. 2b) provide good results, but at the price of a significant computational load and they are prone to important feature matching constraints which often lead to low density depth maps [21].

Moreover, most of the current sensors cannot capture color and depth simultaneously. To overcome these limitations, several attempts have been made to capitalize on the use of the RGB-D Kinect sensor (Fig. 2a, c). The sensor proves to be a simple, fast and cost-effective alternative to collect high density depth maps and the associated color information in a fraction of a second. In spite of the low resolution of the depth map, it generally offers enough precision for most robot manipulation tasks.

Visual data provided by Kinect, has been successfully used for the reconstruction of 3D point clouds of objects by merging data from multiple viewpoints (Fig. 2a), for rigid [7] and non-rigid objects [24, 25] as well. A few open-source [26] and commercial solutions [27] are also available. To collect a full 3D model the sensor is turned around the object of interest following a trajectory similar to the one marked by blue arrows in Fig. 2a and integrating the partial collected point

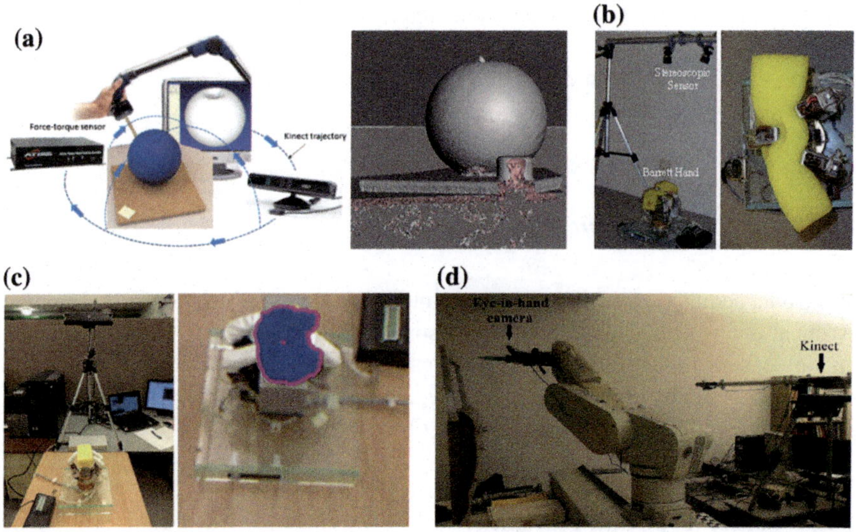

Fig. 2 Multi-sensory vision and tactile data acquisition platforms systems for: **a–c** soft object deformation—**a** Kinect sensor and force-torque sensor collecting 3D data on an object of interest, **b** object handled with a Barrett hand observed by a stereo-system [22], **c** 3D soft object deformation monitoring using a Kinect sensor—and **d** contour following operation [23]

clouds. Alternatively, a partial 2.5D point cloud obtained from a single sensor is sufficient to track contours and detect the object material characteristics (Fig. 2c).

Vision data obtained by Kinect can be also used to locate the object in an unknown environment and to guide the robot arm in proximity to the object. Located behind the robot at a given distance, it can provide the global shape and depth information in complex contour following tasks (Fig. 2d) [23]. This information can, in this case, complement the higher accuracy measurements on the contour location recuperated from an eye-in-hand camera.

4.1.2 Calibration

In the case of large objects, a single Kinect cannot be used to capture the entire surface. When multiple sensors are grouped and operated as a collaborative network of imagers in order to enlarge the overall field of view and allow for modelling large objects, such as automotive vehicles (Fig. 3), a precise mapping between the color and depth components of all the Kinect sensors must be achieved. The internal and external calibration processes proposed in [28] can be used in such situations. The internal calibration corresponds to estimating intrinsic parameters for the color and IR cameras inside a given Kinect, while the extrinsic process provides accurate estimates of the extrinsic parameters in between any respective pair of Kinect devices.

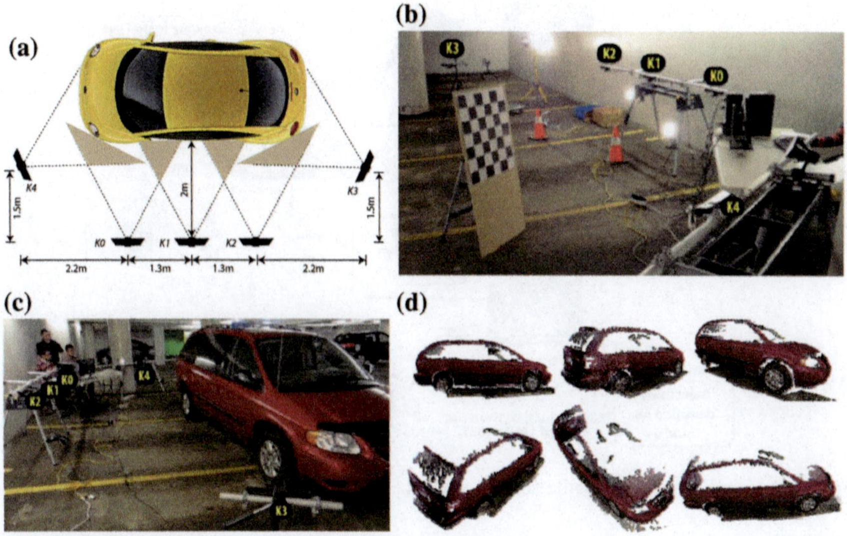

Fig. 3 **a** Sensor system using 5 Kinect sensors K0-K4 for vehicle inspection [28], **b** calibration of sensor using checkerboard, **c** data acquisition over vehicle, and **d** views of reconstructed vehicle

4.1.3 Data Acquisition Schemes

In terms of visual data acquisition schemes, several options are possible. Uniform sampling offers a straightforward solution to ensure complete coverage of a surface. However, in order to achieve adequate sampling density over regions where the local geometry is most likely to vary, the sampling density must be uniformly high over the entire surface and this may lead to inefficiency in certain applications. Each point of the object has an equal chance of being measured in random sampling, but only a lower number of points are actually collected. With an increase in the percentage of sampled points, the cost gets higher to eventually become equal to that of uniform sampling. As well, sampling points randomly might lead to missing important features. In stratified sampling, spaced samples are generated by subdividing the sampling domain into non-overlapping partitions and then by sampling independently from each partition. Such a technique ensures that an adequate sampling is applied to all partitions. It can also be employed in the context of post-processing of large point clouds or meshes [29, 30], where a subdivision of models into grid cells occurs and sample points falling into the same cell are replaced by a common representative. However, all these methods are not meant to be incorporated in the actual sampling procedure, but they rather post-process collected data.

Meant to be incorporated directly in the sampling procedure, a framework to achieve automated selective scanning over large workspaces [31] is illustrated in Fig. 4. A self-organizing neural network architecture, namely a growing neural gas network, adaptively selects regions of interest for further refinement from a cloud of

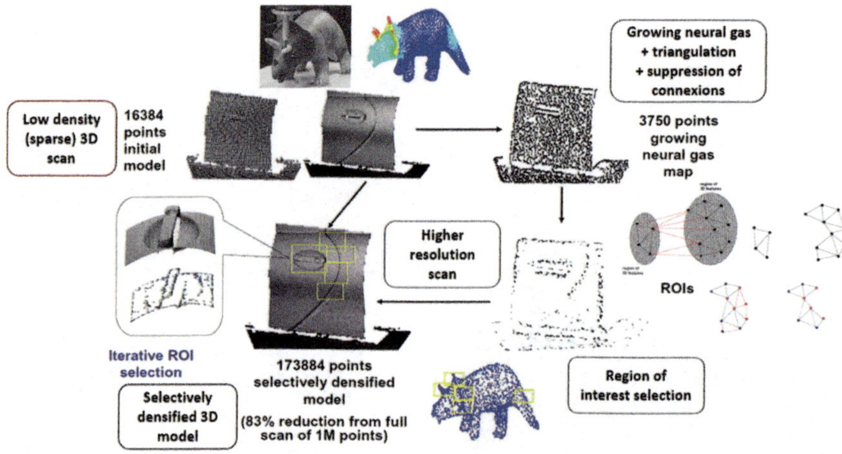

Fig. 4 Selective vision and tactile scanning scheme (adapted from [31])

3D sparsely collected measurements. Starting from an initial low resolution scan of an object, the network is employed to model the resulting point cloud. Those regions that are worth further sampling in order to ensure an accurate model are detected by finding higher density areas in the resulting map. This is achieved by applying a Delaunay tessellation to the resulting growing neural gas output map and by subsequently removing from the tessellation all the triangles that are larger than a set threshold. The latter is automatically computed based on the length of vertices for every triangle in the tessellation.

Rescanning at higher resolution is performed for each identified region (shown in yellow over the car door model in Fig. 4) and a multi-resolution model is then built by augmenting the initial sparse model with the higher resolution data from regions of interest. In this way, a much more compact model can be achieved (i.e. for the car door model in Fig. 4 only 17% of the total number of points that would have resulted from a full-scan) and that contains accurate details only in the regions of interest.

In terms of data acquisition schemes for tactile measurements, current research concerns itself with computer generated objects and their simulation. Conducting strain-stress relationship measurements for objects made of materials that exhibit nonlinear behavior is extremely challenging. Therefore, many applications leave the choice for the selection of elastic parameters to the user, or values are chosen according to some a priori knowledge regarding the deformable object model. This is a subjective process that cannot be applied where accuracy is expected. When measurements of elastic behavior are performed, often a single probing of the object is collected. While this procedure gives satisfactory results for objects made of homogeneous materials, it is unsuitable for objects that are non-homogeneous and have varying elastic properties in different parts of their bodies. Furthermore, the procedure for the acquisition of tactile measurements from each point of an object is extremely time-consuming. These two aspects explain the considerable interest in finding fast sampling procedures for the measurement of the tactile properties of 3D object surfaces. Appropriate sampling control algorithms should be able to minimize the number of the sampling points by selecting only those points that are relevant to the elastic characteristics.

Due to the human vision-touch experience showing the ability of vision in assisting grasping, handling and manipulation tasks, visual information and particularly the regions of interest into visual information can be used for the collection of tactile measurements. This approach is also justified by the fact that changes in the geometry are very often associated with changes in the elastic behaviour of objects. Using the same framework shown in Fig. 4, if the growing neural gas network is applied not only over the geometry data, but is supplemented with compliance information (an approximate measure of elasticity), during the learning procedure, the model contracts asymptotically towards the points in the input space, respecting their density and thus taking the shape of the object encoded in the point cloud. The regions of interest are identified in a similar manner to the one followed for visual data, but removing from the tessellation not only all the triangles that are larger than a set threshold, but also those which have the same compliance. Due to these properties, if tactile measurements are collected over these identified regions of interest (marked with yellow boxes over the triceratops model in the bottom of Fig. 4), the density of the tactile probing points is higher in the regions with more pronounced variations in the geometric shape. The advantage of such a model is not only to identify relevant sampling points, but also to allow for the determination of clusters of sampling points with similar geometric properties, due to its ability to find an optimal finite set that quantizes the given input space. This provides a robust mechanism that can be extended to model non-homogeneous objects.

It is expected that the collection over points of interest inspired from a visual attention mechanism in vision data [32] could also improve the tactile data acquisition process. The consideration of various aspects derived from psychological studies could also be included in advanced intelligent sensing systems to enable the next generation of intelligent autonomous robotic manipulators. For example, the bias of visuo-haptic estimates towards vision, that is the fact that stimuli are judged to be slightly softer under vision-only condition than under touch-only condition and that the haptic softness perception is more reliable with deformable as compared to rigid surfaces [33] can efficiently guide sensing strategies. Additional testing is required before confirming the effectiveness of these procedures.

4.1.4 Data Cleaning and Synchronization

Data collected using the vision sensors, such as Kinect, often contains undesired elements, such as a background or a surface over which an object is placed, some fixed landmarks required by the software to merge 3D data from multiple viewpoints, or the probing tip when a force-torque sensor is used, as it can be noticed in

Fig. 2a. These can be eliminated in part automatically (e.g. supporting surface and landmarks). However, when a tactile sensor probing tip touches the surface and gets acquired as part of the object, a manual intervention might be required to remove the tip and fill the resulting holes. A mesh processing software (e.g. Meshmixer [34]), can be employed for this purpose.

Due to the different sampling rates found in vision (3D data collection on one side and image analysis to recuperate the angle of the probing tip with respect to the surface on the other side) and force-torque sensors (force magnitude measurements), a synchronization process is also required in order to associate the correct surface deformation with the corresponding force magnitude and angle measurements. This can be achieved by calculating a mean of all the recorded force magnitude and also of the angle of measurement over the time it takes for the 3D object model to be collected. The deformed object model can be considered as a result of the application of a force with a magnitude equal to the mean magnitude and applied at an angle equal to the mean angle value.

4.2 Object Modelling and Simulation

4.2.1 Object Position Recuperation and Segmentation

The object of interest is normally selected in the visual environment using user guidance. A user-selected point can guide the segmentation algorithm towards the location of the object of interest. Such user guidance is common in current tracking literature, going from more extreme approaches in which a prior segmentation of the object of interest is assumed to be available in [11], to cases when the user is asked to crop the object in the initial frame [9]. Other solutions capitalize on the automation of the process, by exploiting the fact that the manipulation of objects takes generally place in relatively controlled environments. Therefore the solutions need to be insensitive to smooth changes in lighting, contrast and background, but do not have to deal with multiple moving objects, or with severe changes in the environment. One such solution based on growing neural gas [6] is illustrated in Fig. 5, where the segmentation is treated as a clustering problem based on color information (HSV color components) and spatial features (X and Y coordinates of each pixel in a color image extracted from a video stream). The HSV color space is chosen because it represents better the color similarities and is able to more accurately identify pixels on the same surface in spite of some differences between their colors due to non-uniform illumination or shading effects. A growing neural gas is adapted over the color and spatial information and the resulting map is then classified as one of two categories: object of interest or background based on the mean HSV value computed for the two clusters and making the assumption that generally the background is darker in color than the object of interest. The latter assumption is generally satisfied due to the controlled environment in which the experiments are performed. To identify the color of the object of interest, the mean

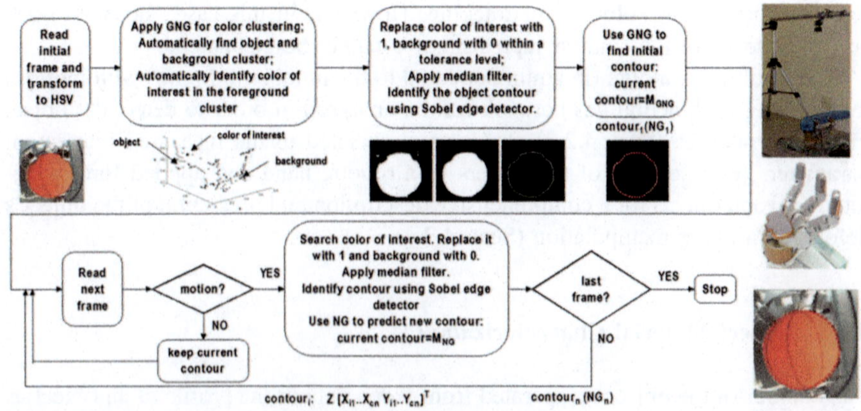

Fig. 5 Object segmentation and tracking (adapted from [6])

is computed for all HSV values in the corresponding cluster. The identified color is then searched in the initial image and over all images in the sequence where movement occurs and all pixels with this color code or a very similar code (within a tolerance level required due to different lighting conditions and due to the fact that the object edges are perceived darker in the image because of shadow effects) are replaced with 1 and the rest with 0 in order to segment the object of interest in subsequent frames. A median filter is finally applied on the result to reduce isolated patches of color and the contour of the object is identified based on the filtered image with the aid of the Sobel edge detector.

4.2.2 Monitoring/Tracking the Object

Once the soft object contour is extracted, it can be tracked over the video sequence as it progressively deforms. To achieve this, a second growing neural gas is initially used to detect the optimum number of points on the contour that accurately represent its geometry. This compact description is employed as an initial configuration for a sequence of neural gas networks that track the contour over each frame in the image sequence in which motion occurs. In each case, a new neural gas network is applied, initialized with the contour of the object in the previous frame, to predict and adjust the position of its neurons to fit the new contour. As illustrated in the flowchart of Fig. 5, this process is repeated until the end of the sequence (i.e. last frame). Due to the choice of a fixed number of nodes used in the neural gas network and to the proposed learning mechanism, the nodes in the contour retain their correspondence with specific points throughout the deformation. This one-to-one correspondence of the points during tracking helps to avoid their mismatch during deformation and ensures a unified description of the contour throughout the frames. Methods such as fast level sets [35] are also an interesting alternative to the proposed neural gas solution for tracking. However, in this case, the one-to-one correspondence of the contour representation cannot be guaranteed.

The resulting contours (in a number equal to the number of frames with motion) representing each neural gas network can be analyzed in order to detect the object material properties (Sect. 4.2.3) or further associated to the measured interaction parameters (e.g. position of the fingers of a robotic hand and applied force magnitude at each finger) for a comprehensive description and prediction of the object's deformation under manipulation (Sect. 4.2.4).

4.2.3 Object Material Characterization

The contour of the object recuperated from video data or the profile of an object as recuperated from a laser scanner can be used to characterize the object elastic properties based on the following observations: elastic objects return to their initial shape or profile once the interaction with them stops (Fig. 6). Therefore, in order to detect if the object is elastic, the final deformation profile, after the interaction stops is compared to the initial deformation profile, collected in the beginning of the measurement procedure, before any force is applied. If the two contours are almost identical, within a certain tolerated noise margin, the object is elastic. The comparison between the initial profile, the profile under force and final profile after force removal can also be exploited to detect plastic and elasto-plastic deformations. If these are different (more than a threshold to cover the noise in the recuperated profiles), it means that either a plastic or an elasto-plastic behavior occurred. The distinction between the plastic and elasto-plastic behaviors can be made by comparing the final deformation contour with the contour while force is applied. If they are identical, it means that a plastic deformation occurred.

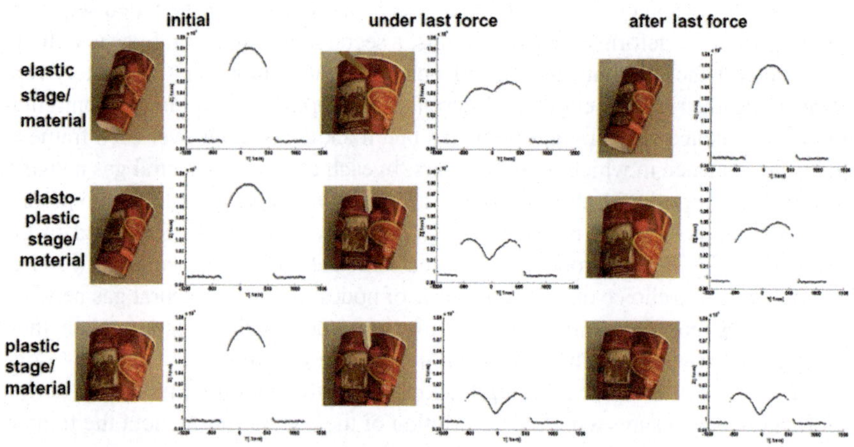

Fig. 6 Object material characterization [31]

If they are different, the material exhibits elasto-plastic properties or the object is within its elasto-plastic deformation stage. If the three profiles are identical, the object is rigid. If a one-to-one correspondence is maintained in the tracked contours, as it is the case of the neural gas solution in Sect. 4.2.2, this comparison is trivial. If one has to deal with profiles or contours of different lengths, an efficient solution to automatically compare them is dynamic time warping [36].

4.2.4 Data Fusion and Deformation Prediction for Multisensory Vision and Tactile Models

Capitalizing on the automated selective scanning framework in Sect. 4.1.3, data selectively collected over regions of interest in terms of vision and tactile measurements can be fused in a representation based on tactile patches (Fig. 7). Such a representation is coherent with psychological studies that have shown that the synthesis of a complex shape is based on the geometric properties of simpler primitives and that this phenomenon occurs in human vision and tactile sensing as well [37]. In such a model, regions of interest from vision are probed at higher resolution and the geometric component of the multi-resolution object is based on the sparse collected data enhanced with these regions. The identification of regions where changes occur in the elastic behavior can lead into the separation of the object in "tactile patches" each exhibiting different elastic properties (i.e. the bottle cap and the bottle body in Fig. 7). A feedforward neural architecture is then used for each patch to capture the relationship between the measured parameters (force magnitude, angle of application, point of interaction of the probe with the object, and object pose with respect to measurement equipment) and the object surface

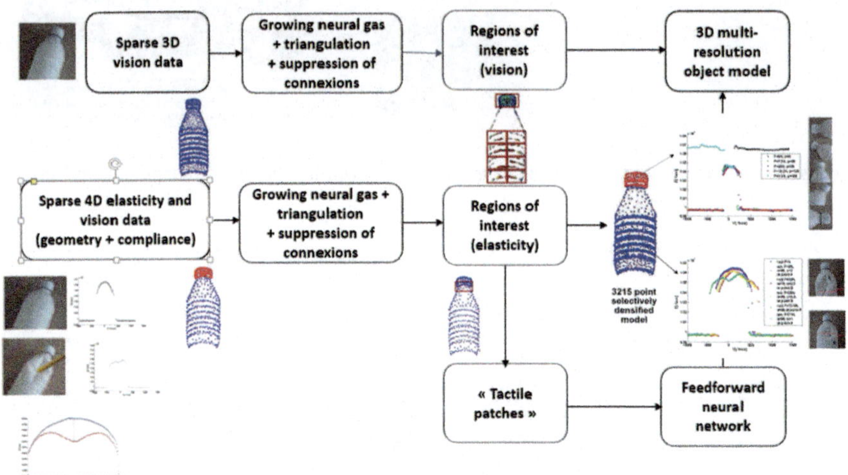

Fig. 7 Modelling of 3D objects as tactile patches (adapted from [31])

deformation. Due to its properties, any of the neural networks is able to provide real-time estimates of the elastic behavior (providing the deformation profile) for those points where the behavior was not probed, therefore eliminating the need for any interpolation of values that normally occurs in any classical model for deformable objects.

The use of neural networks also avoids the problem of recuperating explicitly elastic parameters, which is almost impossible to solve for highly nonlinear elastic materials. The proposed scheme deals easily with piecewise homogeneous materials, due to the existence of tactile patches.

In a similar manner, feedforward architectures can capture and predict the local deformations when the deformed contour is recuperated from a sequence of images of an object under interaction with a robot hand (Fig. 8). If instead of tracking the contour or the profile of an object, the deformation of a 3D object is monitored using a Kinect sensor turning around an object of interest under the interaction of forces exerted with an ATI force-torque sensor (Fig. 2a), a solution to capture implicitly the object behavior capitalizing on a stratified sampling procedure based on the deformation depth, followed by a neural gas-tuned simplification [39] is illustrated in Fig. 9.

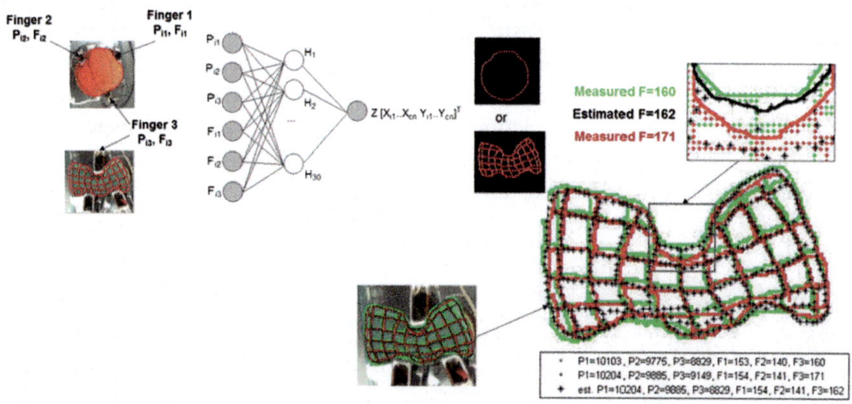

Fig. 8 Prediction of object shape under manipulation with a robot hand (adapted from [38])

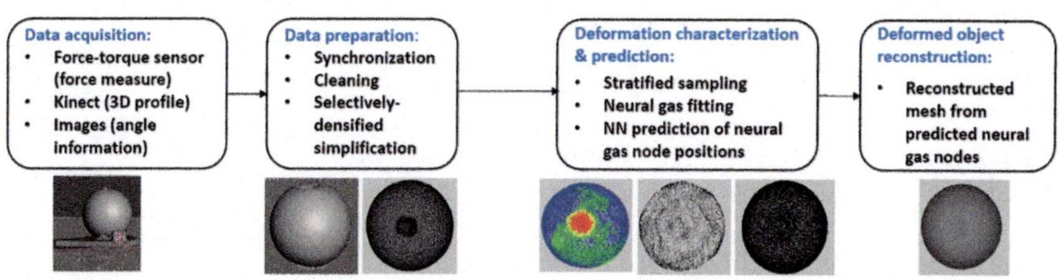

Fig. 9 Data-driven representation of deformable objects under interaction with force-torque sensor

After the cleaning and synchronization of collected data, in order to better characterize the deformation around the probing tip, instead of using the entire collected point cloud, a selectively-densified mesh is first constructed, in which the area around the point of interaction between the probing tip and the object surface is preserved at higher resolution, while the other areas are simplified. This ensures the small deformed area around the probing tip has a higher density of points with respect to the rest of the object's surface. Each deformed mesh is then clustered according to the distance to the initial non-deformed mesh (i.e. blue in Fig. 9 is the closest, and progressing to green, yellow, orange and red as the distance gets higher). Points are sampled randomly but in various proportions from each cluster to identify the adequate amount of data to be used by monitoring the evolution of errors (see Sect. 4.2.5). These proportions are varied by taking into consideration the fact that a good representation is desired specifically in the deformed area and therefore more samples are desired for regions in which the deformation is larger. But this stratified sampling is not sufficient, as the fine differences around the deformed zone might not be appropriately represented, which is the reason why a neural gas-tuned mesh simplification is also applied. The latter is important in order to ensure that fine differences around the deformed zone can be captured in the model. This fitting allows a redistribution of triangles over the mesh such that the fine details are accurately reproduced. The type of model obtained is denser in the region of the deformation (i.e. an average of 97% perceptual similarity with the collected data in the deformed area), while still preserving the object overall shape (i.e. average of 71% similarity over the entire surface) and only using on average 30% of the number of vertices in the mesh. If desired, a feedforward neural network can then be trained to predict the position of the vertices in the neural gas fitted mesh representing each deformed shape of an object based on an applied force magnitude at a given angle.

4.2.5 3D Model Quality Assessment

The quality of a 3D geometrical model can be evaluated from the quantitative and qualitative points of view. In terms of quantitative approaches, Metro [40] allows comparing two models based on the computation of the Hausdorff distance and returns the maximum and mean distance as well as the variance. The second category of quantitative errors can be a form of perceptual error, such as the normalized Laplacian pyramid-based image quality assessment error [41] that takes into account human perceptual quality judgments. As this error is meant to be used on images, images have to be collected over the models of objects from multiple viewpoints and these images can be used pairwise to compute the error. The error measures for each object are then to be reported as an average over the viewpoints. A qualitative evaluation of the results is obtained using Cloud Compare [42] that allows visualizing in an intuitive, color-coded manner the regions most affected by error with respect to its original version.

If the model is a predictive one, in the sense that it is able to predict the deformation of an object for unknown force measurements, the model can be validated qualitatively and quantitatively by using the same metrics as above, but by comparing the predicted mesh with a limited number of real measurements, in which the prescribed forces are exerted on the object.

4.3 Control Schemes for Object Interaction

Combining vision sensors with touch sensing has been explored in recent research to determine the appropriate forces to be applied by the fingers of a robotic hand on a deformable object under manipulation. The goal is to ensure that the hand adapts its behavior to the type of object and to the interaction scenario to achieve an intelligent autonomous manipulation. The stereoscopic vision system depicted in Fig. 2b provides global information by detecting and tracking the deformation of the object in three dimensions. Force and tactile sensors embedded in a Barrett robotic hand are used to provide local information about the deformation at contact points. This knowledge is then used to estimate an object's elastic characteristics and a corresponding control law is defined to maintain a stable and stationary grasp.

From a control system point of view, hand grasping and manipulation processes are carried out by controlling interaction forces at the contact points with an object. Most of the developed control algorithms follow either one of two classical control strategies to solve the force control problem. These are respectively the hybrid position/force control scheme [43] and the impedance control scheme [44]. However, there still exist only limited solutions to the control of robotic manipulation of deformable objects, as the classical approaches require in-depth a priori knowledge of the manipulated object dynamics. The various vision and touch sensing strategies explored in the previous sections can better support the control process by providing live and more comprehensive information about the behavior of soft objects under manipulation.

In general, the typical sequence to grasp and manipulate an object with a robotic hand involves a sequence of logical steps: (i) estimate the object's pose and geometry using 2D or 3D vision sensors; (ii) safely approach the hand with position control to perform the grasp; (iii) determine the contact points and required forces to ensure stable grasp and prevent damages; and (iv) perform manipulation process under force feedback. Vision sensors generally remain involved but mainly to monitor the overall process and detect possible failures. In the initial research that we performed, the focus has been placed on testing the object with a robotic hand immediately after the first contact and the initial grasp is established in order to automatically determine its elasticity characteristics and fine tune the grasping and manipulation process.

Closed-loop manipulation with a robotic hand typically involves a certain form of compliance. Conventional PID controllers have proved successful to ensure stable contact against a compliant surface. However, such an approach requires the modeling of the hand dynamics through which torques, or forces, at joints are determined to drive the fingers' motion during the manipulation. In a real implementation, estimating the dynamic system's parameters with full accuracy proves very difficult, even impossible, which compromises the tuning of a PID controller. Consequently, adaptive control [45] offers a promising alternative to tackle the modeling and control issues for robotic hand control. Considering parameter estimation and adaptive dynamic control simultaneously copes with varying operating conditions, non-nonlinearities, or un-modeled dynamics, as those characterizing deformable objects. Therefore an adaptive feedback control algorithm is being developed for the purpose of dexterous robotic hand manipulation. While the detailed development of the control scheme remains beyond the scope of this chapter, the multisensory systems presented in Fig. 2b, c, that combine stereoscopic and 2.5D RGB-D imaging with force and tactile sensing, along with the characterization process for deformable objects reported earlier, are being put to advantage to control in real-time the motion of a robotic hand with specific determination of the amount of force to be applied at the fingertips. Fine tuning the interaction at each contact point is critical when manipulating deformable objects in order to ensure stable grasp, integrity of the object, or achieving desired shape forming.

5 Conclusion

The chapter discussed some recent research achievements related to sensing issues and interfacing techniques to enable safe interaction of commercial-grade robot manipulators with objects exhibiting rigid or soft surfaces, as developed over years in the authors' research groups. The main challenges of such systems are described, and recent trends in the literature are presented. The main objective being the development of autonomous robotic systems able to handle a wide variety of objects, the solution is decomposed in a series of interconnected challenges to be solved, starting with the data acquisition, data modelling and simulation of objects, and the definition and tuning of control schemes to handle safely the manipulation or the interaction with the object. Within each of these problems potential solutions are proposed and exemplified in the context of practical applications such as: adaptive surface and contour following, object characteristics identification from video and RGB-D data and dexterous robot hand manipulation of soft objects using the Barrett hand.

Acknowledgements The authors wish to acknowledge the contribution of numerous graduate students and collaborators to the research projects summarized in this chapter, and more specifically: A. Chavez-Aragon, F. Hui, A. Huot, F.F. Khalil, P. Laferrière, R. Macknojia, D. Nakhaeinia, E.M. Petriu, B. Tawbe, and R. Toledo. This research has been supported in part by the Natural Sciences and Engineering Research Council of Canada (NSERC), the Canada Foundation for Innovation (CFI), the Ontario Innovation Trust (OIT), the Ontario Centres of Excellence (OCE), and the Fonds de recherche du Québec - Nature et Technologies (FRQNT).

References

1. M. Seitz, Towards autonomous robotic servicing: using an integrated hand-arm-eye system for manipulating unknown objects. J. Robot. Auton. Syst. **26**, 23–42 (1999)
2. D. Henrich, H. Worn (eds.), *Robot Manipulation of Deformable Objects* (Springer, London, 2000)
3. M. Saadat, P. Nan, Industrial applications of automatic manipulation of flexible materials. Int. J. Ind. Robots **29**, 434–442 (2002)
4. F.F. Khalil, P. Payeur, Dexterous robotic manipulation of deformable objects with multi-sensory feedback—a review. Robot Manipulators, Trends and Development, ed. by A. Jimenez, B.M. Al Hadithi (Vukovar, Croatia, In-Tech, 2010), pp 587–619
5. L. Zaidi, B. Bouzgarrou, L. Sabourin, Y. Menzouar, Interaction modeling in the grasping and manipulation of 3D deformable objects, in *Proceedings of IEEE International Conference on Advanced Robotics, Istanbul, Turkey* (2015), pp. 504–509
6. A.-M. Cretu, P. Payeur, E.M. Petriu, Learning and prediction of soft object deformation using visual analysis of robot interactions, in *Proceedings of International Symposium on Visual Computing, Las Vegas, Nevada, US*, ed. by G. Bebis et al. LNCS 6454 (2010), pp. 232–241
7. M. Krainin, P. Henry, X. Ren, D. Fox, Manipulator and object tracking for in-hand 3D object modeling. Int. J. Robot. Res. 1311–1327 (2011)
8. D. Navarro-Alarcon, H.M. Yip, Z. Wang, Y.-H. Liu, F. Zhong, T. Zhang, P. Li, Automatic 3-D manipulation of soft object by robotic arms with an adaptive deformation model. IEEE Trans. Robot. **32**(2), 429–441 (2016)
9. M.-H. Choi, S.C. Wilber, M. Hong, Estimating material properties of deformable objects by considering global object behavior in video streams. Multimed. Tools Appl. **74**, 3361–3375 (2015)
10. A.S. Prabuwono, S. Said, R. Sulaiman, Performance evaluation of autonomous contour following algorithms for industrial robot, in *Robot Manipulators Trends and Development*, ed. by A. Jimenez, B.M. Al Hadithi. InTech (2010), pp 377–398
11. A. Petit, V. Lippiello, B. Siciliano, Real-time tracking of 3D elastic objects with an RGB-D sensor, in *Proceedings of IEEE International Conference on Intelligent Robots and Systems, Hamburg, Germany* (2015), pp. 3914–3921
12. L. Zaidi, B. Bouzgarrou, L. Sabourin, Y. Menzouar, Modeling and analysis of 3D deformable object grasping, in *Proceedings of International Conference on Robotics in Alpe-Adria-Danube Region, Smolenice Castle, Slovakia* (2014), pp. 504–509
13. S. Hirai, T. Tsuboi, T. Wada, Robust grasping manipulation of deformable objects, in *Proceedings of IEEE Symposium on Assembly and Task Planning* (2001), pp 411–416
14. J. Stuckler, S. Behnke, Perception of deformable objects and compliant manipulation for service robots, in *Soft Robotics*, ed. by A. Verl et al. (Springer, 2015), pp. 69–80
15. C.M. Mateo, P. Gil, D. Mira, F. Torres, Analysis of shapes to measure surfaces, in *Proceedings of IEEE Conference on Informatics in Control, Automation and Robotics, Colmar* (2012) pp. 60–65
16. C. Elbrechter, R. Haschke, H. Ritter, Folding paper with anthropomorphic robot hands using real-time physics-based modeling, in *Proceedings of IEEE International Conference on Humanoid Robots, Osaka* (2012), pp. 210–215
17. Q. Pan, G. Reitmayr, T. Drummond, ProFORMA: probabilistic feature based on-line rapid model acquisition, in *Proceedings of 20th British Machine Vision Conference (BMVC), London* (2009)
18. D. Kraft, N. Pugeault, E. Baseski, M. Popovic, D. Kragic, S. Kalkan, F. Worgotter, N. Kruger, Birth of the object: detection of objectness and extraction of object shape through object action complexes. Int. J. Hum. Robots **5**(2), 247–265 (2008)
19. J. Schulman, A. Lee, J. Ho, P. Abbeel, Tracking deformable objects with point clouds, in *Proceedings of IEEE International Conference on Robotics and Automation* (2013)

20. J. Hur, H. Lim, S.C. Ahn, 3D deformable spatial pyramid for dense 3D motion flow of deformable object, in *Proceedings of International Symposium on Visual Computing, Las Vegas, Nevada, US*, ed. by G. Bebis et al. LNCS 8887 (2014), pp. 118–127
21. D. Nakhaeinia, R. Fareh, P. Payeur, R. Laganière, Trajectory planning for surface following with manipulator under RGB-D visual guidance, in *Proceedings of Safety, Security and Rescue Robotics, Linkoping* (2013), pp. 1–6
22. F.F. Khalil, P. Payeur, A.-M. Cretu, Integrated multisensory robotic hand system for deformable object manipulation, in *Proceedings of IASTED International Conference Robotics and Applications, Cambridge, Massachusetts, USA* (2010), pp. 159–166
23. D. Nakhaeinia, P. Payeur, R. Laganière, Adaptive robotic contour following from low accuracy RGB-D surface profiling and visual servoing, in *Proceedings of Canadian Conference on Computer and Robot Vision* (2014), pp. 48–55
24. M. Zollhofer, et al., Real-time non-rigid reconstruction using and RGB-D camera. ACM Trans. Graph. **33**(4), 156:1–156:12 (2014)
25. M. Dou, J. Taylor, H. Fuchs, A. Fitzgibbon, S. Izadi, 3D scanning deformable object with a single RGBD sensor, in *Proceedings of IEEE Computer Vision and Pattern Recognition* (2015), pp. 493–501
26. Microsoft Kinect Fusion. https://msdn.microsoft.com/en-us/library/dn188670.aspx
27. Skanect. http://skanect.occipital.com/
28. R. Macknojia, A. Chavez-Aragon, P. Payeur, R. Laganière, Calibration of a network of Kinect sensors for robotic inspection over a large workspace, in: *Proceedings of IEEE Workshop on Robot Vision* (2013), pp. 184–190
29. D. Nehab, P. Shilane, Stratified point sampling of 3D models, in *Proceedings of Eurographics Symposium Point-Based Graphics*, ed. by M. Alexa, S. Rusinkiewicz (2004), pp. 49–56
30. H. Song, H.-Y. Feng, A point cloud simplification algorithm for mechanical part inspection, in *Information technology for balanced manufacturing systems*, ed. by W. Shen (Springer, 2006), pp. 461–468
31. A.M. Cretu, Experimental data acquisition and modeling of 3D deformable objects, Ph.D. thesis, University of Ottawa (2009)
32. M. Chagnon-Forget, G. Rouhafzay, A.-M. Cretu, S. Bouchard, Enhanced visual-attention model for perceptually-improved 3d object modeling in virtual environments. 3D Res. **7**(4), 1–18 (2016)
33. K. Drewing, A. Ramisch, F. Bayer, Haptic, visual and visuo-haptic softness judgements for objects with deformable surfaces, in *Proceedings of Eurohaptics Conference on Haptic Interfaces for Virtual Environments and Teleoperator Systems* (2009), pp 640–645
34. Meshmixer. http://www.meshmixer.com/
35. Y. Shi, W.C. Karl, A real-time algorithm for the approximation of level-set-based curve evolution. IEEE Trans. Image Process. **17**(5), 645–656 (2008)
36. M. Muller, *Information Retrieval for Music and Motion* (Springer, 2007)
37. J.M. Ehrich, M. Flanders, J.F. Soechting, Factors influencing haptic perception of complex shapes. IEEE Trans. Haptics **1**(1), 19–26 (2008)
38. A.-M. Cretu, P. Payeur, E.M. Petriu, Soft object deformation monitoring and learning for model-based robotic hand manipulation. IEEE Trans. Syst. Man Cybern. Part B **42**(3), 740–753 (2012)
39. B. Tawbe, A.-M. Cretu, Data-driven representation of soft deformable eobjects based on force-torque data and 3D vision measurements, in *Proceedings of 3rd International Conference on Sensors and Applications* (2016) (in press)
40. P. Cignoni, C. Rocchini, R. Scopigno, Metro: measuring error on simplified surfaces. Comput. Graph. Forum **17**(2), 167–174 (1998)
41. V. Laparra, J. Balle, A. Berardino, E.P. Simoncelli, Perceptual image quality assessment using a normalized Laplacian pyramid, in *Proceedings of Human Vision and Electronic Imaging*, vol. 16 (2016)
42. CloudCompare—3D Point Cloud and Mesh Processing Software. http://www.danielgm.net/cc/. Accessed 1 Aug 2016

43. M.H. Raibert, J.J. Craig, Hybrid position/force control of manipulators. ASME J. Dyn. Syst. Meas. Control **102**, 126–132 (1981)
44. N. Hogan, Stable execution of contact tasks using impedance control, in *Proceedings of IEEE International Conference on Robotics and Automation* (1987), pp 1047–1054
45. G. Tao, *Adaptive Control Design and Analysis* (Wiley-IEEE Press, 2003)

IEEE1451 Smart Sensors Architectures for Vital Signs and Motor Activity Monitoring

Octavian Postolache, Vítor Viegas, João Freire, José Miguel Dias Pereira and Pedro Girão

Abstract The chapter presents a set of smart sensing systems, compatible with IEEE1451 standard, intended to unobtrusively monitor the vital signs and motor activity of a wheelchair user. The IEEE 1451 standard was considered to add smartness to transducers: clause 1451.4 adds plug-and-play and self-identification capabilities to the sensors, and clause 1451.1 provides the information model used to share data and functionalities across the network. The .NET Framework was considered to implement the 1451.1 information model as it provides a rich development environment. The .NET's implementation of the 1451.1 information model, including the customization efforts to meet application-specific requirements, is described. Referring to the measurement channels considered in the present approach, photo-plethysmography (PPG), skin conductivity (SKC), ballistocardiography (BCG), and electrocardiography (ECG) were chosen to monitor cardiorespiratory activity, while a simple 3D accelerometer was used to monitor motor activity.

1 Introduction

In the last years the necessity to reduce the hospitalization costs and to allow proactive and preventative care led to a set of developments in the field of home health monitoring [1–3]. Such systems, denominated vital signs monitors, can include measurement units for blood pressure, heart rate and oxygen saturation [4].

O. Postolache (✉)
Instituto de Telecomunicações/ISCTE-IUL, Lisboa, Portugal
e-mail: opostolache@lx.it.pt

V. Viegas · J.M.D. Pereira
Instituto de Telecomunicações/ESTSetúbal-IPS, Lisboa, Portugal

J. Freire · P. Girão
Instituto de Telecomunicações/DEEC-IST, Lisboa, Portugal

B. George et al. (eds.), *Advanced Interfacing Techniques for Sensors*,
Smart Sensors, Measurement and Instrumentation 25,
DOI 10.1007/978-3-319-55369-6_10

Usually the data from the monitors are sent to a clinical server for early identification of clinical needs, adjustments to the treatment plan and medications, thus reducing emergency room visits and unplanned hospitalizations. In order to reduce the number of clinical staff needs, but also to diminish the stress induced on the patient by the measurement setup (e.g. deployment of ECG electrodes), unobtrusive embedded vital signs monitors associated with daily used objects represent interesting alternative solutions [5]. Our group has been developing a set of smart objects for vital signs and motor activity assessment of elderly people, or related to rehabilitation and mobility assessment, as part of different prototypes of smart wheelchairs and smart walkers [6–9]. To assure modular and flexible solutions, sensors are assembled in plug-and-play modules characterized by appropriate conditioning circuits, processing units make use of low power microcontrollers and communications are supported by wireless links. Several solutions regarding "plug-and-play" smart sensors are reported in the literature [10–13].

The IEEE 1451 standard [14] provides guidelines to build open, high performance and smart transducer systems. It defines a set of neutral communication interfaces for connecting transducers to computers, instrumentation systems and networks. The IEEE 1451 for smart sensors is composed by the following clauses:

- IEEE 1451.0: Introduces the concept of Transducer Interface Module (TIM) as a peripheral that interfaces one or more transducers. The TIM acts as a Data Acquisition (DAQ) board with support for transducer self-identification, self-diagnosis and self-calibration. The clause defines common functions, communication protocols and Transducer Electronic Data Sheet (TEDS) formats to support these advanced features.
- IEEE 1451.2/3/5: Specify the TIM concept for the cases where the TIM connects to the computer through a set of different communication interfaces.
- IEEE 1451.4: Defines digital wiring, communication protocols and TEDS structures to make analog transducers smarter.
- IEEE 1451.7: Defines communication protocols and TEDS structures to interact with RFID tags.
- IEEE 1451.1 [15]: Defines a generic information model to represent the functionalities of a smart transducer system. The processor that runs the application (and hence implements the information model) is called "Network Capable Application Processor" (NCAP). The information model is composed by three classes of objects: Blocks (intended to process data), Components (intended to store block-related data), and Services (intended to support communications).

The use of commercial development platforms to implement smart transducer interfaces helps to improve market acceptance of the IEEE 1451 standard. A good example is the .NET Framework [16, 17], which can be used to implement the 1451.1 information model. This well-known and well-supported framework provides last-generation programming languages (C# and VB.NET), a high-performance execution environment, and a set of highly productive software libraries. Among these is included the Windows Communication Foundation

(WCF) [18, 19], which contains all the stuff needed to develop distributed applications (raw sockets, protocol stacks, web services, etc.).

Following this idea, we implemented a software library—named "Service eXtensions for Instrumentation" (SXI)—that combines the better of two worlds: the *communication model* provided by WCF and the *information model* provided by the IEEE 1451.1 clause.

Several implementations of the IEEE 1451 standard for smart transducers are considered, with special attention being granted to clauses 1451.1 and 1451.4. The effectiveness of the SXI library was proven through its usage in the particular field of healthcare monitoring, together with IEEE1451.4 hardware implementations that are following described.

2 IEEE 1451.4 Hardware Implementation

The IEEE 1451.4 standard for smart sensors defines a mechanism for adding self-describing behavior to the traditional transducers characterized by analog output signals [20, 21]. This is the case of the biomedical sensors used for vital signs (PPG, SKC, BCG, ECG) and motor activity monitoring. According to this standard the implemented hardware is based on a microcontroller platform (uCtrl) that allows the acquisition of both analog and digital data from vital signs sensors (Fig. 1). The connectivity between sensors interface modules (SIMi) is assured by a connection board (CB) characterized by a set of 8 sensors and a network controlled circuit (NCC). The SIMi are powered through the power supply line (PWR).

In the present case, the identification is based on the use of an EEPROM memory that stores the IEEE 1451.4 Basic TEDS (Transducer Electronic Data Sheet) information for each vital sign transducer characterized by analog output (e.g. photoplethysmography transducer). The information stored in the memory is transmitted to the microcontroller through a 1-wire interface multi-drop bus. According to the IEEE 1451.4 class 2 architecture, the analog signals are acquired by the ADC of the microcontroller after signal conditioning performed at the CCi level. The communication between the microcontroller platform and the host PC is performed using IEEE 802.15.4 wireless communication protocol. The used topology was the star network topology, thus Digi XBee ZigBee transceivers were used for each ZigBee end nodes and a USB to XBee wireless adapter was installed in a client PC to materialize the ZigBee coordinator. The client PC is an important component of the smart sensing system as it performs advanced data processing, data logging and data publishing. The main components of the implemented IEEE1451.4 system are detailed in the following sections.

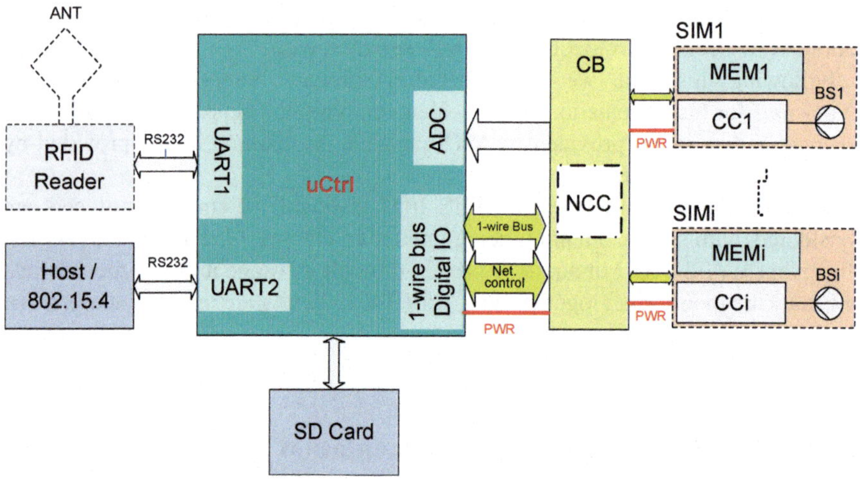

Fig. 1 The block diagram of a microcontroller platform compatible with IEEE 1451.4 standard for smart sensors embedded in wheelchair (BSi biomedical sensors, CCi—conditioning circuit, ANT-RFID antenna, SIMi sensor interface module, PWR-power supply lines, CB—connection board, MEMi—1-wire EEPROM, NCC—network control circuit)

2.1 *The Smart Gateway*

The prototype of the implemented smart gateway (Fig. 2) is based on a PIC 24F128GA010 microcontroller and is responsible for the management of the smart sensor network. The smart gateway can be considered as a reduced version of NCAP (network capable application processor) defined by clause IEEE 1451.2 [14].

The platform assures acquisition of biomedical sensors (BSi) analog output signals using the 10-bit A/D converter of the microcontroller. The microcontroller operates at a frequency of 8 MHz with the A/D block converter allowing a sample rate of up to 10 kS/s for each channel. Considering the ADC range, the conditioning circuit (CCi) must assure signal amplitudes between 0 V and 3.3 V. The TEDS memories communicate through a 1-wire interface allowing a multi-drop bus configuration. In this way, all the memories share the same bus hence reducing the microcontroller digital ports requirement, since the 1-wire protocol bus only uses one single pin. Additionally, two universal asynchronous receiver/transmitter interfaces of the microcontroller (UART1 and UART2) are used to establish communication with the client PC based on IEEE802.15.4 protocol and provide connectivity with a 125 kHz RFID reader.

The RFID reader is added to the system to provide identification functionalities for the wheelchair or walker user through the usage of passive LF RFID tags. Additionally, the LF RFID technology can be used for indoor walker or wheelchair localization where the LF RFID reader detects a set of distributed tags [22].

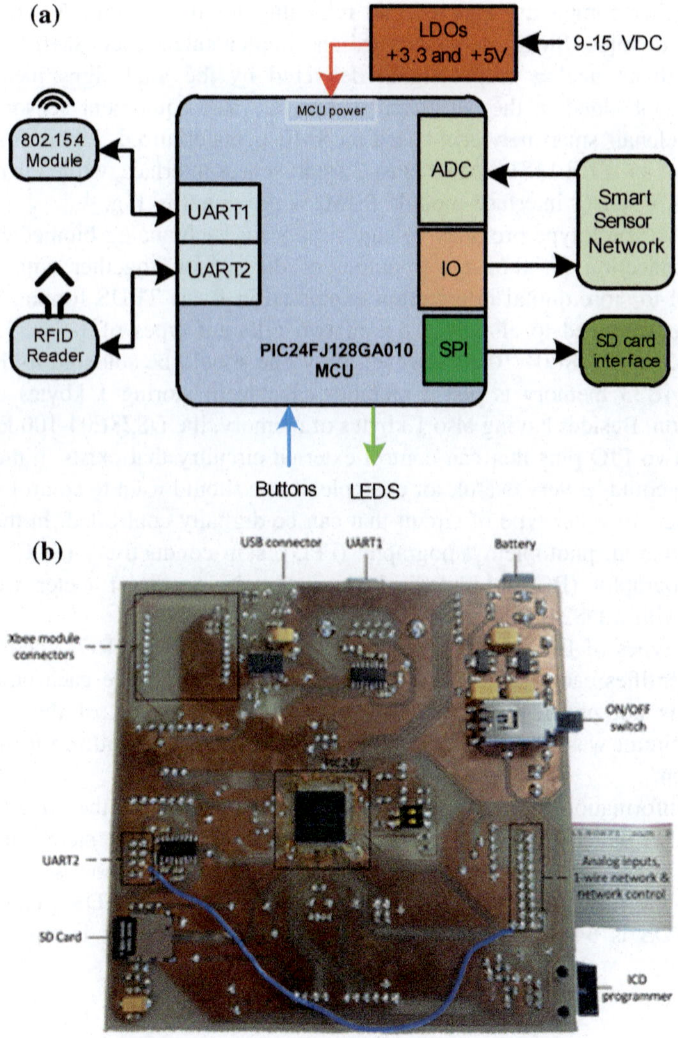

Fig. 2 IEEE1451.4 Smart Gatteway: **a** block diagram; **b** implemented prototype

2.2 *IEEE 1451.4 Sensor Interface Module (SIM)*

All transducers for vital signs monitoring were designed to fulfill the IEEE 1451.4 mixed mode interface (MMI) class 2 specifications having two separate connections for analog and digital signals. The use of class 2 was considered taking into account that the class 1 requires switching between the analog and digital signals, which means higher complexity and reduced reliability for the system. A set of sensor interface modules (SMIi) were designed and implemented, each SMIi being associated with an analog output signal delivered by the vital signs measurement channels embedded in the wheelchair. Thus, a smart equipment sensor network (e.g. wheelchair smart network) based on SMIi is materialized. Figure 3 shows an example of an IEEE 1451.4 MMI class 2 smart sensor interface, while the prototype IEEE1451.4 sensor interface module (SIM) is presented in Fig. 4.

The SMIi prototype provides voltage supply for each analog biomedical sensor and a connection for the analog output of the sensor, together with a 1-wire EEPROM to store digital information expressed in Basic TEDS format. The SIM board was designed to allow the use of two different types of 1-wire EEPROM (DS24B33 or DS28E04-100), however, only one should be soldered to the board. The DS24B33 memory is just a memory capable of storing 1 kbytes of digital information. Besides having also 1 kbytes of memory, the DS28E04-100 EEPROM has also two PIO pins that can control external circuitry that exists in the BSn or CCi. This could be very useful, for example, if one should want to control switches, multiplexers or other type of circuit that can be digitally controlled. In the current implementation, photoplethysmography (PPG), skin conductivity (SKC) and ballistocardiography (BCG) biomedical sensors and an accelerometer were used together with a DS24B33 1-wire EEPROM.

These types of EEPROMs have a unique 64-bit (8 bytes) ROM code address which identifies each node in the smart sensor network. Since each node of the network is not physically associated with the analog output of the sensor, an external circuit was developed to allow the microcontroller platform to make this association.

The information stored into the EEPROMs contains the IEEE 1451.4 Basic TEDS which is the minimum of information that can be stored in order to fulfill the IEEE 1451.4 requirements. Additional information about sensors can be stored in the host computer and accessed using the Basic TEDS elements. The Basic TEDS is 64-bit (8 bytes) long and has a structure that is represented in Table 1.

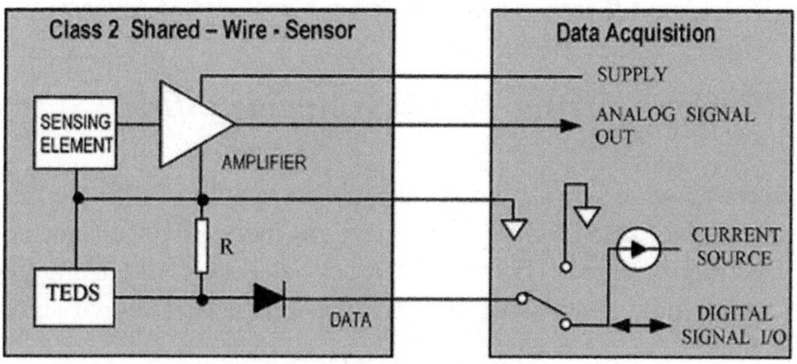

Fig. 3 Example of IEEE 1451.4 class 2 multi-wire interface

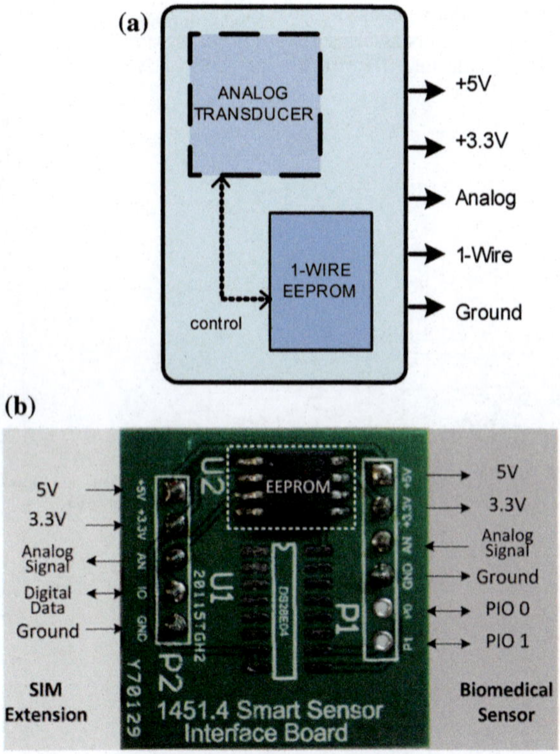

Fig. 4 Sensor Interface Module: **a** block diagram; **b** implemented prototype based on 1-wire EEPROM DS24B33

Table 1 Basic TEDS content

Basic TEDS				
Manufacturer ID	Model number	Version letter	Version number	Serial number
14 bits (17–16381)	15 bits (0–32767)	5 bits Chr5 (A–Z)	6 bits (0–63)	24 bits ($0–2^{24}–1$)

By using a smart sensor database, the Manufacturer ID and Model Number will be responsible for identifying the type of biomedical sensor (e.g. PPG, SKC or BCG) or motor activity sensor. The remaining Basic TEDS fields will serve as upgrade and version control data of the sensors. As an example, the Manufacturer ID "107" was used, since all Man. IDs from 0 to 106 are reserved or already taken by other manufactures [23]. To distinguish between the types of sensors that can be used, Model Number based identification was made as it is presented in Table 2.

Table 2 Model numbers for smart sensors associated with Manufacturer ID "107"

Model number	Type of sensor
1	Photoplethysmography
2	Skin conductivity
3	Electrocardiography
4	Ballistocardiography
5	Acceleration

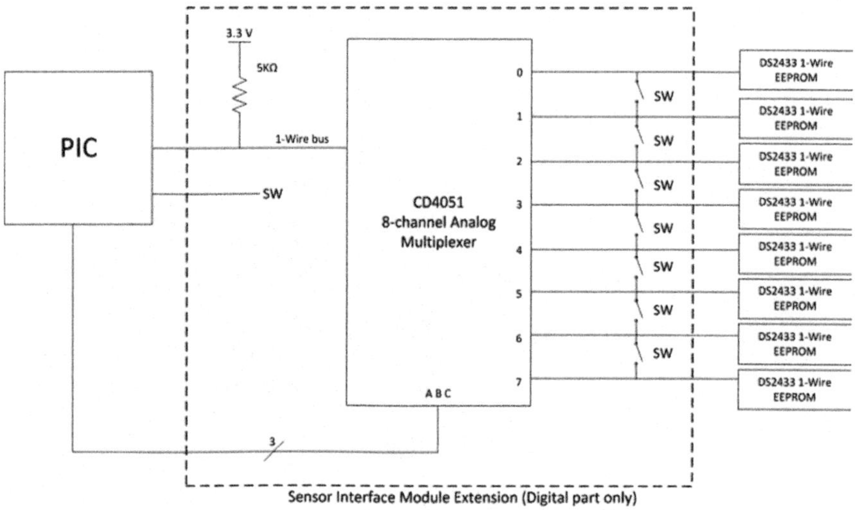

Fig. 5 The block diagram of 1-wire network control system embedded in the sensor interface module extension board

2.3 *Sensor Interface Module Connection Board*

Since the smart gateway based on microcontroller is mounted on a wheelchair, an additional NCAP-SIM external connection board was developed. A NCAP-SIM extension board turns the system modular and more flexible because the sensor input connectors are not physically attached to the microcontroller board. The circuit designed to control the 1-wire smart sensors network is also on the sensor interface module extension board (Fig. 5).

It consists of two quad-analog switches that sets up the network as a 1-Wire multi-drop bus or as a bus of 8 single point-to-point connections to 8 channel analog multiplexer. The system needs to switch between two buses typologies so the smart gateway based on microcontroller can identify which of the EEPROMS is associated with which analog output, i.e., get the unique 64-bit ROM code of each memory that will be associated with an analog channel. This way, it is possible to link the information of the Basic TEDS to an analog channel without having to previously program the TEDS with the measurement location of the sensor or to use an address programmable EEPROM (like a I^2C memory type used in the previous implementation [4]). This requires four additional digital pins from the microcontroller: one pin to control the analog switches (SW) and three pins (ABC) to control the multiplexer selection input/output (IO) bits from 000 to 111 (0 to 7). Each multiplexer IOs is virtually associated with an analog channel. To find the 64-bit ROM from each memory, SW must be zero (point-to-point connections), and ABC should go from 000 to 111 to go through all IOs of the multiplexer. Once the ROM codes are retrieved from the memories, SW should go high (multi-drop bus) and the last multiplexer ABC combination that has an EEPROM available should be selected by the microcontroller. This multiplexer channel will be responsible to transmit and receive all the multi-drop bus information, thus all the other channels from the multiplexer go into a high impedance state.

The connection between the smart gateway based on microcontroller and the NCAP-SIM connection board is made through a 20-pin cable. There are 8 pins to all 8 analog channels, with 3.3, 5 V voltage, 1 ground, and 4 remaining pins for desired future connections. Voltages 3.3 V and 5 V will be used to power the biomedical sensors, conditioning circuit and the 1-wire EEPROM DS28E04-100, if used. The 1-wire EEPROM DS24B33 is powered by the data line. The connection board includes a set of 8 connectors with 5 pins associated with the maximum 8 SIMs of the system.

Each SIM board connects to the NCAP-SIM connection board through a 5 pin connector, which has pins for analog input, digital data, 3, 5 V and ground.

2.4 *Communication and Data Logging*

The host computer performs the remote control of the microcontroller platform using a set of commands. Several of the implemented commands are listed in Table 3. It should be noted that all command bytes are followed by a length byte to indicate the number of bytes to be read by the host.

Considering the possibility of adding or replacing new sensors, the start host connection becomes important as it gives the number of available sensors in the network. The Read Basic TEDS command retrieves the Basic TEDS from each network node in the format ‘0x02 + length (two bytes) +[channel number + basic teds bytes (8 bytes)] × *number of channels available*’. The real-time data from the sensors is transmitted to the PC host in the format ‘–0x0000 0x1111 0x2222 0x3333 0x4444 0x5555 0x6666 0x7777\n’ if all sensors are available. The previous message represents the maximum size of a real-time data message and has a total size of 25 bytes (space characters included). The 10-bit ADC data from each

Table 3 Microprocessor platform supported commands

Command	Function
0x01	Start host connection and get number of available channels
0x02	Read Basic TEDS from smart sensors memories
0x03	Write Basic TEDS into smart sensors memories. Ex: **0x03** + packet length + channel ID + Basic TEDS bytes
'a'	Toggle ADC data acquisition at a specified sample rate. E.g.: **'a'** + length + 0x03E8 (for 1 KS/s)

channel is represented by two bytes. To get data from each channel, the microcontroller scans the available analog channels in a polling fashion.

The communication to a client PC can be made either through wires or wirelessly by using the wireless transceiver connected to one of the available microcontrollers UART ports. In the case of wired communication to the client PC, in order to convert the signals coming from the microcontroller's UART bus to client PC USB port, a FTDI RS232-to-USB converter is used for wired communication. For wireless communication, a Maxstream XBee 802.15.4 module is used and configured to work in transparent mode at a baud rate of 57600 bps.

Even though the sampling can be up to 10 kS/s for each analog input, transmitting the data in real-time has its limitations due to the use of the UART transceiver. This happens in both wired and wireless communications since the interface used is the same. Since the microcontroller is working at 8 MHz, the communication rate is limited to 57600 bits/s. If the UART is setup with one stop bit, transferring 1 byte implies sending also one start and one stop bit, i.e. 10 bits. Thus, the data transferred will be 57600/10 = 5760 bytes/s. Since each message transmitting real time data is 25 bytes long, there is only the possibility of sending 5760/25 bytes = 230 messages frames per second, which means the maximum real-time transmission rate will be of 230 S/s per channel. In the implemented system the sending rate of 100 S/s per channel was used.

When there is no connection available to client PC, the acquired data is saved into an SD Card through an available SPI bus. To manage the reading and writing to the SD Card, the Microchip's Memory Disk Drive File System library was used, and in this way can be carried out the write and read files operation from FAT32 file system. When the data will be saved in the memory card faster sample rate of the signals can be reached, since writing to the SD Card takes about 0.5 ms between messages. Since scanning 8 analog channels takes about 0.25 ms, only after 0.25 + 0.5 = 0.75 ms is possible to sample again. Thus, the maximum allowed sampling rate is about 1333 Hz/channel, which is acceptable for the bandwidth of the biomedical signals that were considered in the present implementation.

2.5 Prototype Tests

The benefits offered by the adoption of the IEEE 1451.4 standard for vital signs and motor activity sensing was evaluated for the particular case of a smart wheelchair. Thus, the signals of the sensors embedded in the wheelchair and the TEDS information stored in the 1-wire memories are acquired by the microcontroller platform that transmits the data to the host computer running 'SmartSwheel' software developed in LabVIEW. This software is able to perform TEDS operations associated with smart sensors and wheelchair user identification, digital signal processing of the acquired signals, and display the processed signals and computed vital signs parameters (e.g. heart rate). Referring to the signal processing component of the developed LabVIEW software, it includes digital filtering, peak detection, and the calculation of statistic parameters related with motor activity performed by the wheelchair user, who is identified using the RFID also controlled by the implemented software.

Taking into account the complexity of the software designed for configuration and for data reading, different software functionalities are active according to the type of the user. Thus, the administrator can access, after identification, all the software functionalities including TEDS, Measurement and Data Base Management. However, a physician, nurse or therapist can access only the Measurement and Data Base Management functionalities (Fig. 6). Regarding to the wheelchair

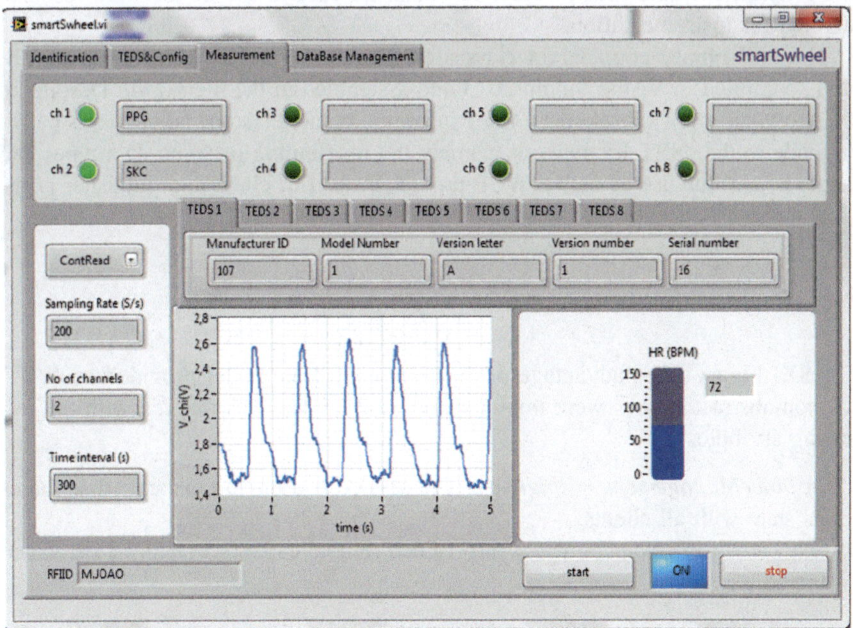

Fig. 6 The graphical user interface of smart sensors for wheelchair (SmartSwheel) LabVIEW application

user, the identification provides him the possibility of reading the data of the different vital signs channels and of calculated parameters (e.g. heart rate variability) and of motor activity.

In Fig. 6 can be observed the capability of the system to highlight the active sensors (PPG and SKC in the present case) using a set of software LEDs in accordance with the detected active channels. At the same time, for each detected active channel, the Basic TEDS information is presented and also the signal associated with the active channel (PPG associated with ch1 active measurement channel). Several vital signs parameters, such as Heart Rate (HR), are calculated and also presented on the user interface level. Using the RFID reader control capabilities and a stored table of the wheelchair users the name or the ID of the identified user is also presented in the GUI's digital display.

Taking into account the requirement of IEEE1451's NCAP fast software prototyping, the following paragraph focus on the developed Service eXtension for Instrumentation library (SXI) that merges the strengths of the 1451.1 information model with the advantages provided by the .NET Framework and the Windows Communication Foundation (WCF).

3 Service eXtensions for Instrumentation

The SXI library is expressed as a materialization of the IEEE1451.1 model using last generation framework in order to obtain an easy-to-use middleware resource targeted for instrumentation.

The SXI library contains 14 classes (see Table 4), which were coded in VB. NET, compiled in Visual Studio 2013 and assembled in the file *sxi.dll*. Despite all the efforts to follow closely the 1451.1 information model, some adjustments had to be made to the .NET Framework, namely the redefinition of some data types and methods, and the use of native WCF proxies instead of client and publisher ports.

3.1 Communication Services

The SXI library takes advantage of the communication model provided by WCF. All non-abstract classes were implemented as WCF services marked with the following attributes:

- *InstanceManagement* = *singleton*: The service is a unique instance that shares its state with all clients.
- *UseSynchronizationContext* = *false*: Each remote call is served by a dedicated thread.

Table 4 SXI information model

Class name	Description
Root	Root of all classes
Entity	Adds functionalities to identify and localize objects.
Block	Base class for blocks. Adds functionalities to retrieve information about the block, change its execution state and interact with its owned objects.
PBlock	Top-level block that represents the application as whole.
FBlock	Base class for function blocks.
Component	Base class for components.
Parameter	Represents a network-visible variable.
Parameter with update	Adds functionalities to synchronize the parameter value with its owning block.
Physical parameter	Adds metadata to describe the contents of the parameter.
Scalar parameter	Represents a physical quantity well represented by a scalar.
Service	Base class for services.
Client	Dynamic proxy useful to make remote calls.
Publisher	Issues publications on a UDP multicast address.
Subscriber	Listens to publications on a UDP multicast address.

Notes
(1) Classes in bold are non-abstract

- *ConcurrencyMode* = *single*: The remote calls pass by a built-in arbitration mechanism that guarantees exclusive access to the service. In other words, the remote calls are served one at a time.

The WCF follows the client/server communication model: each service, during instantiation, registers itself on a client/server endpoint and exposes its methods on the network. Any client, having the dispatch address of the service, can create a proxy, call a given method and collect the results (if any). Client/server endpoints can use one of the following bindings:

- *BasicHttpBinding*: Binding fully compatible with standard web services (WS). Messages are formatted according the Simple Object Application Protocol (SOAP) and transported over HTTP. It promotes interoperability over performance, making it suitable for communications between cross-platform applications across the internet.
- *WSHttpBinding*: Adds support for web services enhancements (WS-*), making it more versatile but also less interoperable. Well-suited for communications between uni-platform applications across the internet.
- *NetTcpBinding*: Messages are formatted according a proprietary binary protocol and transported over TCP. It promotes performance over interoperability, making it suitable for communications between uni-platform applications inside an intranet.

The publish/subscribe communication model was implemented taking advantage of the multicast capabilities of UDP. Publications were implemented as methods with no return values registered on a given multicast address. A call to such a method is forward to all subscribers registered on the multicast address, thus acting as a "one-way" message from one to many.

The SXI library is quite flexible in terms communications: remote calls can be made using the client/server model (for one-to-one communications) or the publish/subscribe model (for one-to-many communications); data can be formatted using SOAP or a proprietary binary protocol; messages can be transported over HTTP (using port 80) or directly over TCP (using a user-defined port number). With so many options available, communications can easily be tuned for performance (by transmitting raw binary messages over TCP) or for interoperability (by transmitting SOAP messages over HTTP).

3.2 Control Stations

The objects contained in the SXI library are the bricks used to build high-level applications. If those applications are meant to control physical processes they are referred as "control stations".

A typical control station works as a Programmable Logic Controller (PLC): it loops continuously acquiring data from sensors, executing control algorithms and updating actuators (see Fig. 7a). Function blocks are used to implement the control strategy, while parameters are used to store network-visible variables. Remote calls can be made to change the behavior of functions blocks and to read/write variables.

Control stations can issue publications to inform the network about significant events. During start-up, periodically as a beacon, or on demand, the station can issue publications of type *PSK_NCAPBLOCK_ANNOUNCEMENT* and *PSK_NETWORK_VISIBLE_SERVER_OBJECT_PROPERTIES* to say that is alive and to announce the services it exposes. During runtime, the station can issue publications of type *PSK_PHYSICAL_PARAMETRIC_DATA* whenever the value of a parameter changes.

Figure 7b shows a proposal of graphical interface for control stations. The interface is composed by a tree list that is populated by all the services the application exposes. The value of each parameter can be read or written provided that is scalar.

3.3 Engineering Stations

"Engineering stations" are SXI-based applications intended to take care of control stations. A single engineering station can manage all the control stations attached to the same intranet. The management work consists of two main tasks:

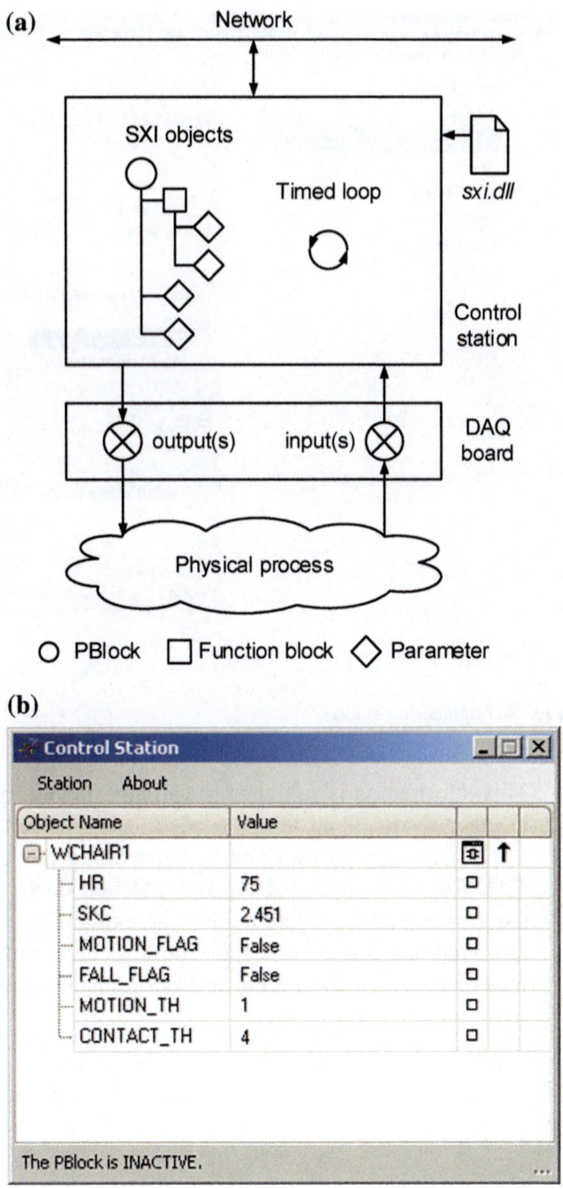

Fig. 7 Control station: **a** architecture; **b** graphical interface

- System configuration: Task done by subscribing publications of type *PSK_NCAPBLOCK_ANNOUNCEMENT* and *PSK_NETWORK_VISIBLE_SERVER_OBJECT_PROPERTIES*. The attached data lists all the control stations visible on the network and the services they expose. Each service is described in terms of type, name, unique IDentifier (ID), dispatch address and owning relation.

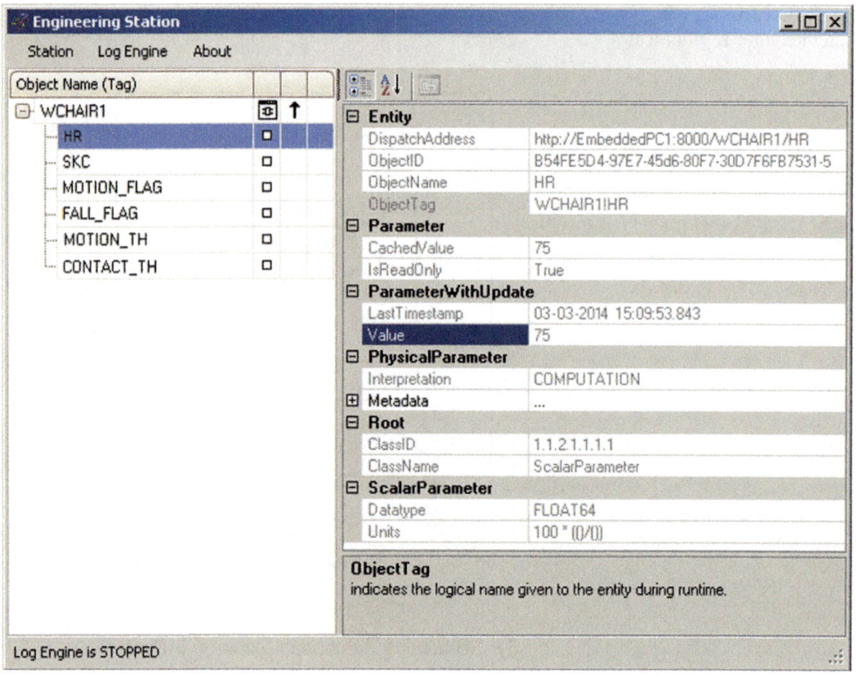

Fig. 8 Engineering station

This information is sufficient to call any method on any service on any station; in other words, is all that is needed to configure the entire system.

- Data monitoring: Task done by subscribing publications of type *PSK_PHYSICAL_PARAMETRIC_DATA*. The attached data contains the updated value of the parameter and meta-fields that explain its meaning. The value can be extracted and logged to a file. The log engine makes use of fast-access binary files of type TDMS (Technical Data Management Streaming).

Figure 8 shows a proposal of graphical interface for engineering stations. The interface is composed by two panes: on the left pane, the user can navigate through all the services available on the network; on the right pane, the user can read/write the properties of a given service.

4 SXI Support for Smart Wheelchair

An example of SXI support for a measurement system associated with healthcare monitoring is following presented. Thus, the considered measurement system is expressed by a wheelchair characterized by vital signs and motor activity measurement channels. The implemented measurement channels provides information

Fig. 9 System layout

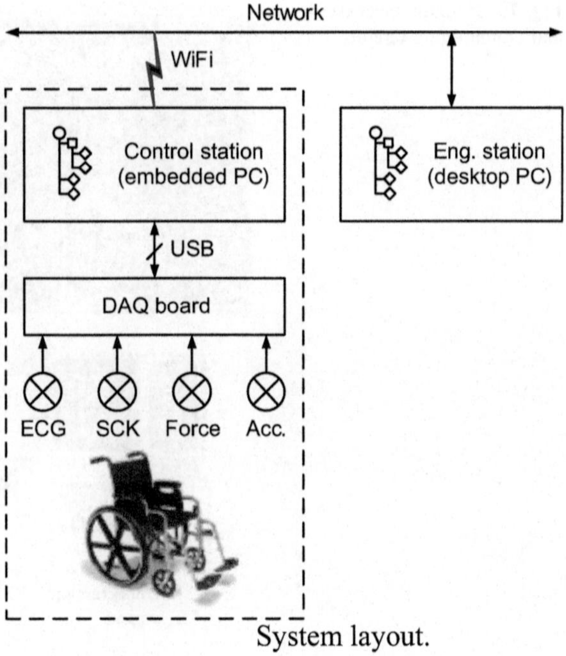

about electrocardiogram (ECG), heart rate (HR), skin conductivity (SKC) and motion activity. The measurement channels, including the sensors and conditioning circuits are connected to the analog inputs of data acquisition module USB connected to an embedded computer (see Fig. 9).

4.1 *Vital Signs Measurement Channels*

The ECG and SKC signals are sensed by dry electrodes made of e-textile. The electrodes were sewn to the lining of the wheelchair arms. As shown in Fig. 10, the ECG electrodes are in contact with both hands, while the SKC electrodes are in contact with one hand only.

The ECG signal is conditioned by an instrumentation amplifier (INA114 with gain = 107), followed by a 2nd order Butterworth band-pass filter (with gain = 10 and cutoff frequencies = 0.3 and 20 Hz).

The SKC signal is conditioned by a voltage divider (formed by the skin resistance and a 2 MΩ fixed resistor), followed by a buffer.

Fig. 10 E-textile electrodes and conditioning circuits

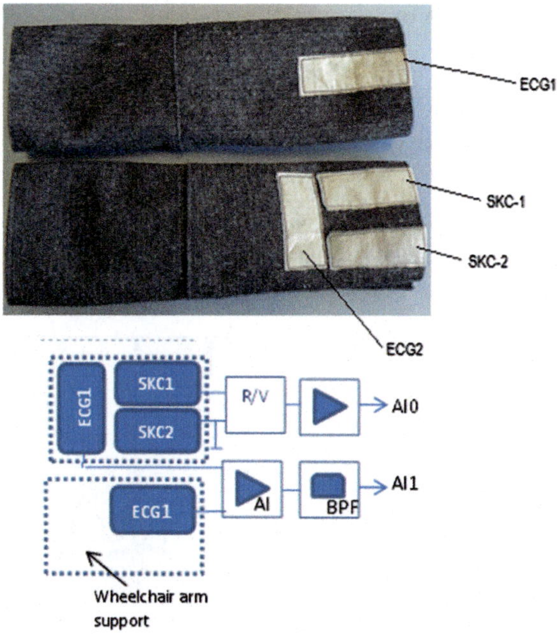

4.2 Acceleration and Force Measurement Channels

The acceleration signals are provided by a 3D micro-machined accelerometer, model MMA7260QR from Freescale. The sensor was properly positioned to sense motion on the XY plane and to detect freefalls on the Z axis. The device is characterized by built-in signal conditioning, low power consumption and selectable sensitivity (1.5g, 2g, 4g or 6g).

It is known that the forces applied to the e-textile electrodes affect the contact resistance. To avoid this problem (i.e. to increase the repeatability of the measurements), two force sensors were placed under the electrodes. Only if both forces are above a pre-defined threshold, the ECG and SKC signals are considered to be valid.

The forces are measured by two piezoresistive sensors, model FlexiForce A201-25 from Tekscan. The sensors are characterized by good linearity (in terms of conductance $G = 1/R$) and high sensitivity (11 kgf full scale). The conditioning circuit is done by an inverter amplifier excited by a constant voltage.

4.3 Data Acquisition

The conditioned signals (ECG + SKC + 2 forces + 3 accelerations) are acquired by a general-purpose, low-cost DAQ board, model USB-6009 from National

Instruments (NI). The acquisition is done at a fixed sampling rate of 1 kS/s, in records of 4000 samples, covering time intervals of 4 s.

The configuration of the analog inputs was facilitated by the use of virtual TEDS compliant with the IEEE 1451.4 clause [10–13]. Each virtual TEDS is composed by two sections: the basic TEDS, which uniquely identifies the sensing channel; and the template TEDS, which contains the technical characteristics of the sensing channel (such as signal type, range, coupling, excitation, etc.). In the present case, the template ID = 30 was used because all the sensing channels are classified as "high-level voltage output". The virtual TEDS were built using the TEDS Library for LabVIEW (provided by NI) having resulted in four files with extension **.ted*.

4.4 Data Processing

The acquired samples are processed by a control station hosted by the embedded PC, which is characterized by an i5 processor, 4 GB RAM and 240 GB SSD disk. The station, which runs on top of Windows 7 Professional and .NET Framework v4.5, performs the following tasks (for each record of samples):

- The physiological signals are valid if the user is quiet and its hands are effectively in contact with the e-textile electrodes. The first condition is met if the variance of the XY acceleration (in g^2) stays below the threshold *MOTION_TH*. The second condition is met if both forces stay above the threshold *CONTACT_TH*.
- The ECG waveform is applied to a Hilbert-based peak detector to extract the HR. The extraction algorithm consists of an input filter (to remove motion artefacts), an adaptive thresholding stage (based on the Hilbert transform) and an R-peak detector.
- The SKC is extracted from the mean value of the corresponding waveform.
- If the variance of the XY acceleration (in g^2) exceeds the threshold *MOTION_TH*, then the Boolean variable *MOTION_FLAG* is turned on (otherwise is turned off). This variable is useful to track the motion activity of the wheelchair.
- If the Z acceleration falls below the threshold *FALL_TH*, then the Boolean variable *FALL_FLAG* is turned on (otherwise is turned off). The default value of the threshold is 0.5 g, which means a slope of 60º. The Boolean variable is useful to detect eventual falls of the wheelchair (and its user).

The control station was coded in VB.NET and compiled in Visual Studio 2013. The application exposes the following services (here identified by their tags):

- *WCHAIR1* (type = *PBlock*): Top-level block that owns all the services hosted by the station.

- *WCHAIR1!MOTION_TH* and *WCHAIR1!CONTACT_TH* (type = *Parameter*, access = read and write): Represent the thresholds of motion activity and fall alert, respectively. Their values can be accessed only through remote procedure calls (following the client/server communication model).
- *WCHAIR1!HR* and *WCHAIR1!SKC* (type = *ScalarParameter*, access = read only): Represent the HR and SKC variables, respectively. Their values can be accessed by making remote procedure calls or by intercepting publications of type *PSK_PHYSICAL_PARAMETRIC_DATA*.
- *WCHAIR1!MOTION_FLAG* and *WCHAIR1!FALL_FLAG* (type = *ScalarParameter*, access = read only): Represent the motion and fall flags, respectively. As in the previous case, their values are published on the network whenever they are updated.

4.5 Data Monitoring

An engineering station was added to the network to perform system configuration and data monitoring. The station runs on a desktop PC characterized by an i7 processor, 8 GB RAM and 500 GB hard risk. The station can handle multiple wheelchairs (i.e. control stations) although, in this case, only one has been used.

5 System Tests

Tests were done to characterize the sensing channels and to verify the communication capabilities of the SXI library.

5.1 Heart Rate Estimation

The accuracy of the HR sensing channel was evaluated as follows:

- A patient sat on the wheelchair and put its hands over the e-textile electrodes. At the same time, additional wet-electrodes were placed on the wrists and the right leg to sense cardiac activity using the Medlab P-OX100, a commercial pulse oximeter and ECG monitor. The Medlab P-OX100 outputs two analog signals representing the ECG waveform and the corresponding HR estimate.
- The signals provided by the Medlab P-OX100 were acquired by the DAQ board and processed side-by-side with those provided by the wheelchair.

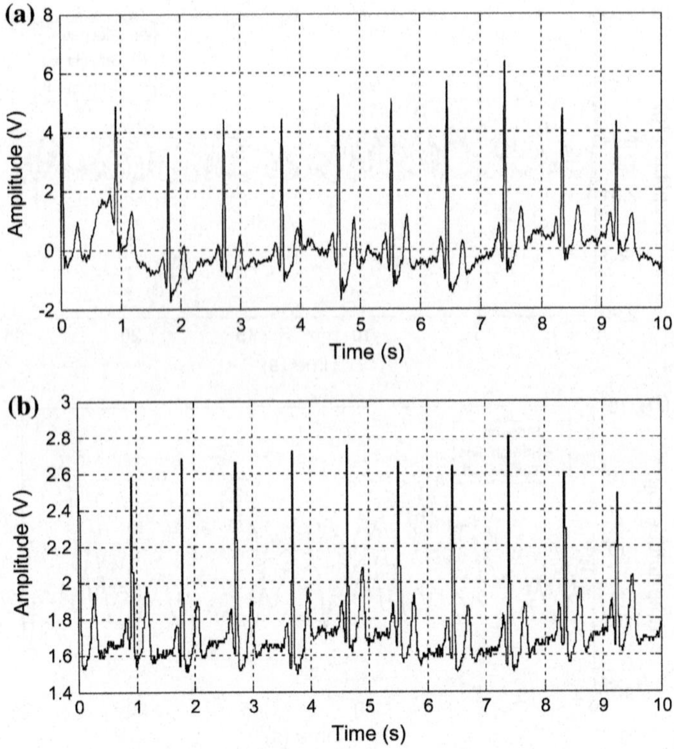

Fig. 11 ECG waveforms: **a** from the e-textile electrodes (under test); **b** from Medlab P-OX100 (used as reference)

- Both ECG waveforms, the one from the e-textile electrodes (under test), and the one from Medlab P-OX100 (used as reference), were logged to a file and plotted on Fig. 11.

5.2 *Motion Activity*

Acceleration measurements were made to help defining a threshold value for motion activity. Figure 12a plots the amplitude of the XY acceleration vector (in g) for several motion profiles: wheelchair stopped (black line); wheelchair moving in a straight line (blue line); and wheelchair turning left (green line). Figure 12b shows the corresponding variances (in $\mu g_{RMS}{}^{\wedge 2}$), calculated using 0.5 s time windows with 0.3 s of overlap. Analyzing the results obtained, a variance of 1 $\mu g_{RMS}{}^{\wedge 2}$ seems like a good value to distinguish between motion and no motion.

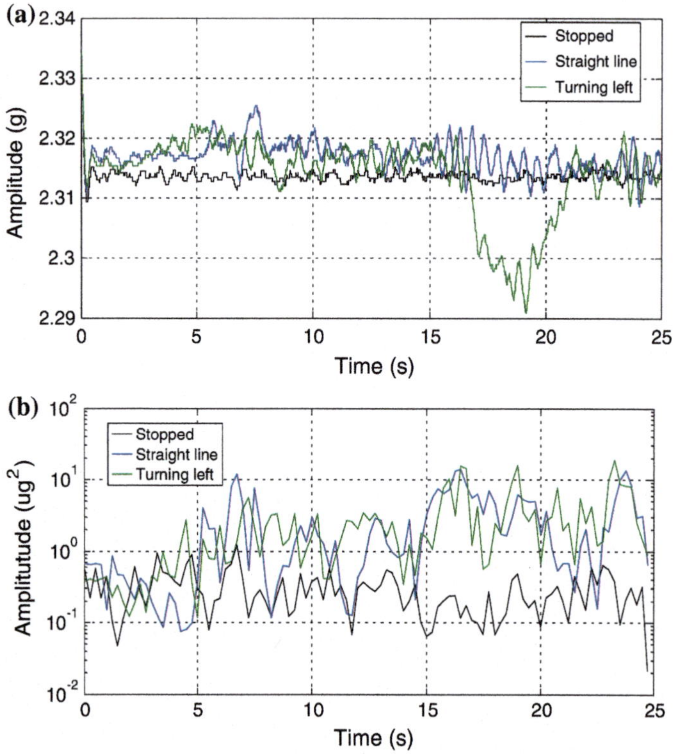

Fig. 12 XY acceleration for different motion profiles: **a** amplitude; **b** variance

6 Conclusion

A set of architectures based on IEEE1451 including hardware and software for vital signs and motor activity of the wheelchairs' users was presented in the chapter. A detailed description of the design and implementation of a microcontroller platform for IEEE 1451.4 is carried out highlighting in particularly elements of hardware expressed by eight measurement channels where the digital information is accessed through 1-wire protocol. The presented IEEE1451 implementations mainly focus on wheelchair user physiological parameters monitoring based on the information acquired through the ECG, PPG, BCG, SKG measurement channels, special attention being granted to kinematic variables (motion activity) monitoring. In one of the hardware implementation physiological variables are sensed by e-textile electrodes embedded on the wheelchair arms, while motion activity is sensed by a 3D MEMS accelerometer. Data acquisition and processing is done onboard the wheelchair by a control station that runs on an embedded PC with a WiFi link. Data monitoring is done remotely by an engineering station installed on a desktop PC. A service extension for instrumentation library is described and the usage for control and engineering stations which implements the 1451.1 information model using WCF services is exemplified. Additionally, smart biomed LabVIEW software, including graphical user interface adapted to the user type was presented, as so as the test results obtained with the presented IEEE1451 implementations. Some tests were done to characterize the sensing channels and to verify the communication capabilities of the SXI library. The obtained results, when compared to reference instruments, showed high performance and good robustness.

References

1. T. Tamura, M. Sekine, Z. Tang, M. Yoshida, Y. Takeuchi, M. Imai, Preliminary study of a new home healthcare monitoring to prevent the recurrence of stroke, in *Proceedings of 37th Annual International Conference of the IEEE Engineering in Medicine and Biology Society (EMBC)*, pp. 5489–5492, Milan, Italy, August 2015
2. S. Hossain, Patient status monitoring for smart home healthcare, in *Proceedings of IEEE International Conference on Multimedia & Expo Workshops (ICMEW)*, pp. 1–6, Seattle, USA, July 2016
3. J. Ma, M. Dong, R&D of versatile distributed e-home healthcare system for cardiovascular disease monitoring and diagnosis, in *Proceedings of International Conference on Biomedical and Health Informatics (BHI)*, pp. 444–447, Valencia, Spain, June 2014
4. O. Postolache, P.M. Girão, H. Ijaz, IEEE 1451.4 embedded smart sensors architecture for wheelchair user monitoring, in *Proceedings of IEEE International Workshop on Medical Measurements and Applications*, vol. 1, pp. 1–5, Budapest, Hungary, May 2012)
5. I. Korhonen, J.E. Bardram, Guest editorial introduction to the special section on pervasive healthcare. IEEE Trans. Inf. Technol. Biomed. **8**(3), 229–234 (2004)
6. O. Postolache, P.M. Girão, E.C. Pinheiro, J.M. Pereira, R. Madeira, G. Postolache, J.G. Mendes, M.C. Moura, Multi-usage of microwave Doppler radar in pervasive healthcare systems for elderly, in *Proceedings of Instrumentation and Measurement Technology Conference (I2MTC)*, pp. 1–5, Hangzhou, China, May 2011
7. O. Postolache, P. Girão, J.M. Joaquim, E. Pinheiro, G. Postolache, Physiological parameters measurement based on wheelchair embedded sensors and advanced signal processing. IEEE Trans. Instrum. Meas. **59**(10), 2564–2574 (2010)
8. M.R. Ribeiro, O. Postolache, P.M. Girão, Modular platform architecture for fast prototyping of vital signs and motor activity monitors, in *Proceedings of IEEE International Instrumentation and Technology Conference (I2MTC)*, pp. 1–6, Minneapolis, US, May 2013
9. O. Postolache, P. Girao, J.M. Dias Pereira, J. Pincho, C. Maia Moura, G. Postolache, Smart walker for pervasive healthcare, in *Proceedings of International Conference on Sensing Technology (ICST)*, pp. 1–5, Parlmerston North, New Zealand, December 2011
10. N. Ulivieria, C. Distanteb, T. Lucaa, S. Rocchia, P. Sicilianob, IEEE1451.4: A way to standardize gas sensor. Sens. Actuators B Chem. 114(1), 141–151, 2006. (30 March 2006)
11. National Instruments, An Overview of IEEE 1451.4 Transducer Electronic Data Sheets (TEDS). http://standards.ieee.org/develop/regauth/tut/teds.pdf (accessed March 2017)
12. Transducers Techniques, TEDS IEEE 1451.4 Plug & Play Smart Load Cell. https://www.transducertechniques.com/teds-ieee-1451-4-plug-and-play-smart-load-cell.aspx (accessed March 2017)
13. R. Wang, K. Deevy, Reconfigurable adaptive wireless sensor node technology using IEEE 1451.4 standard, in *Proceedings of 39th Annual Conference of the IEEE Industrial Electronics Society (IECON)*, pp. 3988–3993, November 2013

14. E.Y. Song, K. Lee, Understanding IEEE 1451—networked smart transducer interface standard. IEEE Instrum. Meas. Mag. 11(2), 11–17 (2008)
15. V. Viegas, J.M. Dias Pereira, P. Silva Girão, A brief tutorial on the IEEE 1451.1 standard. IEEE Instrum. Meas. Mag. 11(2), 38–46 (2008)
16. D.S. Platt, *Introducing Microsoft .NET*, 3rd ed. (Microsoft Press, WA, USA, 2003). ISBN 9780735619180
17. J. Ritcher, *Applied Microsoft .NET Framework Programming* (Microsoft Press, WA, USA, 2002). ISBN 9780735614222
18. J. Lowy, *Programming WCF Services* (O'Reilly, USA, 2007). ISBN 9780596526993
19. J. Smith, *Inside Windows Communication Foundation* (Microsoft Press, WA, USA, 2007). ISBN 0780735623064
20. K. Lee, Sensor networking and interface standardization, in Proceedings IEEE Instrumentation and Measurement Technology Conference (IMTC), pp. 147–152, Budapest, Hungary, May 2001
21. P. Doyle, D. Heffernan, D. Duma, A time-triggered transducer network based on an enhanced IEEE 1451 model. Microprocess. Microsyst. 28, 1–12 (2004)
22. R. Madeira, O. Postolache, G. Postolache, Development of therapeutic serious game for a pervasive assistive environment, in *Proceedings of IEEE International Conference on Serious Games and Applications for Health (SEGAH)*, pp. 1–11, Braga, Portugal, November 2011
23. IEEE Standards, IEEE 1451.4TM Manufacturer ID Public Listing. http://standards.ieee.org/develop/regauth/manid/public.html) (accessed March 2017)

Printed by Printforce, the Netherlands